Praise for *The Physics of Superheroes:*

"[Kakalios] uses Superman, Spider-Man and less well-known characters such as Ant-Man—yes, there was a superhero called Ant-Man—to explain physics principles from gravity and the laws of motion to quantum mechanics."
— *Star-Tribune* (Minneapolis)

"The physics, even when Kakalios points out where the comics got it wrong, is drawn out sympathetically and with good humor."
— *Nature*

"Delightful."
— *Richmond Times-Dispatch*

"A droll but sincere look at what Superman and Spider-Man can teach about physics....Entertaining....His explanations are lucid and smooth."
— *Science*

"With passion, general affability, and a penchant for bad (truly bad) jokes, Kakalios ably relates the most baffling of theorems."
— *Kirkus Reviews*

"A fascinating seminar on just how the superpowers of various heroes would and wouldn't work, scientifically."
— *Starlog*

"Entertaining, and often quite funny."
— *The Post and Courier* (Charleston, SC)

"What could superheroes and physics possibly have in common? More than you'd think. [Kakalios] takes a number of famed comic-book heroes and analyzes their abilities and actions in terms of how plausible they are from a physics standpoint."
— *Calgary Sun*

"Knowledgeable. . . . Cleverly informative reading."
—*Comic Shop News*

"A physics textbook that will be interesting. . . . Very enter-taining."
—About.com

"Kakalios draws on the Atom, Iron Man, X-Men, the Ant-Man, and the Hulk, among others, to cover topics as diverse as electromagnetism, quantum mechanics, string theory, and thermodynamics. That all of this is accomplished with enough humor to make you laugh aloud is an added bonus."
—*Publishers Weekly*

"[Kakalios] describes the powers of comic-book superheroes and uses those traits as an opportunity to launch into interesting physics."
—*Physics Today*

"A treat for anyone interested in physical science and can be enjoyed readily by math phobes and those with little science education, since Kakalios explains it all with clear detail and a good measure of fun. Highly recommended."
—*Library Journal*

"Clearly, Kakalios is a man who loves both physics and comics, and it really shines through."
—*SFX* magazine

"Kakalios's use of such stories to elucidate the finer points of impulse and momentum is extremely readable."
—*The Guardian*

"After reading so many comic books, Kakalios is a bit of a comic himself, and he makes this a thoroughly entertaining read. Why can't all physics professors be like this?"
—Science a GoGo

"[An] engaging and informative look at the surprising role of physics in comic books."
—*Science News*

"*The Physics of Superheroes* is clear, rapid, funny, and endlessly informative—as if Stan Lee and George Gamow had teamed up to battle the nefarious forces of ignorance."
—Gerard Jones, author of *Men of Tomorrow: Geeks, Gangsters, and the Birth of the Comic Book*

"Author James Kakalios is a scientific genius who could put Lex Luthor and Dr. Doom to shame. Superman should have him on retainer. I do—because *The Physics of Superheroes* is this comic-book writer's newest favorite indispensable resource."
—Mark Waid, writer of *Spider-Man, Superman,* and the *Fantastic Four*

James Kakalios is a professor in the School of Physics and Astronomy at the University of Minnesota, where he has taught since 1988, and where his class "Everything I Needed to Know About Physics I Learned from Reading Comic Books" is a popular freshman seminar. He received his Ph.D. in 1985 from the University of Chicago, and has been reading comic books for much longer. He lives in Minneapolis with his wife and three children.

THE PHYSICS OF SUPERHEROES

SPECTACULAR SECOND EDITION

James Kakalios

AVERY
an imprint of Penguin Random House
New York

an imprint of Penguin Random House LLC
375 Hudson Street
New York, New York 10014

Previously published in paperback by Gotham Books, November 2009
Copyright © 2009 by James Kakalios

Most Avery books are available at special quantity discounts for bulk purchase for sales promotions, premiums, fund-raising, and educational needs. Special books or book excerpts also can be created to fit specific needs. For details, write SpecialMarkets@penguinrandom house.com.

ISBN 978-1-592-40508-4

Printed in the United States of America
20 19 18 17 16 15

Book design by Sabrina Bowers

TO THERESE

CONTENTS

SECTION 2–ENERGY–HEAT AND LIGHT

SECTION 3—MODERN PHYSICS

SECTION 4—WHAT HAVE WE LEARNED?

FOREWORD

ALTHOUGH WILE E. COYOTE is by no stretch of the imagination a superhero, I have to admit that it was this hapless villain—who escaped death episode after episode while continuing to fruitlessly chase the Road Runner with almost Sisyphean intensity day in and day out—who first got me thinking about the physics of illustrated characters. Even as a relatively young boy hooked on television, I suspected there was something fishy whenever I saw Wile E. run off a cliff and hover indefinitely until the moment he realized there was no solid ground underneath. Somehow it seemed to me even then that gravity should continue to work, whether or not one was conscious of it.

I bring this example up, in spite of the fact that it involves no superheroes, and in fact involves a television cartoon rather than a comic-book figure, because it illustrates a point that has become central to the way I think about teaching physics: Few things are more memorable than confronting one's own misconceptions. Indeed, some of us who study "physics education" for a living suggest that it is only by directly encouraging students to run up against their own misconceptions that you can help them internalize what you're teaching them. I don't know whether this is true or not, but I do know that if you want to reach out to understand popular misconceptions, then exploiting where we get our cultural perspectives from is a good place to start. And if that means borrowing from Superman, or *Star Trek*, I am all for it!

Now, I don't want you to think that I bring up comic books and popular misconceptions in the same paragraph because I want to denigrate the former. Far from it! Indeed, the comics sometimes actually get it right, and as James Kakalios describes in his introduction to this far-reaching journey from the gravity of Krypton to the quantum mechanics of the X-Men, students often seem to grumble about how the standard examples from his introductory physics class have nothing to do with the real world they will encounter upon graduation. But when they are instead introduced to the physics of superheroes, this complaint never arises!

One might initially wonder whether Superman might seem more real to students than pulleys, ropes, and inclined planes. But the real reason students don't complain is undoubtedly that the comic-book examples are fun, while inclined planes aren't. And that is perhaps one of the most useful reasons for thinking about the physics of superheroes. Not only can you imagine, and be introduced to, lots of interesting physics, from everyday phenomena to esoteric modern subjects, but it is actually fun to think about. Moreover, while subjects like quantum mechanics might seem intimidating, who could be intimidated by cute Kitty Pryde?

Some who remember the comic books that enthralled them as young people might also recall a sense of wistfulness in pondering whether our own world could ever capture the excitement and drama of the worlds of comic-book superheroes. In truth, however, it is far more interesting and exciting to open up our minds to the hidden wonders of nature that science has revealed to us over the past four hundred or so years. Truth is far stranger than fiction, even comic-book fiction. And finding out why is all part of the fun.

Lawrence M. Krauss
Cleveland, Ohio

PREFACE

I WAS A COMIC-BOOK FAN as a kid, but like many who have come before and after me, I abandoned the hobby in high school upon discovering girls. My mother, following the standard script, used this opportunity to throw my collection away. I renewed my comic-book-reading habit years later, in graduate school, as a way to relieve the stress of working on my dissertation. Now, as an adult, I've rebuilt much of my comic-book collection (or, as my wife refers to it, "the fire hazard"), but as a precaution, my mother is not allowed near it.

Back in 1998, the University of Minnesota, where I am a physics professor, introduced a new type of class termed "freshman seminars." These are small, seminar-type classes open to entering students, and while they are for credit, they are not tied to any particular curriculum. Professors are encouraged to develop classes on unconventional topics, and freshman seminars on Bio-Ethics and the Human Genome; The Color Red (a chemistry class); Trade and the Global Economy; and Complex Systems: From Sandpiles to Wall Street, are among the many offered. In 2001 I introduced a class originally entitled Everything I Know About Physics I Learned from Reading Comic Books. This is an actual physics class, covering most of the topics traditionally covered in an introductory course, but rather than employing illustrations of masses on springs, or blocks sliding down inclined planes, all of the examples came from the four-color adventures of costumed superheroes, and focused in particular on those situations where the comic books got their physics right.

The present book, while inspired by this class, is not a textbook per se. It is written for the nonspecialist who is interested in a relatively pain-free way to learn about the basic physics concepts underlying our modern technological lifestyle. Topics such as forces and motion, conservation of energy, thermodynamics, electricity and magnetism, quantum mechanics, solid-state physics and materials science are discussed, and real-world applications such as automobile airbags, transistors, and microwave ovens are explained. I hope you will be so busy enjoying this superhero ice-cream sundae that you won't realize that I am sneakily getting you to eat your spinach at the same time.

This book is intended for both longtime comic-book fans and those who can't tell Batman from Man-Bat. In order to describe the physics connected with certain superheroes or story lines, I have had to summarize key plot points in various comic books. Therefore, for those who have not yet read these classics, the following two words apply to this entire book: "Spoiler Alert."

Readers interested in consulting the source materials considered here will find citations of the comic books discussed in the text at the end of the book. I have listed the original comic-book information and, whenever possible, where the issue can be found in a reprint volume. The dates listed for a given comic book, printed on its cover, are not when the issue first appeared on the newsstands. To extend shelf life, the date listed indicated when the comic was to be returned to the publisher for credit, and not when it became available for purchase. In an effort to attract new readers who desire first issues as collectors' items, comic books will occasionally restart their numbering scheme while keeping the name of the comic unchanged. If not otherwise noted, issue numbers refer to the first volume of a comic. I have listed, where known, the writer and artist for each comic listed in the endnotes. My omission of the inkers should not be construed as denigrating their contribution to the finished comic (I most certainly do not believe that such a job is equivalent to "tracing"), but rather a reflection of the fact that the artist, along with the writer, typically have the primary responsibility for the physics in a given comic-book scene.

Any discussion of physics in comic books naturally invites the scrutiny of physicists as well as comic-book fans, both of whom are known for their, let us say, attention to detail. Each of the incidents I've selected illustrates a particular physics principle. Sometimes the very next issue would contain a scene contradicting the physically plausible manifestation of a superpower described here. When considering characters that have starred in multiple comic books for more than half a century, it is a sure bet that there will be counter-examples for any statement I make. Consequently, while examining the physics associated with a superhero's powers will, in many cases, provide a better appreciation for their talents, my comrades in fandom are advised that this book is not intended to provide definitive accounts of any character's power or adventures.

Similarly, my physics colleagues are warned that this book is for a nonexpert audience. I have attempted to keep things simple, while acknowledging the rough edges and complications of the real world. A complete discussion of many of the topics considered here could easily expand to fill several volumes and would provide a concrete illustration of Dr. Manhattan's final words at the conclusion of Alan Moore's and Dave Gibbons' *Watchmen*: "Nothing ends, Adrian. Nothing ever ends."

The language that describes the physical world is mathematical in nature. Why this should be so is a deep, philosophical question (the physicist Eugene Wigner referred to the "unreasonable effectiveness of mathematics" in accounting for nature's properties) that has puzzled and stirred all those who have studied it. It is tempting, in a book involving comic-book superheroes, to avoid even the slightest whiff of mathematics. However, that would be a cheat, worse than omitting any reproductions of artwork in a book about Picasso, or not providing a CD of recordings for a book about the history of jazz, because mathematics is necessary in any thorough discussion of physics.

The reader may protest that they do not understand math, or cannot think mathematically. But for this book, all that is required is that one recognizes that $\frac{1}{2} + \frac{1}{2} = 1$. That's it, that two halves equal a whole. If you are comfortable with $\frac{1}{2} + \frac{1}{2} = 1$, then writing it as $2 \times (\frac{1}{2}) = \frac{2}{2} = 1$ (that is, two multiplied by one half) should cause no concern, because obviously two halves equal one. It seems so simple that you may be surprised to discover that we have already been doing algebra (and you thought you'd never use that again after high school!).

As many students have long suspected, there is a trick to algebra, and the trick is the following: If one has an equation describing a true statement, such as $1 = 1$, then one can add, subtract, multiply, or divide (excepting division by zero) the equation by any number we wish, and as long as we do it to both the left and right sides of the equation, the correctness of the equation is unchanged. So if we add 2 to both sides of $1 = 1$, we obtain $1 + 2 = 1 + 2$ or $3 = 3$, which is still a true statement. Dividing both sides of $1 = 1$ by 2 gives us $\frac{1}{2} = \frac{1}{2}$. Since $1 = 1$, then $\frac{1}{2} + \frac{1}{2} = 1$, which in turn can be written as $\frac{2}{2} = 1$. I'll make a deal with you, Fearless Reader: I won't use any mathematics more sophisticated than what's described in this paragraph if you'll refrain from panicking when a mathematical equation appears. You can always glide right by the math, and your understanding will be no worse for it. But if you take a notion to calculate a velocity or a force for a situation other than considered here, you'll have the tools to do so. In any event, I promise you that there will be no quizzes at the end of the book!

PREFACE TO THE SECOND EDITION

IN GRADUATE SCHOOL, as the writing of my physics dissertation neared completion, my thesis adviser impressed upon me that no scientific research is ever "finished." Regardless of the answers you have obtained to your problem, there are always at least a few open questions—loose threads waiting to be pulled. Having written *The Physics of Superheroes*, I have discovered that, similarly, no book is ever finished. The manuscript is delivered to the editor, galley proofs are checked (though the typographical errors that inevitably slip through are guaranteed to be the first thing the proud author spots upon cracking open the newly published book), the book shows up on the shelves of bookstores, and yet threads remain loose. There is always more that could be said, or said better.

Consequently, I was pleased when Gotham Books offered me the opportunity to revise *Superheroes* for a second edition. The feedback I received from readers on the 2005 first edition suggested many topics that warranted additional attention. My estimation of the time Superman would spend pushing his weight down against the ground as he prepared to leap a tall building in a single bound in particular provoked many friendly discussions with readers, and I am happy to have a chance to elaborate on this point in this second edition.

Additionally, soon after publication, I came across superhero illustrations of various physics principles that, had I been aware of earlier, I certainly would have included. This second edition also includes new chapters on fluid mechanics, angular momentum, and materials science, as reflected in discussions of the heroes and villains Aquaman, the Top, and Whirlwind; and the entire Justice League of America and Avengers, respectively. Some chapters in the Mechanics section have been reordered, in order to improve the flow of the topics discussed. Every chapter has been edited, some extensively. Moreover, mathematical analysis indicates that in this second edition, the jokes are now 12.7 percent cornier, in some cases by the utilization of the anti "rule of three" (two punch lines followed by a straight line). I hope that you will find this second edition to be a fun and accessible way to learn the basic principles of physics.

In addition to all of those I thanked in the first edition, there are a few new names whose assistance with the second edition I wish to acknowledge. Patrick Mulligan at Gotham Books has been a wonderful and insightful editor of this edition. I am proud to thank my son Thomas for his careful reading of the new chapters and his suggestions for improvement and clarification. Laura Gjovaag, who runs The Unofficial Aquaman website, and Rob Kelly at The Aquaman Shrine website, were both great resources for all things Aquaman-related and provided indispensable information and images of the Aquatic Ace. I appreciate their generous time and effort. Once again, Jenny Allen's assistance preparing the new figures for the second edition is greatly appreciated.

THE PHYSICS OF
SUPERHEROES

INTRODUCTION

SECRET ORIGINS:
HOW SCIENCE SAVED
SUPERHERO COMIC BOOKS

IF I HAD EVER WONDERED whether or not my students found studying physics to be a waste of time, all doubt was removed several years ago. I was returning from lunch to the physics building at my university when I overheard two students as they were leaving. From their expressions and the snippet of conversation I caught, it seemed that they had just had a graded exam returned to them. I'll quote here what I heard (but in the interest of decorum, I'll clean it up).

The taller student complained to his friend, "I'm going to bleeping buy low, and bleeping sell high. I don't need to know about no bleeping balls thrown off no bleeping cliffs."

There are three nuggets of wisdom contained in this statement: (1) the secret to financial success; (2) as scholars of English grammar can attest, "bleeping" can be both an adjective *and* an adverb; and (3) that the examples used in traditional physics classes strike many students as divorced from their everyday concerns.

The real world is a complicated place. In order to provide illustrations in a physics lesson that emphasize only a single concept, such as Newton's Second Law of Motion or the principle of Conservation of Energy, physics teachers have developed over the decades an arsenal of overly stylized scenarios involving projectile motion, weights on pulleys, or oscillating masses on springs. These situations seem so artificial that students inevitably lament, "When am I ever going to use this stuff in my real life?"

One trick I've hit upon in teaching physics involves using examples culled from superhero comic books that correctly illustrate various applications of physics principles. Interestingly enough, whenever I cite examples from superhero comic books in a lecture, my students never wonder when they will use this information in their "real life." Apparently they all have plans, post-graduation, that involve protecting the City from all threats while wearing spandex. As a law-abiding citizen, this notion fills me with a great sense of security, knowing as I do how many of my scientist colleagues could charitably be termed "mad."

* * *

I first made the connection between comic books and college education back in 1965, when for the princely sum of twelve cents I purchased *Action Comics # 333*, which featured the adventures of Superman. While not a huge fan of the Man of Steel at the time, I was seduced by the comic's cover (see fig. 1), which promised a glimpse into the inner workings of our institutions of higher learning. As a kid I was deeply curious as to what college life would be like. Now that I am a university professor, I realize that this was a premonition that once I entered college I would never get out, and that my matriculation would turn into some sort of life sentence.

A story in *Action # 333*, titled "Superman's Super Boo-Boos," featured a scene in which, as a "tribute to all he's done for humanity," Superman was to be granted an "honorary degree of Doctor of Super-Science" by Metropolis Engineering College. (I should point out that such a degree option was not offered when I enrolled in graduate school.) On the cover of this issue, Superman was in a large auditorium on a college campus and was "writing" his name on a bronze honor scroll using his heat vision. The aged faculty in attendance, wearing commencement dress, were aghast, not due to the fact that searing beams of energy were emanating from Superman's eyes but rather because the professors saw a fire-breathing dragon up on the stage instead of Superman, due to an illusion cast by Superman's archnemesis, Lex Luthor. This was part of Luthor's scheme to constantly confound Superman's expectations until he was paralyzed by indecision and hence unable to stop Luthor's evil schemes.* Even as

---

* To this end, Lex Luthor publicly came to Superman's aid on several occasions, with the goal of making Superman doubt his judgment regarding Luthor's intentions. So committed was Luthor to this plan that he even directly saved Superman's life when he was threatened by another crook

Fig. 1.
Cover to
Action
Comics # 333,
featuring a
scene from
Superman's
ill-fated visit
to Metropolis
Engineering
College.

a grade-school kid, I realized that this depiction of college life was probably not too realistic. Nevertheless, the cover did provide two insights that, in the fullness of time, have turned out to be fairly accurate. The first is that all college professors, at all times, always wear caps and gowns. The second is that all college professors are eight-hundred-year-old white men.

---

wielding a Kryptonian sword capable of killing him. You might think that it would be simpler at this point to abandon his plan to "gaslight" Superman and just let the other crook kill him and be done with it, but who can truly understand the thought processes of a criminal mastermind like Lex Luthor?

While this may have been my first indication that comic books and college could coexist, it would not be my last. Over the years, I have continued to enjoy reading and collecting comic books. (This is not a "guilty pleasure" of mine, simply because I don't believe in "guilty" pleasures. Snobbery is just the public face of insecurity.) And in my reading I've noted that the writers and artists creating superhero comic-book stories get their science right more times than you might expect. Those not overly familiar with superhero comic books may be surprised to learn that anything in comic books could be scientifically correct, and one can learn a lot of science from reading comic books.

A typical example is shown in fig. 2, featuring a scene from *World's Finest # 93*, from April 1958. The big superstars at National Comics (which later became Detective Comics and is today known as DC Comics) are Superman, Batman, and Robin and each issue of World's Finest contained a team-up adventure of the Man of Steel and the Dynamic Duo. In this story, a crook, Victor Danning, has his intelligence accidentally increased to "genius level" during a botched attempt to steal a "brain amplifier" machine. Using his enhanced mental powers, he proceeds to commit a series of "super-crimes," drawing the attention of Batman and Robin and Superman. After several of his schemes are nar-

Fig. 2. *A scene from* World's Finest # 93. *In this scene, a crook whose intelligence has been artificially enhanced describes his plan to utilize the dispersion of underground shock waves in order to determine the hidden location of the Batcave.*

rowly foiled by our heroes, Danning decides to take a proactive approach and attempts to discover the hidden headquarters of Batman and Robin, the Batcave. (It's not really explained how this would defeat Batman and Robin, and it is presented as a given that if you are a crook, you want to learn the Batcave's address.)

Danning instructs his henchmen to plant sticks of dynamite along the perimeter of Gotham City. Monitoring the resulting shock waves on his "radar seismograph," Danning explains that the waves traveling through a cave will have different speeds than those moving through solid rock, and in this way the location of the Batcave can be discerned.

In this example, the evil genius Victor Danning is on solid scientific ground, for it is indeed true that the velocity of sound or a shock wave will depend on the density of the material through which it moves. Sound waves need a medium in which to propagate. In a dilute medium, such as air, there are large spaces between molecules, which makes it harder for pressure waves to propagate, compared with water, steel, or the thin walls of an apartment. Roughly speaking, the denser the medium, the faster the sound travels, which is why in old Western movies a character would put his ears to the railroad tracks to determine whether a train, too distant to see, was approaching. He could hear the train's vibrations through the steel rail much sooner than if he waited for the same noise to reach him through the air. Indeed, geologists make use of this variation of the speed of sound waves in order to locate underground pockets of oil or natural gas. The depiction in comic books of actual scientists and how they work, on the other hand, often leaves much to be desired. In the very same issue of *World's Finest*, as shown in fig. 3., the inventor of the "brain amplifier," Dr. John Carr, describes his latest experiment at a scientific conference. He boasts that his device will "increase any man's mental power 100 times." Unfortunately, Carr points out that there's one catch, saying: "There's one ingredient that's still missing [before his machine will work], and I haven't found out yet what it is!" This is equivalent to inventing a machine that turns lead into gold, or tap water into gasoline—but needs one key element (and who knows if it even exists) to work! Rarely are presentations made at physics conferences of work in such an unfinished state (at least not intentionally). Watching this presentation

Fig. 3. *Another scene from* World's Finest # 93, *where "crooked ex-scientist" Victor Danning first learns of the "brain amplifier" device, along with its only drawback.*

is Victor Danning (this is where he was inspired to steal the brain amplifier, despite its intrinsic design flaw), who is labeled in the caption box as a "crooked ex-scientist." This part rings true, for I speak not just as a physicist but for all scientists when I say that once you become "crooked," we kick you out of the club and strip you of your "scientist" title, ripping the epaulets off your lab coat and tossing you to the curb.

* * *

The incorporation of scientific principles into superhero adventures is only occasionally found in stories from the 1940s (referred to by fans as the "Golden Age" of comic books) but is much more common in comics from the late 1950s and 1960s (known as the "Silver Age"). Between these two epochs lies the "Dark Age" of comic books, when sales plummeted and the very concept of superheroes came under attack by psychiatrists, educators, and congressmen. Those circumstances that led to the existence of two "ages" of superhero comic books are also responsible, it can be argued, for the "scientific" tone in Silver Age comics published in the post-Sputnik era. Since we will be relying on superheroes to illustrate scientific concepts for the rest of this book, it is useful to first take a moment to consider the early roots of these mystery men.

A BRIEF HISTORY OF SUPERHERO COMIC BOOKS

Before there were comic books, there were comic strips.* Weekly broadsheets in Victorian England, nicknamed "penny dreadfuls," featured humorous strips in the music-hall tradition. Their popularity with poor workers was a great offense to middle-class sensibilities. A fierce newspaper rivalry in the 1890s between Joseph Pulitzer and William Randolph Hearst spurred the creation of the newspaper comic strips that proved extremely popular with newly arrived immigrants who were only semi-conversant in English. These newspaper strips became highly effective weapons in the circulation wars of the time. In fact, it is claimed that the association of Hearst papers with Richard F. Outcault's visually striking and popular comic strip character, the Yellow Kid, who even appeared on editorial pages, inspired critics of sensationalistic newspapers to denigrate them with the sobriquet "yellow journalism."

Despite occasional forays into presenting comic strips in a magazine format (such as a Buster Brown comic book published in 1903, Little Nemo in 1906, and Mutt and Jeff in 1910), comic books did not become firmly established until 1933. At the time, newsstands were filled with extremely popular pulp magazines, which contained an entire original novel for just ten cents. Costs were kept down, in part, by printing on the poor-quality paper that gave these journals their name. There were popular titles devoted to mystery stories, such as *Detective Fiction Weekly* and *Black Mask* (which first published Dashiell Hammett and Raymond Chandler); science fiction, such as *Amazing Stories* and *Astounding Stories* (wherein Theodore Sturgeon, Isaac Asimov, and Ray Bradbury got their start); horror and fantasy, such as *Unknown* and *Weird Tales* (the home of H. P. Lovecraft, Robert E. Howard, and even the playwright Tennessee Williams); and action/adventure titles like *The Shadow, The Spider, G-8 and His Battle Aces, The Mysterious Wu Fang,* and *Doc Savage*. At the

---

* The concept of using words and drawings to tell a story is more than five hundred years old. There are examples of woodcut "broadsheets" from the Middle Ages that use paneled borders, speedlines, and word balloons.

height of their popularity, some pulp titles sold several hundred thousand issues per month, which even at ten cents a copy was big money during the Depression. In this highly competitive environment, George Janosik, George Delacorte, Harry Wildenberg, and Maxwell C. Gaines (a schoolteacher before he became a comic-book publisher) decided to take a chance and reprint the color newspaper comic strips from the Sunday supplements onto standard tabloid-size sheets of newsprint, folding them into a 6 ⅝" × 10 ⅛" pamphlet (thereby establishing the standard format for comic books, which remains unchanged to this day). The comic book *Funnies on Parade* was distributed with coupons for Procter & Gamble products and similar promotional giveaways, and the popularity of the first print run of 10,000 copies inspired them to stick a ten-cents price tag on another issue and sell them on newsstands. The speed with which these newsstand comics sold out (despite the fact that they contained only reprints of material that had been previously available in Sunday newspapers) demonstrated that there was a future in these "funny books."

Newspaper comic strips were leased to regional papers through "distribution syndicates," which controlled the reprint rights to these strips. In order to satisfy the demand for material by comic-book publishers who couldn't secure (or didn't want to pay for) reprint rights for newspaper strips, Maj. Malcolm Wheeler-Nicholson hired a group of work-hungry young artists and writers and commissioned original comic material. These brand-new stories were published as New Fun Comics by National Allied Publications. The drawn, inked, lettered, and ready-for-printing comic pages that came out of Wheeler-Nicholson's studio enabled the publishers to avoid the high rates charged by the then-powerful typesetters unions. Soon, the stories that had occupied the pulp-fiction magazines were being told in graphic form, and comics featuring detective and police stories, horror, funny animals and straight gags, adventure heroes, secret agents, and crime fighters with mystical powers filled the newsstands. These ranks were joined in 1938 by a strange visitor from another planet with powers and abilities far beyond those of mortal men.

Superman was the brainchild of Jerry Siegel and Joseph Shuster, two teenagers from Cleveland who dreamed of earning riches through the creation of a popular newspaper adventure strip. Combining attributes from two of Edgar Rice Burroughs's characters, Tarzan and John Carter of Mars, Siegel and Shuster turned the conventional science-fiction adventure story on its head. Instead of an ordinary Earthman traveling to a strange new planet or the far future (as in the stories featuring Flash Gordon or Buck Rogers), a citizen of a distant world with strange powers came to Earth. This innovation, together with the colorful uniform the hero wore (inspired, perhaps, by the costumed strongmen who performed in circuses at the time) and the then-novel concept of a secret identity for the heroic adventurer, made their strip so original that it was swiftly turned down by every newspaper distribution syndicate they approached. After four years of constant rejection, Siegel and Shuster were desperate enough to try to sell their Superman concept to the decidedly lesser market of comic books. They eventually found a receptive audience in Sheldon Mayer, a young editor who saw potential in the rough, early strips of Siegel and Shuster. Mayer convinced fellow editor Vin Sullivan that this character was exactly what was needed for his new comic-book title that was due at the printer but lacked a lead story. With no time to redo the strip to fit the comic-book format, panels from the two weeks of sample newspaper strips were hastily cut and pasted into a thirteen-page story. With a cover adapted from one of the panels in the strip showing Superman lifting an auto sedan over his head while crooks fled in panic, *Action Comics # 1*, cover price ten cents, appeared on newsstands in June 1938. The rest, to coin a phrase, is history.

* * *

Evolutionary biology teaches that random mutations can lead to the creation of new species. When these new species exhibit superior adaptation to a changing environment, they can quickly dominate an ecological niche. Similarly, the superhero comic book struck a resonant chord with Depression-era readers, and was an instant success. Soon the newsstands were full of superhero comics, featuring characters possessing a dazzling array of powers and abilities.

All of these new characters shared one crucial attribute: differing sufficiently from Superman in order to avoid joining Fawcett Publications' Captain Marvel in a lawsuit over copyright infringement by National Publications (which owned the legal rights to Siegel and Shuster's creation). Many of these newborn heroes had a single superpower, such as superspeed (the Flash, Johnny Quick), flight (Hawkman, Black Condor), superstrength (Hourman, Captain America), or none of the above (Batman). Some of these heroes gained their superpowers via "scientific means." The Flash of the 1940s, for example, became superfast after a chemical laboratory accident in which he inhaled some "hard water."* Chemist Rex Tyler developed a pill that provided enhanced strength and speed for sixty minutes, enabling him to fight crime as Hourman. The 4F-army-reject Steve Rogers became the superhero Captain America, following a series of injections with a "super soldier" serum (nowadays, these would probably be described as "steroids").

Much more common, though, was a mystical or supernatural origin for the characters' abilities, due to the acquisition of, or exposure to, a magical object from some hidden corner of the globe. In this way, as they would throughout their existence, comic books merely reflected the popular cultural zeitgeist. For example, in the 1940s the Green Lantern was a hero who had come into possession of a mystical lantern originally from ancient China, from which he fashioned a ring that endowed the wearer with a wide variety of powers but was ineffective against wood. Viewed in the cultural context of the times, the world was a bigger place back in 1940. To the adolescent imagination, the "Far East" and the "Congo" were still vast repositories of powerful secrets and mysterious artifacts. When the Green Lantern character was reinvented in 1959, he was given a new costume and origin, and his lantern and ring became extraterrestrial artifacts. The ring's vulnerability to objects that were colored yellow was now attributed to a chemical impurity in the ring's composition, which could not be removed without making the ring ineffective.† Similarly, the Green Lantern's colleague, Hawkman, was a reincarnated Egyptian prince in 1940, while the 1960s version of

---

* Water is characterized as "heavy" when the two hydrogen atoms in H_2O each have an extra neutron, while water with a large mineral content is termed "hard." Fortunately for the Golden Age Flash, water softeners were not common in the 1940s.
† Hence, an effective anti–Green Lantern weapon, regardless of the period, would be a yellow, wooden baseball bat.

the same hero was an intergalactic policeman from the planet Thanagar.

This origin transition continues today. In 1962, Peter Parker gained the powers of Spider-Man when he was bitten by a spider that had accidentally become radioactive in a physics lab demonstration, while in a 2000 reinterpretation of the same character (as well as in the 2002 film version), the fateful bite was from a genetically engineered superspider that escapes during a molecular-biology lab demonstration. Thus the one constant would appear to be that the creation of the superhero is a way of binding the cultural anxiety of the day, whether of the "distant other" in the 1940s, radioactivity in the 1960s, or genetic manipulation today.

The original incarnations of various superheroes in the late 1930s and 1940s were products of their time and reflected life during the Great Depression and World War II. After the war, soldiers who had acquired a comic-book-reading habit while overseas continued to buy comics after returning to the States, and certain publishers catered to this older audience with adult-themed stories featuring more graphic violence. Some of the young comic-book writers and artists had also been drafted into the armed services, and their wartime experiences resulted in a more serious, and in some cases darker, tone to their postwar work. In 1945, Maxwell Gaines ended his association with National Comics and started a new publishing firm called Educational Comics, printing such titles as *Picture Stories from Science*, *Picture Stories from American History*, and *Picture Stories from the Bible*. After his untimely passing in 1947, his son William Gaines changed the firm's name to Entertaining Comics (EC) and shifted their inventory to such comics as *Tales from the Crypt*, *Crime SuspenStories*, *Weird Science-Fantasy*, and *The Vault of Horror*. These comics were neither suitable nor intended for the same young audience as Captain Marvel. It was just a matter of time before someone noticed and complained.

Dr. Fredric Wertham's 1953 bestselling book, *Seduction of the Innocent*, forcefully argued that such lurid stories corrupted the minds of young children, leading them directly to careers as juvenile delinquents. In a cycle that appears to repeat itself in every generation, there was a growing concern among parents and authority figures in the post–World War II era over the coarsening effects of popular culture on the attitudes and mores of teenagers.

The U.S. Senate Subcommittee on Juvenile Delinquency, headed by the ambitious Sen. Estes Kefauver, held hearings on the connection between comic books and teenage crime. Initially, the committee intended to focus solely on horror and crime comics, but Wertham, a consultant to the subcommittee, brought superhero comics to the senators' attention. Seeking to avoid the imposition of federal oversight and regulation, the major comic-book publishers created a self-regulatory agency called the Comics Code Authority (CCA). The publishers developed a set of rules governing acceptable comic-book content, with explicit instructions that gore, lewdness, drug use, zombies, and vampires were prohibited in any comic book bearing the Comics Code Authority seal of approval on its cover. Many of the guidelines created by the CCA seemed designed solely to ensure the destruction of nearly the entire EC line of comics (the only survivor being a relatively new comic satire magazine by the name of *MAD*). All comic-book stories had to be submitted to the CCA (whose staff was funded by the publishers) for approval before being published, similar to the current ratings board that vets motion pictures.

While it played an important role in the 1950s and 1960s in convincing parents that comic books could be viewed as "wholesome" entertainment, the influence of the Comics Code Authority has waned over the years as the average age of the typical comic-book reader has increased. This is reflected in the decreasing size of the CCA seal on comic-book covers. In 1964, the seal took up an area roughly that of postage stamp, or two thirds of a square inch (as it was a prominent marketing tool to assure parents that the story contained within was acceptable for their children), while it was less than a quarter of a square inch in 1984 and is a barely detectible tenth of a square inch in 2004 (for DC Comics, that is; Marvel Comics quit its participation in the CCA in 2001 and employs a self-determined labeling system roughly equivalent to the PG, PG-13, and R ratings used by movies).

Declining sales from the loss of a major distribution network and the competition from television led to the near collapse of the comic-book industry, and from 1953 to 1956, only about a half-dozen superhero comics continued to be published, a dramatic reduction from the 130 different superhero titles available at newsstands at the peak of the Golden Age. Funny animal stories, Westerns, and young-romance comics were safer alternatives for the few companies that persevered in publishing comics during this period.

In 1956 National Comics decided to test the superhero waters with the reintroduction of the Flash in *Showcase # 4*. The sales figures for each issue of *Showcase* that featured the Flash indicated that the market for superheroes had returned, and over the next few years, National brought back new versions of the Green Lantern, the Atom, Hawkman, and others. The Silver Age of superhero comic books had begun, and superheroes have remained a mainstay of comic books ever since.

From the very beginning in *Showcase # 4*, examples of correctly applied physics principles appeared in these stories. With the launching of the Soviet satellite Sputnik in 1957 at the height of the Cold War, there was considerable anxiety over the quality of science education that American schoolchildren were receiving. The Comics Code Authority seal on their covers plus the inclusion of science concepts may have convinced wary parents that there was a net positive benefit to these four-colored adventures.

In addition to employing accurate science, comics from the Silver Age often had scholarly nuggets from other learned disciplines buried within their stories. For example, the plot of "The Adventure of the Cancelled Birthday" in *The Atom # 21* (written by Gardner Fox, who was both a lawyer and a writer for science-fiction pulp magazines) revolved around the obscure fact that in 1752, when Great Britain adopted the Gregorian calendar to replace the Julian calendar, eleven days were omitted during the transition. That is, September 2, 1752, was followed the next day by September 14, in order to regularize the British calendar with other parts of Europe. Two issues later, the letters column in *The Atom* printed a complaint from one such fan, arguing over the poor choice of historical characters, such as the obscure Justice Fielding. The editor of Atom comics, Julius "Julie" Schwartz, responsible for reintroducing the Flash in 1956, defended the story in the letters column, pointing out that it was high time that the reader become acquainted, as had the Atom, with Henry Fielding, the author of *Tom Jones*.

Even if they were not woven into the plot, bits of historical or scientific trivia would occasionally pop up in comic-book adventures through the appearance of a caption box containing a fact that had no direct bearing on the story at hand. For example, in *The Brave and the Bold # 28* featuring the first appearance of the Justice League of America, an alliance of National Comics' super-heroes, Aquaman swims by a puffer fish and has a brief conversation with him using his "fish telepathy." The puffer fish relates some crucial information gleaned while floating on the surface of the ocean. A caption in this panel informs us that "by swallowing air into a special sac beneath its throat, the puffer fish becomes inflated like a football—whereupon it rises to the surface and floats upside down."

Why take the time to include these educational captions? They may have been a consequence of the habits of the former pulp-fiction writers penning these tales. Prior to editing comic books at National, Mort Weisinger and Julie Schwartz, lifelong fans of science fiction, had been literary agents for science-fiction and fantasy writers, including Ray Bradbury, Robert Bloch, and H. P. Lovecraft. Certain comic-book writers had previously made their living as pulp-science-fiction writers. As such, they were walking storehouses of obscure historical and natural knowledge. The Hugo Award winner Alfred Bester, author of science-fiction classics *The Demolished Man* and *The Stars, My Destination*, also wrote comics during the 1940s and penned the original Green Lantern oath. In an autobiographical essay, Bester tells of spending hours browsing through reference books in the New York Public Library searching for odd historical tidbits around which he could construct a story. Knowing a lot of trivia could also help these pulp-fiction writers' financial bottom line, as these authors were paid by the word. Consequently they would frequently pad their work with all sorts of barely relevant tangents, as reflected in this joke:

Q: How many pulp-fiction writers does it take to change a lightbulb?

A: The history of the lightbulb is a long and interesting tale, beginning in 1879 in the quiet town of Menlo Park, New Jersey, and continuing on to the present day.

While the Silver Age comic-book writers may have had an economic incentive to be verbose, it is also likely that they were motivated by considerations of self-preservation to inject educational elements into their stories. As mentioned above, the introduction of science facts and principles into these stories may have arisen from a genuine desire on the part of the writers and editors to educate, or perhaps simply to avoid any further congressional attention.

A PHYSICIST READS A COMIC BOOK

Reading classic and contemporary superhero comic books now, with the benefit of a Ph.D. in physics, I have found many examples of the correct description and application of physics concepts. Of course, nearly without exception, the use of superpowers themselves involves direct violations of the known laws of physics, requiring a deliberate and willful suspension of disbelief. However, many comics needed only a single "miracle exception"—one extraordinary thing you have to buy into—and the rest that follows as the hero and villain square off would be consistent with the principles of science. While the intent of these stories has always primarily been to entertain, if at the same time the reader was also educated, either deliberately or accidentally, this was a happy bonus.

It is these happy bonuses, such as the one illustrated in fig. 2, that I wish to consider here. In this book, I'll present an overview of certain scientific principles, using examples of their correct application as found in comic books. I will describe characters and situations that illuminate various physics concepts, rather than systematically considering the physics underlying an array of superheroes. (Consequently, it is conceivable that your favorite superhero may not be discussed. I know that several of my own favorites didn't make the cut.) By the end of this book you will have been exposed to the key concepts in an introductory physics class, with a little bit of upper-level quantum mechanics and solid-state physics thrown in for fun. By examining the physical principles underlying certain comic-book adventures, we will at the same time gain an understanding of the mechanisms behind many real-world applications, from television to telephones to stellar nucleosynthesis of the elements.

I will focus primarily, but not exclusively, on the Silver Age period in comic-book history (from the reintroduction of the Flash in *Showcase # 4* in 1956 to the death of Gwen Stacy in *The Amazing Spider-Man # 121* in 1973) because the writers of this period made more of an effort than those in the Golden Age to incorporate scientific principles into their stories. In addition, the Silver Age characters have demonstrated lasting popularity, and their iconic status will make it easier to refer to their exploits without forcing the reader to constantly consult the back-issue bins of their local comic-book shop. It is all too easy to find flaws and errors in the science referenced in comic-book stories, and this is not the aim of this book. In addition to being unsporting and uncharitable (after all, these stories were never intended to function as science textbooks, despite the occasional student's attempts at surreptitious substitution), it is more difficult to make a point when the only illustrative examples are negative ones. Nevertheless, sometimes we will find that some scenes in comic books are simply not physically plausible, even with the granting of a "miracle exception."

Before I begin I would like to say a few words about a common misconception concerning physicists. Despite the impression gleaned from popular movies, being a physicist does not require an encyclopedic knowledge of equations and fundamental constants, coupled with the ability to perform complex arithmetic in one's head with robotic speed and precision. Physics is not about having memorized all the answers, but rather about asking the right questions. For when the right question is posed of a phenomenon, either the answer becomes clear or at least a path to further and more fruitful questioning is revealed.

To illustrate that asking one right question can be more important than a bushel full of correct answers, consider the simple physics experiment of tossing a ball in an arc. There are many questions we may ask, such as: How high does the ball travel? How far to the right does it move? How long is it in the air? How fast is it going? What is the geometric shape of its path? However, I would argue that there is one simple question that implies all of the above questions and gets to the heart of the issues concerning the ball's motion. That one single question is the following: Does the ball have any choice? If the ball does not have any choice in its motion, if it lacks free will, then its trajectory is completely determined by forces external to itself. Once we determine the nature of these forces and how they influence the ball's motion, we may then calculate the path of the ball for a given initial velocity imparted by the thrower. This calculated trajectory would then contain any and all information we may desire regarding how high the ball rises, how far it moves, its time in flight, what its velocity is, and so on. If we then repeat the toss with exactly the same initial position and velocity as before, then the ball must exactly and faithfully trace out the calculated trajectory, for the ball does not have any choice in the matter.

This is the beauty and attraction of physics, at least for those of us lucky enough to make our living from its study. The promise and potential is that if we can determine the forces acting on an object and how these forces influence the object's motion, we will then be able to predict the development of future events. By performing careful experiments, these predictions can be empirically tested and, if correct, confirm our understanding of how nature operates. On the other hand, if the experiment contradicts our model (a far more likely outcome, initially), we modify our equations and try again, using the failed test as an important clue as to what was missing from the initial calculation.* In this way, our understanding of nature progresses until we have a valid model, which is then termed a theory. To dismiss any idea that survives this exhaustive vetting as "just a theory" is equivalent to describing the Hope Diamond as "just a crystal."

Scientific knowledge comes only at the price of increased doubt: The more we learn, the more clearly we see all that remains uncertain. Doubt is to be embraced in science, for the only answers we can trust are those that survive the crucible of questioning and experimental testing. I hope to share with you in this book the true pleasure that comes from seeing how the asking

---

* The accumulated body of knowledge about the world is now so vast that physicists are able to make continued progress only by specializing in either experimental or theoretical research. Experimentalists work in laboratories and carry out measurements, while theoreticians perform calculations and computer simulations. I am an experimentalist, while Stephen Hawking is a theorist (the differences begin there). One of the last physicists who truly excelled at both experimental and theoretical research was Enrico Fermi.

of a few key questions can lead to a wealth of answers about the world we live in.

I begin, as do all standard textbooks in freshman physics, with the fundamental laws of motion as first described by Isaac Newton. Fitting such an original and profound contribution to modern thought, our first comic-book example involves an equally seminal contribution to Western civilization. I refer, of course, to the first true comic-book superhero, faster than a speeding bullet, more powerful than a locomotive, and, most relevant to our next discussion, able to leap tall buildings in a single bound.

SECTION
1

MECHANICS

UP, UP, AND AWAY—

FORCES AND MOTION

AS DESCRIBED IN *Superman # 1*, Jor-El, a scientist on the distant planet Krypton, discovers that his world is about to explode and kill its entire population. Possessing only a small prototype rocket ship, he and his wife elect to save their infant son, Kal-El, sending him to Earth so that he will not share their fate.* After traveling great distances through the vastness of space, the rocket crash-lands on Earth with its sole passenger none the worse for wear. Discovered by the childless Kansas farmers the Kents, Kal-El is immediately given up to an orphanage. Following a change of heart, the Kents return to the orphanage (where the superbaby has been wreaking havoc) whereupon they adopt Kal-El, name him Clark, and raise him as their own human son. As Kal/Clark Kent grows into adulthood, he develops a series of extraordinary abilities with which he fights the never-ending battle for Truth, Justice, and the American Way.

In his first Golden Age incarnation, Superman's powers differed significantly from those we associate with him today. He could lift a car over his head, for example, but not a continent. He was fast, able to outrace a "streamlined train," but not a light beam. And he could not fly, but simply leap great distances (one eighth of a mile was his originally stated range).

---

* As a father myself, I can certainly empathize with Jor-El. Many are the times I've been tempted to put my own kids in a rocket ship and send them off into deep space.

Jerry Siegel and Joseph Shuster's original conception of Superman was that of a pulp action hero with a liberal dose of science fiction added to lend an air of plausibility for their hero's great strength. The source of Superman's powers on Earth was credited in the Golden Age to his Kryptonian heritage, specifically the fact that his home planet had a far stronger gravity than Earth's. For example, the moon's much smaller size compared with Earth results in a weaker gravitational field, so objects on the moon weigh less than they do on Earth. Consequently, an Earthman, whose muscles and bones are adapted to Earth's gravity, is able to lift moon cars overhead and leap moon buildings in a single bound. Similarly Superman's great strength ("more powerful than a locomotive") and tougher skin ("nothing less than a bursting shell" could pierce it) resulted from his relocating to a planet with a far weaker gravity than Krypton's. (Even though Superman was sent to Earth as an infant, presumably his Kryptonian DNA had been encoded for the development of muscles and bones that were suited to a stronger gravitational field.)

By the late 1940s, Superman would gain the power of true flight, able to choose and alter his trajectory after leaving the ground.[*] At this point, Superman can be considered to have gained free will over the laws of physics. Over time, he acquired a host of other abilities that could not be reasonably accounted for by the stronger gravity of his home planet. These powers included various visions (heat, X-ray, and others), super-hearing, super-breath, and even super-hypnotism.[†]

The origin of Superman's powers was subsequently revised, beginning in the March 1960 issue of *Action Comics # 262*, to

---

[*] Superman first flew in his cartoon adventures, animated by the Fleischer Studios and then Famous Studios from 1941 to 1943. It took too long and was too costly to animate Superman constantly crouching and leaping, and the animators petitioned to allow Superman to just fly. Slowly and somewhat inconsistently, this power migrated over to the comic book adventures of the Man of Tomorrow.

[†] This last power was introduced to explain why only a simple pair of eyeglasses created such a perfect disguise that no one ever realized that mild-mannered reporter Clark Kent and the world-famous Superman were the same person. As described in *Superman # 330*, Superman apparently subconsciously hypnotizes everyone who sees him into believing that his face is markedly different from Clark Kent's.

claim that Superman's fantastic abilities derived from the fact that the Earth orbited a yellow sun, as opposed to the red sun of Krypton. The color of a sun is a function of both its surface temperature and the atmosphere through which it is viewed. The blue portion of the solar spectrum is strongly scattered by the atmosphere, which is why the sky looks blue. Viewed straight on, our sun appears yellow because the atmosphere is also more absorptive toward the blue end of the spectrum, except at dawn or sunset when the position of the sun is low on the horizon and sunlight must travel a greater distance through the atmosphere. Nearly all wavelengths are then absorbed, except for lower-energy red light, which gives sunsets their characteristic hues (the greater number of particulates in the air at the end of the day compared with the beginning also contributes to the difference in shading between sunset and dawn). These spectral features are for the most part independent of the chemical composition of the gases making up the Earth's atmosphere. There is no physical mechanism by which a shift in the primary wavelength of sunlight from yellow (a wavelength of 570 nanometers, or 570 billionths of a meter) to red light (650 nanometers) would endow someone with the ability to bend steel in his bare hands. Consequently, at this stage in his history, *Superman* ceased being a science-fiction strip and became a comic book about a fantasy hero. Changing a superhero's origin in order to accommodate new powers or circumstances occurs so frequently in comic books that comic-book fans have coined a term, "retconning," to describe this retroactive continuity repair.

Interestingly, Superman's foes went through a similar evolution around this same time. In the early years of Action and Superman comics, Siegel and Shuster gave voice to the revenge fantasies of their young and economically disadvantaged Depression-era readers. Superman first used his powers to fight corrupt slumlords, coal-mine owners, munitions manufacturers, and Washington lobbyists. In his very first story, he psychologically tormented a lobbyist by holding him as they both fell from a tall building. At this early stage of his career, the story lines indicated that only a few people knew of Superman's existence, and the lobbyist believed that the fall would be fatal. He willingly divulged the information Superman was after rather than risk another such fall. In the 1940s and 1950s, in addition to selling millions of comics per month, Superman had become a star of radio serials, movie shorts (both animated and live action), and a popular television program. His adversaries subsequently morphed into criminal masterminds with colorful personas and costumes, such as the Toyman, the Prankster, and Lex Luthor, whose schemes for grand larceny or world domination Superman would foil while keeping the corporate power structure safely undisturbed. As befits the escalating capabilities of the villains he faced, Superman entered a superpower arms race, eventually growing so powerful that it became difficult for writers to concoct credible threats to challenge his godlike abilities. Radioactive fragments of his home planet, known as Kryptonite, became a frequent device to extend any given story beyond the first page of the comic.*

It is the simpler, original Superman of the Golden Age, the last son of Krypton, that I wish to consider here.

THE PHYSICS OF JUMPING AND ALL OTHER MOTION

In the first years of his comic-book history, Superman was unable to fly but could simply "leap tall buildings in a single bound," thanks to Earth's weaker gravity.

Well, how high could he leap? According to his origin story in *Superman # 1*, Superman's range was given as one-eighth of a mile, or 660 feet. Assuming he could jump this high straight up, this is approximately equivalent to the height of a thirty- to forty-story building, which in 1938 would have been considered quite tall. So our question can be rephrased as: What initial velocity would Superman need, lifting off from the sidewalk, to vertically rise 660 feet?

---

* How a radioactive element from his native planet would affect Superman so strongly, while he remained immune to radioactive isotopes found on Earth, is more an issue of contract negotiations than physical plausibility. Kryptonite was first introduced in the radio serial *Adventures of Superman* in 1943, or so the story goes, when the overworked radio actor portraying the Man of Steel wanted a vacation. The radio scriptwriters created Superman's mineral nemesis so that another actor could portray the stricken superhero by groaning into the microphone. Several years later, the comic-book writers adopted and adapted this creative device, and a rainbow of Kryptonite (green, red, gold, silver, and others) with a broad range of effects on Superman was introduced into the comic-book universe.

Whether we wish to describe the trajectory of a leaping Man of Steel or of the tossed ball from our earlier example in the Introduction, we employ the three laws of motion as first elucidated by Isaac Newton in the mid-1600s. These laws are frequently expressed as: (1) an object at rest remains at rest or, if moving, keeps moving in a straight line if no external forces act upon it; (2) if an external force is applied, the object's motion will change in either magnitude or direction, and the rate of change of the motion (its acceleration) when multiplied by the object's mass is equal to the applied force; and (3) for every force applied to an object, there is an equal and opposite force exerted back by the object. The first two laws can be expressed succinctly through one simple mathematical equation:

$$\textbf{FORCE} = (\textbf{MASS}) \times (\textbf{ACCELERATION})$$

That is, the force F applied to an object is equal to the resulting rate of change in the object's velocity (its acceleration a) when multiplied by the object's mass m, or $\textbf{F} = \textbf{ma}$.

Acceleration is a measure of the rate of change of the velocity of an object. A car starting from rest (velocity = 0) and accelerating to 60 mph would have a change in velocity of 60 mph − 0 mph = 60 mph. The acceleration is found by dividing this change in velocity by the time needed to make the change. The longer the time, the lower the acceleration needed for a given change in speed. An automobile speeding up from 0 to 60 mph in six seconds will have a much larger acceleration than if it does so in six hours or six days.* The final speed will be the same for all three cases, namely, 60 mph, but the accelerations will be radically different owing to the different times needed to affect this change in velocity. From Newton's $\textbf{F} = \textbf{ma}$, the force needed to create the former, faster acceleration is obviously much larger than for the latter, slower case.

When the acceleration is zero, there is no change in the motion. In that case, a moving object keeps moving in a straight line, or if sitting still, remains so. From the expression $\textbf{F} = \textbf{ma}$, when a = 0, then the force F = 0, which is the whole point of Newton's first law of motion.

---

* You never forget your first car!

While this may be straightforward from a mathematical point of view, from a common-sense perspective it is nothing short of revolutionary. Newton is saying (correctly) that if an object is moving, and there is no net outside force acting on it, then the object will simply continue moving in a straight line. However, you and I, and Isaac Newton for that matter, have never seen this occur! Our everyday experiences tell us that to keep something moving, we must always keep pulling it or pushing it with an external force. A car in motion does not remain in motion unless we keep pressing the accelerator pedal, which ultimately provides a force. Of course, the reason that moving objects slow down and come to rest when we stop pushing or pulling them is that there are forces of friction and air resistance that oppose the object's motion. Just because we stop pulling or pushing does not mean, in the real world, that there are no forces acting on the object. There's nothing wrong with Newton's laws—we just have to make sure we account for friction and air resistance when applying them. It is these unseen "drag forces" that we must overcome in order to maintain uniform motion. Once our pulling or pushing exactly balances the friction or air drag, then the net force on the object is zero, and the object will then continue in straight-line motion. Increasing the push or pull further will yield a net nonzero force in the direction of our push or pull. In this case, there will be an acceleration proportional to the net force. The constant of proportionality connecting the force to the acceleration is the mass, **m**, reflecting how much the object resists changing its motion.

It is worth pointing out here that mass is not the same as weight. "Weight" is another term for "force on an object due to gravity." Mass, on the other hand, is a measure of how much stuff ("atoms" for you specialists) an object contains. The mass of the atoms in an object is what gives it its "inertia," a fancy term to describe its resistance to change when a force is applied. In outer space, an object's mass is the same as it is on the Earth's surface, because the number and type of atoms it contains does not change. An object in outer space may be "weightless," in that it is subject to a negligible attractive force from nearby planets, but it still resists changes in motion, due to its mass. A space-walking astronaut in deep space cannot just pick up and toss a space station around (assuming she had a platform on which to stand), even though the station and everyone on it is "weightless." The mass of the space station is so large that the force the astronaut's muscles can apply produces only a negligible acceleration.

For objects on the Earth's surface (or that of any other planet, for that matter), the acceleration due to gravity is represented by the letter g (we'll discuss this more in a moment). The force that gravity exerts on the object of mass, m, is then referred to as its Weight. That is, **Weight = (mass) × (acceleration due to gravity)** or **W = mg**, which is just a restatement of **F = ma** when **a = g**. Mass is an intrinsic property of any object, and is measured in kilograms in the metric system, while Weight represents the force exerted on the object due to gravity, and is measured in pounds in the United States. In Europe, Weight is commonly expressed in units of kilograms, which is not strictly correct, but easier to say than "kilogram–meter/sec^2," the unit of force in the metric system (also known as a "Newton"). When a weight in the metric system is compared to one in the United States, the conversion ratio is 1 kilogram is equivalent to 2.2 pounds. I say "equivalent" and not "equal" because a pound is a unit of force, while kilograms measure mass. An object will weigh less than 2.2 pounds on the moon and more than 2.2 pounds on Jupiter, but its mass will always be 1 kilogram. When calculating forces in the metric system, we'll stick with kg-meter/sec^2 rather than "Newtons," in order to remind ourselves that any force can always be described by **F = ma**.

To recap, Superman's mass at any given moment is a constant, because it reflects how many atoms are in his body. His weight, however, is a function of the gravitational attraction between him and whatever large mass he is standing on. Superman has a larger weight on the surface of Jupiter, or a lesser weight on the moon, compared with his weight on Earth, but his mass remains unchanged. The gravitational attraction of a planet or moon decreases the farther away one moves from the planet, though technically it is never exactly zero unless one were infinitely far from the planet. It is tempting to equate mass with weight, and easy to do so when dealing only with objects on Earth for which the acceleration due to gravity is always the same. As we will soon be comparing Superman's weight on Krypton to that on Earth, we will resist this temptation.

Finally, the third law of motion simply makes explicit the commonsense notion that when you press on something, that thing presses back on you. This is sometimes expressed as "For every action, there is an equal and opposite reaction." You can only support yourself by leaning on the wall if the wall resists you—that is, pushes back with an equal and opposite force. If the force were not exactly equal and in the opposite direction, then there would be a net nonzero force, which would lead to an acceleration and you crashing through the wall. When the astronaut mentioned above pushes on the space station, the force her muscles exert provides a very small acceleration to the station, but the station pushes back on her, and her acceleration is much larger (since her mass is much smaller).

Imagine Superman and the Hulk holding bathroom scales against each other (which are simply devices to measure a force, namely, your weight due to gravity). When they press against each other's scale, no matter how hard Superman pushes on the left, if they remain stationary, then the Hulk's scale on the right will read exactly the same force. Moreover, no matter how hard Superman is pushing, his scale will read zero pounds of force if the Hulk offers no resistance and just moves his scale out of the way and steps aside.* Forces always come in pairs, and you cannot push or pull on something unless it pushes or pulls back. When you stand on the sidewalk, your feet exert a force on the ground due to gravity pulling you toward the center of the Earth. People on the opposite side of the planet do not fall off because gravity pulls everyone in toward the center of the planet, regardless of where they are located. You do not accelerate while standing; the ground provides an equal and opposite force exactly equal to your weight. During the brief moment when Superman jumps, his legs exert a force greater than just his normal standing weight. Because forces come in pairs, his pushing down on the pavement causes the pavement to push back on him. Thus, he experiences an upward force lifting him up and away.

And that's it—all of Newton's laws of motion can be summarized in two simple ideas: that any change in motion can only

---

* The Hulk is brighter than everyone gives him credit for (his alter ego is a physicist, after all).

result from an external force ($\mathbf{F} = \mathbf{ma}$), and that forces always come in pairs. This will be all we need to describe all motions, from the simple to the complex, from a tossed ball to the orbits of the planets. In fact, we already have enough physics in hand to figure out the initial velocity Superman needs to leap a tall building.

IN A SINGLE BOUND

Superman starts off with some large initial velocity (fig. 4). At the top of his leap, a height h = 660 feet above the ground, his final velocity must be zero, or else this wouldn't be the highest point of his jump, and he would in fact keep rising. The reason Superman slows down is that an external force, namely gravity, acts on him. This force acts downward, toward the surface of the Earth, and opposes his rise. Hence, the acceleration is actually a deceleration, slowing him down, until at 660 feet, he comes to rest.

Imagine ice-skating into a strong, stiff wind. Initially you push off from the ice and start moving quickly into the wind. But the wind provides a steady force opposing your motion. If you do not push off again, then this steady wind slows you down until you are no longer moving and you come to rest. But the wind is still

Fig. 4. Panel from Superman # 1 (*June 1939*) *showing Superman in the process of leaping a...well, you know.*

pushing you, so you still have an acceleration and now start sliding backward the way you came, with the wind. By the time you
reach your initial starting position, you are moving as fast as when
you began, only now in the opposite direction. This constant wind
in the horizontal direction affects an ice-skater the same way gravity acts on Superman as he jumps. The force of gravity is the same
at the start, middle, and highest points of his leap. Since $F = ma$,
his acceleration is the same at all times as well. In order to determine what starting speed Superman needs to jump 660 feet, we
have to figure out how his velocity changes in the presence of a
uniform, constant acceleration g in the downward direction.

As common sense would indicate, the higher one wishes to
leap, the faster the liftoff velocity must be. How, exactly, are the
starting speed and final height connected? Well, when you take a
trip, the distance you travel is just the product of your average
speed and the length of time of the trip. After driving for an hour
at an average speed of 60 mph, you are 60 miles from your starting
point. Because we don't know how long Superman's leap lasts, but
only his final height of h = 660 feet, we perform some algebraic
manipulation of the definition of acceleration as the change in
speed over time and that velocity is the change in distance over
time. When the dust settles, we find that the relationship between
Superman's initial velocity v and the final height h of his leap is
$v \times v = v^2 = 2gh$. That is, the height Superman is able to jump
depends on the square of his liftoff velocity, so if his starting speed
is doubled, he rises a distance four times higher.

Why does the height that Superman can leap depend on the
square of his starting speed? Because the height of his jump is
given by his speed multiplied by his time rising in the air, and the
time he spends rising also depends on his initial velocity. When
you slam on your auto's brakes, the faster you were driving, the
longer it takes to come to a full stop. Similarly, the faster
Superman is going at the beginning of his jump, the longer it takes
gravity to slow him down to a speed of zero (which corresponds to
the top of his jump). Using the fact that the (experimentally
measured) acceleration due to gravity g is 32 feet per second per
second (that is, an object dropped with zero initial velocity has a
speed of 32 feet/sec after the first second, 64 feet/sec after the next
second, and so on) the expression $v^2 = 2gh$ tells us that Superman's
initial velocity must be 205 feet/sec in order to leap a height of

660 feet. That's equivalent to 140 miles per hour! Right away, we can see why we puny Earthlings are unable to jump over skyscrapers, and why I'm lucky to be able to leap a trash can in a single bound.

In the above argument, we have used Superman's average speed, which is simply the sum of his starting speed (v) and his final speed (zero) divided by two. In this case his average speed is v/2, which is where the factor of two in front of the **gh** in $v^2 = 2gh$ came from. In reality, both Superman's velocity and position are constantly decreasing and increasing, respectively, as he rises. To deal with continuously changing quantities, one should employ calculus (don't worry, we won't), whereas so far we have only made use of algebra. In order to apply the laws of motion that he described, Isaac Newton had to first invent calculus before he could carry out his calculations, which certainly puts our difficulties with mathematics into some perspective. Fortunately for us, in this situation, the rigorous, formally correct expression found using calculus turns out to be exactly the same as the one obtained using relatively simpler arguments, that is, $v^2 = 2gh$.

How does Superman achieve this initial velocity of more than 200 feet/sec? As illustrated in fig. 5, he does it through a process that physicists term "jumping." Superman crouches down and applies a large force to the ground, and the ground pushes back (since forces come in pairs, according to Newton's third law). As one would expect, it takes a large force in order to jump up with a starting speed of 140 mph. To find exactly how large a force is needed, we make use of Newton's second law of motion, **F = ma**— that is, Force is equal to mass multiplied by acceleration. If Superman weighs 220 pounds on Earth, he would have a mass of 100 kilograms. So to find the force, we have to figure out his acceleration when he goes from standing still to jumping with a speed of 140 mph. Recall that acceleration describes the change in velocity divided by the time during which the speed changes. If the time Superman spends pushing on the ground using his leg muscles is ¼ second,* then his acceleration will be the change in speed of 200

---

* Experts will argue, correctly, that this value for the time spent pushing against the ground is too long, and should be at least ten times smaller. We will address this point in the next chapter. Have patience, Fearless Reader!

Fig. 5. Panels from Action Comics # 23, *describing in some detail the process by which Superman is able to achieve the high initial velocities necessary for his mighty leaps.*

feet/sec divided by the time of ¼ second, or 800 feet/sec² (approximately 250 meters/sec² in the metric system, because a meter is roughly 39 inches). This acceleration would correspond to an automobile going from 0 to 60 mph in a tenth of a second. Superman's acceleration results from the force applied by his leg muscles to get him airborne. The point of $F = ma$ is that for any change in motion, there must be an applied force and the bigger the change, the bigger the force. If Superman has a mass of 100 kilograms, then the force needed to enable him to vertically leap 660 feet is $F = ma = (100$ kilograms$) \times (250$ meters/sec²$) = 25,000$ kilograms meters/sec², or about 5,600 pounds.

Is it reasonable that Superman's leg muscles could provide a force of 5,600 pounds? Why not, if Krypton's gravity is stronger than Earth's, and his leg muscles are able to support his weight on Krypton? Suppose that this force of 5,600 pounds is 70 percent larger than the force his legs supply while simply standing still, supporting his weight on Krypton. In this case, Superman on his home planet would weigh 3,300 pounds. His weight on Krypton is determined by his mass and the acceleration due to gravity on Krypton. We assumed that Superman's mass is 100 kilograms, and this is his mass regardless of which planet he happens to stand on. If Superman weighs 220 pounds on Earth and nearly 3,300 pounds on Krypton, then the acceleration due to gravity on Krypton must have been fifteen times larger than that on Earth.

So, just by knowing that $F = ma$, making use of the definitions "distance = speed × time" and "acceleration is the change in speed over time," and the experimental observation that Superman can "leap a tall building in a single bound," we have figured out that the gravity on Krypton must have been fifteen times greater than on Earth.

Congratulations. You've just done a physics calculation!

2

DECONSTRUCTING KRYPTON—

NEWTON'S LAW OF GRAVITY

NOW THAT WE HAVE DETERMINED THAT in order for Superman to leap a tall building, he must have come from a planet with a gravitational attraction fifteen times that of Earth, we next ask: How would we go about building such a planet? To answer this, we must understand the nature of a planet's gravitational pull, and here again we rely on Newton's genius. What follows involves more math, but bear with me for a moment. There's a beautiful payoff that explains the connection between Newton's apple and gravity.

As if describing the laws of motion previously discussed and inventing calculus weren't enough, Isaac Newton also elucidated the nature of the force that two objects exert on each other owing to their gravitational attraction. In order to account for the orbits of the planets, Newton concluded that the force due to gravity between two masses (let's call them **Mass 1** and **Mass 2**) separated by a distance d is given by:

$$\text{FORCE DUE TO GRAVITY} = (G) \times \frac{[\,(\text{MASS 1}) \times (\text{MASS 2})\,]}{(\text{DISTANCE})^2}$$

where G is the universal gravitational constant. This expression describes the gravitational attraction between any two masses, whether between the Earth and the sun, the earth and the moon or between the Earth and Superman. If one mass is the Earth and the other mass is Superman, then the distance between them is the radius of the Earth (the distance from the center of the Earth to the surface, upon which the Man of Steel is standing). For a spherically symmetric distribution of mass, such as a planet, the attractive force behaves as if all of the planet's mass is concentrated at a single point at the planet's core. This is why we can use the radius of the Earth as the distance in Newton's equation separating the two masses (Earth and Superman). The force is just the gravitational pull that Superman (as well as every other person) feels. Using the mass of Superman (100 kilograms), the mass of the Earth, and the distance between Superman and the center of the Earth (the radius of the Earth), along with the measured value of the gravitational constant in the previous equation, gives the force F between Superman and the Earth to be F = 220 pounds.

But this is just Superman's weight on Earth, which is measured when he steps on a bathroom scale on Earth. The cool thing is that these two expressions for the gravitational force on Superman are the same thing! Comparing the two expressions for Superman's **weight = (Mass 1) × g** and the **force of gravity = (Mass 1) × [(G × Mass 2) /(distance)²]**, since the forces are the same and Superman's Mass 1 = 100kg is the same, then the quantities multiplying Mass 1 must be the same; that is, the acceleration due to gravity g is equal to (G × Mass 2)/d². Substituting the mass of the Earth for Mass 2 and the radius of the Earth for d in this expression gives us g = 10 meter/sec² = 32 feet/sec².

The beauty of Newton's formula for gravity is that it tells us why the acceleration due to gravity has the value it does. For the same object on the surface of the moon, which has both a smaller mass and radius, the acceleration due to gravity is calculated to be only 5.3 feet/sec²—about one sixth as large as on Earth.

This is the true meaning of the story of Isaac Newton and the apple. It certainly wasn't the case that in 1665 Newton saw an apple fall from a tree and suddenly realized that gravity existed, nor did he see an apple fall and immediately write down $F = G \, (m1 \times m2)/ \, (d)^2$. Rather, Newton's brilliant insight in the seventeenth century was that the exact same force that pulled the apple toward the Earth pulled the moon toward the Earth, thereby connecting the terrestrial with the celestial. In order for the moon to stay in a circular orbit around the Earth, a force has to pull on it in order to constantly change its direction, keeping it in a closed orbit.

Remember Newton's second law of **F = ma**: If there's no force, there's no change in the motion. When you tie a string to a bucket and swing it in a horizontal circle, you must continually pull on the string. If the tension in the string doesn't change, then the bucket stays in uniform circular motion. The tension in the string is not acting in the direction that the bucket is moving; consequently, it can only change its direction but not its speed. The moment you let go of the string, the bucket will fly away from you.

Back to the case of the moon. If there were no gravity, no force acting on it, then the moon would travel in a straight line right past the Earth. If there were gravity but the moon were stationary, then it would be pulled down and crash into our planet. The moon's distance from the Earth and its speed are such that they exactly balance the gravitational pull, so that it remains in a stable circular orbit. The moon does not fly away from us, because it is pulled by the Earth's gravity, causing it to "fall" toward the Earth, while its speed is great enough to keep the moon from being pulled any closer to us. The same force that causes the moon to "fall" in a circular orbit around the Earth, and causes the Earth to "fall" in an elliptical orbit around the sun, causes the apple to fall toward the Earth from the tree. And that same gravitational force causes Superman to slow in his ascent once he leaps, until he reaches the top of a tall skyscraper. Once we know that in order to make such a powerful leap, his body had to be adapted to an environment where the acceleration due to gravity is fifteen times greater than on Earth, that same gravitational force informs us about Krypton's geology.

One consequence of Newton's law of gravitation—which states that as the distance between two objects increases, the gravitational pull between them becomes weaker by the square of their separation—is that all planets are round. A sphere has a volume that grows with the cube of the radius of the orb, while its surface area increases with the square of the radius. This combination of the square of the radius for the surface area with the inverse square of the gravitational force leads to a sphere being the only stable form that a large gravitational mass can maintain. In fact, to address the astrophysical question of what distinguishes a very large asteroid from a very small planet, one answer is its shape. A small rock that you hold in your hand can have an irregular shape, as its self-gravitational pull is not large enough to deform it into a sphere. However, if the rock were the size of Pluto, then gravity would indeed dominate, and it would be impossible to structure the planetoid so that it had anything other than a spherical profile. Consequently, cubical planets such as the home world of Bizarro must be very small. In fact, the average distance from the center of the Bizarro planet to one of its faces can be no longer than 300 miles, if it is to avoid deforming into a sphere. However, such a small cubical planet would not have sufficient gravity to hold an atmosphere on its surface, and it would be an airless rock. Since we have frequently seen that the sky on the Bizarro world is blue like our own (and shouldn't it be some other color if it is to hold true to the Bizarro concept?), this would imply that there is indeed air on this cubical planet. We must therefore conclude that a Bizarro planet is not physically possible, no matter how many times we may feel in the course of a day that we have been somehow instantly transported to such a world.

Back to normal spherical planets like Krypton. If the acceleration due to gravity on Krypton g_K is fifteen times larger than the acceleration due to gravity on Earth g_E, then the ratio of these accelerations is $g_K/g_E = 15$. We have just shown that the acceleration due to gravity of a planet is $g = Gm/d^2$. The distance d that we'll use is the Radius R of the planet. The mass of a planet (or of anything for that matter) can be written as the product of its density (the Greek letter ρ is traditionally used to represent density) and its volume, which in this case is the volume of a sphere (since planets are round). Since the gravitational constant G must be the same on Krypton as on Earth, the ratio g_K/g_E is given by the following simple expression:

$$g_K / g_E = [\rho_K R_K] / [\rho_E R_E] = 15$$

where ρ_K and R_K represent the density and radius of Krypton and ρ_E and R_E stand for the Earth's density and radius, respectively. When comparing the acceleration due to gravity on Krypton to that on Earth, all we need to know is the product of the density and radius of each planet. If Krypton is the same size as Earth, then it must be fifteen times denser, or if it has the same density, then it will be fifteen times larger.

Now if, as we have argued at the start of this book, the essence of physics is asking the right questions, then it is as true in physics as it is in life that every answer one obtains leads to more questions. We have determined that in order to account for Super-man's ability to leap 660 feet (the height of a tall building) in a single bound on Earth, the product of the density and radius of his home world of Krypton must have been fifteen times greater than that of Earth. We next ask whether it is possible that the size of Krypton is equal to that of Earth ($R_K = R_E$) so that all of the excess gravity of Krypton can be attributed to its being fifteen times denser than Earth. It turns out that if we assume that the laws of physics are the same on Krypton as on Earth (and if we give up on that, then the game is over before we begin and we may as well quit now!), then it is extremely unlikely that Krypton is fifteen times denser than Earth.

We have just made use of the fact that mass is the density mul-tiplied by volume, which is just another way of saying that density is the mass per unit volume of an object. Now, to understand what limits this density, and why we can't easily make the density of Krypton fifteen times greater than Earth's, we have to take a quick trip down to the atomic level. Both the total mass of an object and how much volume it takes up are governed by its atoms. The mass of an object is a function of how many atoms it contains. Atoms are composed of protons and neutrons inside a small nucleus, surrounded by lighter electrons. The number of posi-tively charged protons in an atom is balanced by an equal number of negatively charged electrons. Electrons are very light compared to protons or neutrons, which are electrically uncharged particles that weigh slightly more than protons and reside in a nucleus. (We'll discuss what the neutrons are doing in the nucleus in Chap-ter 16.) Nearly all the mass of an atom is determined by the pro-tons and neutrons in its nucleus, because electrons are nearly two thousand times lighter than protons.

The size of an atom, on the other hand, is determined by the electrons or, more specifically, their quantum mechanical orbits. The diameter of a nucleus is about one trillionth of a centimeter, while the radius of an atom is calculated by how far from the nucleus one is likely to find an electron, and is about ten thousand times bigger than the nucleus. If the nucleus of an atom were the size of a child's marble (a diameter of 1cm) and placed in the end zone of a football field, the radius of the electron's orbit would extend to the opposite end zone, 100 yards away. The spacing between atoms in a solid is governed essentially by the size of the atoms themselves (you can't normally pack them any closer than their size).

Thus, if quantum mechanics is the same on Krypton as on Earth, the space taken up by a given number of atoms in a rock (for example) will not depend significantly upon which planet the rock resides on. The rock will weigh more on a planet with a larger gravity, but the number of atoms it contains—as well as the spacing between the atoms, both of which determine its mass density—will be independent of which planet the rock finds itself on. Because the number of atoms also determines the mass of the rock, it follows that the density of any given object will be the same, regardless of the planet of origin.

Most solid objects have roughly the same density, at least within a factor of ten. For example, the density of water is 1 gram/cm^3 while the density of lead is 11 gram/cm^3 (a gram is one thousandth of a kilogram). In other words, a cube that measures 1cm on each side would have a mass of 1 gram if composed of water and 11 grams if composed of lead. The higher density of lead is due almost entirely to the fact that a lead atom is ten times more massive than a water molecule. While there is a lot of water on the surface of the Earth, there's even more solid rock within the planet, so that Earth's average density is 5 gram/cm^3. In fact, Earth is the densest planet in our solar system, with Mercury and Venus close behind. Even if Krypton were solid uranium, it would have an average density of 19 gram/cm^3, which is not even four times larger than Earth's. In order for Krypton to have a gravity fifteen times greater than Earth's due to a larger density alone, it would have to have a density fifteen times larger than Earth's 5 gm/cm^3—that is, 75 gram/cm^3—and no normal matter is this dense.

If the density of planet Krypton couldn't be much greater than Earth's, perhaps the heavier gravity on Krypton is due to it being a larger planet—one with a radius fifteen times larger than Earth's. While planets in our own solar system come in all sizes—from Pluto, with a radius one fifth as large as Earth's, making it just barely bigger than some moons, to Jupiter, with a radius of more than eleven times Earth's—the geology of the planet is a sensitive function of its size. Planets bigger than Uranus, with a radius four times larger than Earth's, include Neptune, Saturn, and Jupiter. These planets are gas giants, lacking a solid mantle upon which buildings and cities may be constructed, let alone supporting humanoid life. In fact, if Jupiter were ten times larger, it would be the size of our own sun. In this case, the gravitational pressure at Jupiter's core would initiate nuclear fusion, the process that causes our sun to shine. So, if Jupiter were just a bit larger, it would no longer be a giant planet but rather a small star.

Big planets are gaseous because if you're going to build a very big planet, you are going to need a lot of atoms, and when you go to the great celestial stockroom, nearly all of the raw materials available are either hydrogen or helium gas. To be precise, 73 percent of the elemental mass in the universe is hydrogen and 25 percent is helium. Everything else that you would use to make a solid planet—such as carbon, silicon, copper, nitrogen, and so on—comprises only 2 percent of the elemental mass in the known universe. So big planets are almost always gas giants, which tend to have orbits far from a star, where the weaker solar radiation cannot boil away the gaseous surfaces they have accreted. The concentration of heavier elements with which solid planets can form is much lower, so they will tend to be smaller and closer to a star. If these inner solid planets got too large, the gravitational tidal forces* from their sun would quickly tear them apart. Krypton's advanced civilization, with scientists capable of constructing a rocket ship, couldn't arise on a gas giant with a radius fifteen times that of Earth's.

So, is that it? Is the story of Superman and Krypton, with an Earth-like surface and a gravity fifteen times that of Earth, totally bogus? Not necessarily. Remember that earlier it was stressed that no *normal* matter could be fifteen times denser than matter on Earth. However, astronomers have discovered exotic matter, with exceedingly high densities, formed from the remnants of supernova explosions. As mentioned, when the size of a gaseous planet exceeds a certain threshold, the gravitational compression at its center is so large that the nuclei of different atoms literally fuse together, creating larger nuclei and releasing excess energy in the process. The source of this energy is expressed in Einstein's famous equation, $E = mc^2$ or Energy E is equivalent to mass m

---

* This is the unbalanced force across the diameter of a planet arising from one side being closer to the sun than the other.

multiplied by the speed of light c squared. The mass of the fused-product nucleus is actually a tiny bit smaller than that of the two initial separate nuclei. The small difference in mass, when multiplied by the speed of light squared (a very big number) yields a large amount of energy. This energy radiates outward from the star's center, producing an outward flow that balances the inward attractive gravitational force, keeping the radius of the star stable.

When all the hydrogen nuclei have been fused into helium nuclei, some of the helium nuclei are in turn fused into carbon nuclei, some of which in turn are compressed to form nitrogen, oxygen, and all of the heavier elements, up to iron. The fusion process speeds up as the star generates heavier and heavier nuclei, so that all of its iron and nickel are created within the last week of the star's life. As heavier and heavier nuclei are combined, the process becomes less and less efficient, so that the energy released when iron nuclei fuse is insufficient to stably counteract the inward gravitational pressure. At this point gravity wins out, rapidly compressing the material into a much smaller volume. In this brief moment, the pressure at the center of the star is so high that one last gasp of fusion occurs, and heavier elements all the way up to uranium are generated, with a concurrent tremendous release of energy. This last stage in the life of the largest stars is termed the "supernova" phase. With this final blast of energy, the elements that had been synthesized within the star are flung out into space, where gravity may eventually pull them together into clumps that can form planets or other stars. Every single atom in your body, in the chair in which you are sitting, or the paper and ink in *Action Comics # 1*, was synthesized within a star that died and subsequently expelled its contents. We are all composed of stardust or, if you're feeling a tad more cynical, solar excrement.

For really big stars, the gravitational pressure at the center is so great that after the supernova phase, there remains a large remnant core for which gravity compresses the protons and electrons into neutrons, which are squeezed until they touch and become a solid, composed of nuclear matter. The remnants of such massive stars are termed "neutron stars," which are composed solely of neutrons and whose density is exceeded only by that of black holes (which are left over from the death of even bigger stars, whose gravitational attraction is so high that not even light can escape its pull). Compared with the density of lead (11 grams per cubic centimeter), the density of neutron-star material is just a tad higher: one hundred trillion grams per cubic centimeter. That is, a teaspoon of neutron-star material on Earth would weigh more than 100 million tons. This is just the stuff to boost the gravity of Krypton.

If a planet the size of Earth had a small volume of neutron star material within its core, the additional mass would dramatically increase the gravity on the surface of the planet. In fact, it would take only a sphere of neutron star material with a radius of 600 meters (about the length of six football fields) at the center of a planet the size of Earth to create an acceleration due to gravity on its surface of 150 meters/sec^2, whereas the acceleration due to gravity on Earth is 10 meters/sec^2. So, in order for Krypton's gravity to have been 15 times greater than Earth's, it must have had a core of neutron-star matter at its center.

And thus, we see why Krypton was doomed! For such a super-dense core would produce enormous strains on the surface of the planet, making a stable distribution of matter tenuous at best. At some point in the planet's history, volcanic activity and plate tectonics would result in massive upheavals. Such pre-shock earthquakes would alert an astute scientist that now would be a good time to put their kid into a rocket ship and send him to some other distant planet, preferably one without a neutron-star core.

Let us pause to admire the scientific insight of Jerry Siegel and Joe Shuster. These teenagers from Cleveland, Ohio, either had an understanding of astrophysics and quantum mechanics that exceeded that of many contemporary physics professors in 1938 or they were very lucky guessers. Only five years earlier, the astronomers Walter Baade and Fritz Zwicky had proposed the existence of neutron stars, the definitive evidence of which would not be found until the 1960s. Perhaps if Sheldon Mayer at National Publications had not taken a chance on their Superman strip, Siegel and Shuster might have considered publication in a scientific journal such as the *Physical Review*, and the history of both science and comic books would be very different today.

Now, in the last chapter I alluded to a question concerning Superman's acceleration when he lifted off the ground as he leapt a tall building in a single bound. We had calculated that in order to rise to a height of one eighth of a mile, he needed a liftoff speed of 200 ft/sec, or 140 miles per hour. If the time he spent pushing against the ground was one-quarter of a second, then this corresponded to an acceleration of 800 feet/sec^2, or twenty-five times the acceleration due to gravity.

Careful readers of the first edition have pointed out that I had overestimated the time that Superman spent jumping. The formula employed to determine his liftoff speed, that is, $v^2 = 2gh$, can also be used to find his acceleration as he leaps. The velocity in this case is still 200 feet/sec, but instead of g, we use his jumping acceleration. While before the height h was the eighth of a mile height of a building, now the height is the difference from his low crouch when he prepares to jump and when he is standing straight up as he first leaves the ground. At most, this distance might be three feet, so whereas earlier we estimated his jumping acceleration to be an incredible 800 feet/sec^2, this new calculation finds that his acceleration must be over 6,600 feet/sec^2. This corresponds to a time spent jumping of not one quarter of a second, but a much speedier 0.03 seconds. Which makes him quite super, indeed.

Using the longer jumping time of one quarter of a second led to a lower jumping acceleration and a corresponding force of 5,600 pounds that Superman's legs must supply during this leap. The shorter time of 0.03 seconds and larger jumping acceleration leads to a much larger force of just over 46,000 pounds. The same arguments as employed in Chapter 1 lead us to conclude that the acceleration due to gravity of Krypton g_K is an astounding 125 times greater than on Earth, compared with our earlier estimate of $g_K/g_E = 15$.

I used the longer time of a quarter second in Chapter 1 for two reasons. Informal experiments in my classes, where I asked students to leap and recorded the time they spent pushing against the ground, seemed consistent with a fraction of a second, and one quarter is roughly the average value obtained. Of course, none of my students were able to leap over a tall building. In Chapter 1, the intent was to explain how one finds an acceleration, knowing the time interval over which a velocity changes. More importantly, I wanted to use a lower jumping acceleration and leg force for the discussion in this chapter. If I had used a leaping time of 0.03 seconds and, upon concluding that the acceleration due to gravity of Krypton must have been 125 times that of Earth, I doubt few would be surprised that it is difficult to construct such a planet. A planet with a gravity fifteen times that of Earth's seems more plausible, and to show that this is also unlikely requires a discussion of the limits of the density of matter and the geology of large planets. My goal is to use the superhero's powers to illustrate real-world physics.

Occasionally I have been taken to task by a reader for contending that the acceleration due to gravity on Krypton is "only" fifteen times larger than that of Earth. I would explain my reasons, listing the above arguments. Some fans would protest: "Look, it's just not realistic that the time spent jumping is as long as a quarter of a second!" To which I responded that we were discussing the exploits of the last son of the imaginary planet Krypton, able to bend steel in his bare hands and change the course of mighty rivers, and the part that they found unrealistic was the time he spent pushing off against the ground while leaping a tall building in a single bound? Few would fail to concede the point at this stage. Not that I mind in the least being corrected and in fact am thrilled that readers are motivated enough to pick up pencil and calculator and keep me honest. That's the whole point, after all.

THE DAY GWEN STACY DIED—
IMPULSE AND MOMENTUM

IF A SENATE SUBCOMMITTEE HEARING marked the beginning of the end of the Golden Age of comics, the death blow of the Silver Age was self-inflicted. Viewed from today's perspective, comics from the Silver Age (from the late 1950s to 1960s) seem suffused with an optimistic outlook and a sunny disposition that borders on the Pollyannaish. The Golden Age characters reinvented by Julius Schwartz and colleagues at DC Comics in the late fifties and early sixties, such as the Flash, Green Lantern, or Green Arrow (an amalgam of Batman and Robin Hood, with a quiver full of gadget arrows such as a "boxing-glove arrow" or a "handcuff arrow," whose successful application violated several fundamental principles of aerodynamics), carried on the positive outlooks and righteousness of their antecedents, and their plot-driven twelve- or twenty-two-page-long stories did not leave much room for character development. A typical Silver Age hero in a DC comic book would gain superpowers through some implausible mechanism and then decide, as a matter of course, to use said powers to fight crime and better humanity (after first donning, of course, a colorful costume), never questioning this career choice.

The situation was very different with the superheroes populating DC's main comic-book competitor, Marvel Comics, whose characters such as the Hulk and the X-Men lamented that if they didn't have bad luck, they'd have no luck at all. In 1961, the Marvel Comics (née Timely and then Atlas) Company was on the verge of going out of business. From its Golden Age peak, when it had published the Human Torch, the Sub-Mariner, and Captain

America comics, the company had fallen to the point where it was barely getting by putting out monster comics, Westerns, funny animal stories, and young-romance stories. This all changed with a golf game between Jack Liebowitz, the head of DC Comics, and Martin Goodman, Marvel Comics' publisher. Liebowitz boasted of the success DC was having with a particular title, the *Justice League of America*, that featured a team of superheroes including Wonder Woman, the Flash, Green Lantern, Aquaman, the Martian Manhunter, and others banding together to fight the *supervillain du mois*. Upon returning to the office, Goodman instructed his editor (and his nephew-in-law, the last remaining full-time employee), Stan Lee, to come up with a comic book featuring a team of superheroes. Marvel was not publishing superhero comics at the time; consequently Lee could not assemble a team book by incorporating characters from other comics, as DC had done. Instead, he and Jack Kirby created a new superhero team from whole cloth. The resulting title, the *Fantastic Four*, written by Lee and with art by Kirby, became a sales success, and led to Marvel Comics' reversal of fortune.*

Lee's and Kirby's unique contribution was to add character development and distinctive personalities to their comic-book stories. As a way of distinguishing themselves from the heroes in DC comics, the superheroes in Marvel's stories did not see their superpowers as a blessing, but frequently bemoaned their fate. When cosmic radiation turned Ben Grimm into the large, orange, rock-complexioned Thing in *Fantastic Four # 1*, he did not revel in his newfound superhuman strength but cursed the fact that he had become a walking brick patio, wanting nothing more than to be restored to his human form. But no character in the Marvel universe complained more about his lot in life than Spider-Man.

In *Amazing Fantasy # 15*, written by Lee and drawn by Steve Ditko in 1962, young Peter Parker was a thin, nerdy high-school student who suffered endless teasing by the popular jocks at his school. Parker was an orphan, living with his overprotective,

---

* Those who were involved in publishing DC and Marvel comics at the time deny that such a golf game ever took place. Nevertheless, because this story is considered the fountainhead of Marvel Comics by so many fans, it has become the accepted legend, regardless of whether it has any factual basis.

elderly relatives, Aunt May and Uncle Ben. Excluded from joining the popular students in an after-school activity, Peter indulged his interest in science by attending a physics lab demonstration on radioactivity. As happened with alarming frequency in Silver Age Marvel comics, an accident with radioactivity resulted in the bestowal of superpowers. In this case, a spider was inadvertently irradiated during the demonstration and bit Peter Parker before expiring, leaving him with radioactive spider blood.

Parker found that he had acquired from this spider bite various arachnid attributes, including heightened agility and the ability to adhere to walls. Because a spider can lift several times its own body weight, Peter could now lift several times his own weight, described in the comics as a "proportionate" increase in strength. Peter also gained a "sixth sense" that alerted him to potential dangers, a Spider-Sense, if you will. One can only guess that Stan Lee, stymied in his attempts to kill real spiders in his bathroom, concluded that spiders used ESP to avoid being squished. Presumably, we can thank the protective Comics Code Authority for the fact that Peter did not also gain a spider's ability to spew organic webbing from his anus, but instead used his knowledge of chemistry and mechanics to construct technological web shooters that he wore on his wrists.*

After a lifetime of ridicule and abuse at the hands of his peers, Peter initially sees his newfound powers as a venue to fame and fortune. After testing his skills in professional wrestling, he creates a colorful blue-and-red costume and mask in order to enter show business. Feeling empowered on the eve of his television debut, he arrogantly refuses to help a security guard stop a fleeing robber, though it would have been easy to do so. However, upon returning home, he learns that gentle Uncle Ben has been slain by an intruder. Capturing the killer using his new powers, Peter discovers to his horror that this was the same robber he could have stopped earlier that day. Belatedly realizing "that with great power there must also come great responsibility," Peter Parker dedicates

---

* In the 2002 motion picture *Spider-Man*, the genetically engineered super-spider bite also gave Peter the ability to shoot organic webbing from ducts in his wrists. This freed the filmmakers from having to explain why teenager Peter Parker was able to invent and manufacture a revolutionary adhesive webbing, yet persistently remained in debt.

himself to fighting crime and righting wrongs as the amazing Spider-Man.

Not that he didn't continue to complain about his life at least three times per issue. One of the novelties that Lee and Ditko introduced in the Spider-Man comic book was a host of real-life concerns and difficulties that bedeviled Spider-Man nearly as much as his colorful rogue's gallery of supervillains. Peter Parker would contend with seething high-school romances and jealousies, money problems, anxiety over his aged aunt's health, allergy attacks, even a sprained arm (he spent issues # 44–46 of the *Amazing Spider-Man* with his arm in a sling), all while trying to keep the Vulture, the Sandman, Doctor Octopus, and the Green Goblin at bay. But the greatest intrusion of reality, which would signal the end of the innocent Silver Age, would come in 1973 in *Amazing Spider-Man # 121* with the death of Peter Parker's girlfriend, Gwen Stacy—a death that was demanded, as we will now show, not by the writers and editors or by the readers, but rather by Newton's laws of motion.

The Green Goblin had first appeared in *Amazing Spider-Man # 14* as a mysterious crime over-boss, and grew into one of Spidey's most dangerous foes. In addition to enhanced strength and an array of technological weapons, such as a rocket-propelled glider and pumpkin bombs, the Green Goblin managed to unmask Spider-Man and learn his secret identity in the classic *Amazing Spider-Man # 39*. Knowing that Peter Parker was really Spider-Man gave the Goblin a distinct advantage in his battles. In *Amazing Spider-Man # 121*, the Goblin kidnaps Parker's girlfriend, Gwen Stacy, and brings her to the top of the George Washington Bridge,* using her as bait to lure Spider-Man into battle. At one point in their fight, the Goblin knocks Gwen from the tower, causing her to fall to her apparent doom (see figs. 6 and 7).

At the last possible instant, Spider-Man manages to catch Gwen in his webbing, narrowly preventing her from plummeting

---

* Though the bridge as drawn is clearly the Brooklyn Bridge, it is identified in the story as the George Washington Bridge. In 2004 the editor of this issue, Stan Lee, accepted responsibility (it comes with great power, after all) for the error. In subsequent reprints the bridge is referred to as the Brooklyn Bridge.

Fig. 6. *Gwen Stacy's fatal plunge off the top of the George Washington Bridge, as told in* Amazing Spider-Man *# 121. Note the "SNAP" near her neck in the second to last panel.*

Fig. 7. Continuation of Gwen Stacy's death scene, where Spider-Man receives a harsh physics lesson, and the Green Goblin's scientific "genius" is called into question.

into the river below. And yet, upon reeling her back up to the top of the bridge, Spider-Man is shocked to discover that Gwen is in fact dead, despite his last-second catch. "She was dead before your webbing reached her!" the Goblin taunts. "A fall from that height would kill anyone—before they struck the ground!" Apparently the Green Goblin, creator of such advanced technology as the Goblin Glider and pumpkin bombs, suffers from a basic misunderstanding of the principle of conservation of momentum.

Of course, if it were true that it was "the fall" that killed poor Gwen, then the implication for the fate of all skydivers would suggest a massive conspiracy of silence on the part of the aviation industry, not to mention that this would make the attraction of bungee jumping even more inexplicable. Nevertheless, comic-book fans have long argued over whether it was indeed the fall or the webbing that killed Gwen Stacy. This question was listed as one of the great comic-book controversies (alongside whether the Hulk is stronger than Superman, and who is faster: the Flash or Superman*) in the January 2000 issue of *Wizard* magazine. We now turn to physics to definitively resolve the question of the true cause of the death of Gwen Stacy.

The central question we pose is: How large is the force supplied by Spider-Man's webbing when stopping the falling Gwen Stacy?

PHYSICS AND THE FINAL FATE OF GWEN STACY

To determine the forces that acted upon Gwen Stacy, we first need to know how fast she was falling when the webbing stopped her. In our previous discussion of the velocity required for Superman to leap a tall building in a single bound, we calculated that the necessary initial velocity v was related to the final height h (where his speed is zero) by the expression $v^2 = 2gh$, where g is the acceleration due to gravity. The process of falling from a height h with initial velocity $v = 0$, speeding up due to the constant attractive force of gravity, is the mirror image of the leaping processes that got him to the height h in the first place.

Consequently we can employ the expression $v^2 = 2gh$ to calculate Gwen Stacy's speed right before she is caught in Spider-Man's webbing. Assuming that Spidey's webbing catches her after she has fallen approximately 300 feet, Gwen's velocity turns out to be nearly 95 mph. Again, air resistance will slow her down somewhat, but as indicated in fig. 6, she is falling in a fairly streamlined trajectory. As we are about to discuss, the danger for Gwen is not the speed but the sudden stopping she'll face when she hits the river.

---

* The Flash!

In order to change Gwen Stacy's motion from 95 mph to zero mph, an external force is required, supplied by Spider-Man's webbing. The larger the force, the greater will be the change in Gwen's velocity, or rather, her deceleration. To calculate how large a force is needed in order to bring Gwen to rest before she strikes the water, we once again turn to Newton's second law, $\mathbf{F} = \mathbf{ma}$. Recall that the acceleration is the change in velocity divided by the time during which the speed changes. Multiplying both sides of the expression $\mathbf{F} = \mathbf{ma}$ by the time over which the speed decreases, we can rewrite Newton's second law as:

$$(FORCE) \times (TIME) = (MASS) \times (CHANGE\ IN\ SPEED)$$

The momentum of an object is defined as the product of its mass and its speed (the right-hand side of the above equation). The product of (Force) × (time) on the left-hand side of this equation is called the Impulse. This equation, therefore, tells us that in order to change the momentum of a moving object, an external force F must be applied for a given time. The larger the interval of time, the smaller the force needed to achieve the same change in momentum.

This is the principle behind the air bags in your automobile. As your car travels down the highway at a speed of, say, 60 mph, you as the driver are obviously also moving at this same speed. When your car strikes an obstacle and stops, you continue to move forward at 60 mph, for an object in motion will remain in motion unless acted upon by an external force (that outside force is coming up in an instant). In the days before seat belts and air bags, the steering column typically supplied this external force. The time your head spent in contact with the steering wheel was brief, so consequently, the force needed to bring your head to rest was large. By rapidly inflating an air bag, which is designed to deform under pressure, the time your head remains in contact with the inflated air bag increases, compared with the steering wheel, so the force needed to bring your head to rest decreases. Distributing the force over the larger surface area of the air bag also helps to reduce injuries in a sudden stop. This force is still large enough to often knock the driver unconscious, but the important point is that it is no longer lethal. The right-hand side of the above equation is determined by the crash—you were moving at 60 mph and then you came to rest because you hit something. If you wish to avoid this, you shouldn't have hit anything. Your mass in the expression for the change in momentum does not change (this is the best-case scenario!). Thus the right-hand side of the equation is governed by the collision, and all you can do to minimize the damage is through control of the left-hand side, that is, the impulse. The product of force and time must always be the same, as the net result is the same—namely, the initial large momentum changing to zero. The longer the time in a given impulse, the smaller the force needed to bring about the change in momentum. This is also the physical justification for a boxer rolling with a punch, increasing the time of contact between his face and his opponent's fist, so that the force his face must supply to stop the fist is lessened.

Now, Spider-Man's webbing does have an elastic quality, which is a good thing for Gwen Stacy, but the time that is available to slow her descent is short, which is an awful thing. For a given change in momentum, the shorter the time, the greater the necessary force. For Gwen, her change in speed is 95 mph−0 mph = 95 mph, and assuming she weighs 110 pounds, her mass in the metric system is 50 kilograms. If the webbing brings her to rest in about one half of a second, then the force applied by the webbing to break her fall is 970 pounds. Hence, the webbing applies a force nearly nine times larger than Gwen's weight of 110 pounds. Recalling that an object's weight is simply $W = mg$, where g is the acceleration due to gravity, we can say that the webbing applies a force equivalent to 9 g's in a time span of 0.5 seconds. As indicated in fig. 6, when the webbing brings Gwen to a halt, a simple sound effect drawn near her neck (the "SNAP!" heard round the comic-book world) indicates the probable outcome of such a large force applied in such a short period of time. In contrast, bungee jumpers allow a distance sufficient for the cord to extend for many seconds, in order to keep the braking force below a fatal threshold.

Traveling at such a speed, coming to rest in such a short time interval, there is no real difference between hitting the webbing and hitting the water. However, there have been recorded cases of people surviving forces greater than that experienced by Gwen Stacy. Col. John Stapp rode an experimental rocket sled in 1954 and was subjected to a force of 40 g's during deceleration, yet lived to describe the experience as comparable to "dental extraction without anesthetic." Of course, Col. Stapp was securely strapped

into and supported by the sled in a reinforced position. More typically, suicide victims who jump from bridges die not from drowning, but rather from broken necks. Hitting a body of water at such
a speed has the same effect as hitting solid ground, as the fluid's
resistance to displacement increases the faster you try to move
through it (we'll discuss this further in Chapter 4 when considering the Flash). Tragically for Gwen Stacy, and for Spider-Man, this
is another example of when comic books got their physics right,
and we readers were not required to suspend our disbelief, no matter how much we may have wanted to.

Spider-Man seems to have learned this physics lesson concerning impulse and change in momentum. A story in *Spider-Man
Unlimited # 2*, entitled, appropriately enough, "Tests," finds the
wall-crawler adhering to the top of a skyscraper when an unfortunate window washer plummets past him. Launching himself after
the falling worker, Spider-Man must solve a real-life physics problem under more pressure than you'd find in a typical final exam.
As he closes the gap between the worker (due to the fact that
Spider-Man pushed off from the building with a larger initial
velocity than did the window washer), Spider-Man considers:
"OK, I have to do this right. Can't snag him with a web-line, or the
whiplash will get him." As shown in fig. 8, Spider-Man recognizes
that his best solution is to match his speed to that of the worker
and then grab hold of him when they are barely moving relative to
each other. (I'm not sure how Spider-Man slows himself down to
match the worker's velocity—perhaps by dragging his feet against
the side of the building?) Then Spidey shoots out a web-line,
where his arm, endowed with spider-strength, is able to withstand
the large impulse associated with their upcoming change of
momentum.

This solution was also employed in the 2002 motion picture
Spider-Man. When the Green Goblin drops Mary Jane Watson
from a tower on the Queensboro Bridge, in a clear homage to the
storyline from *Amazing Spider-Man # 121*, Spider-Man this time
does not stop her rapid descent with his webbing. Rather he dives
after her, and only after catching her does he employ his webbing
to swing them to (relative) safety, using the same procedure as in
fig. 8. One hallmark of a hero, it appears, is the ability to learn
from experience.

While certainly no hero, the above arguments have also made

© 2004 Marvel Comics

Fig. 8. *Scene from the story "Tests"* in Spider-Man Unlimited # 2 *(May 2004),*
in which the caption boxes reveal Spider-Man's thought process as he faces a
practical application of Newton's second law of motion.

an impression on the Green Goblin. As mentioned earlier, the January 2000 issue of *Wizard* magazine described the controversy surrounding the death of Gwen Stacy as a classic open question in
comic-book fandom. This prompted my letter to the editor of *Wizard*, published a few months later, which summarized the above
physics discussion. Two years later, the August 2002 issue of *Peter
Parker: Spider-Man # 45* (written by Paul Jenkins and drawn by
Humberto Ramos) featured a story line in which the Green Goblin
demonstrated that he had also finally learned this physics lesson.
In this issue the Goblin had sent a videotape of Gwen Stacy's
death to the news media in order to psychologically torment
Spider-Man. Portraying himself as the reluctant hero of this
tragedy, the Goblin narrates in the tape:

"Realizing the girl had fallen, I naturally made a course correc-
tion on my glider in an attempt to save her. I began an immediate

descent. But before I had a chance to reach her, Spider-Man did something incredibly stupid: Despite the speed of her fall, he chose to catch her in that rubber webbing of his. In the next instant, her neck was snapped like a rotten twig."

It may have taken the Goblin nearly thirty years, but apparently he at last understands that it wasn't "the fall" that killed Gwen Stacy, but the sudden stopping. If a twisted, evil maniac like the Green Goblin can learn his physics, then there is hope for us all.

4

FLASH FACTS—

FRICTION, DRAG, AND SOUND

IT WAS A DARK and stormy night in Central City as police scientist Barry Allen locked up for the night. Pausing by the chemical storeroom, he marveled at the large collection of chemicals that the CCPD possessed. Despite his scientific training, Allen was standing near an open window during the gathering storm and bore the full brunt of a lightning strike that entered the room. The lightning bolt shattered the chemical containers, dousing him while the electrical current passed through his body.

But the simultaneous exposure to lethal voltages and hazardous chemicals somehow only dazed Allen, knocking him off his feet. Later that evening, he was surprised to discover that he could easily outrace a departing taxicab and catch and restore a spilled plate of food in a diner in the blink of an eye. Realizing that the lab accident had somehow endowed him with superspeed, he naturally adopted a simple yet elegant red-and-yellow costume and used his newfound powers to fight crime as the Flash.*

There is a broad range of physical phenomena associated with speed, and John Broome, Robert Kanigher, and Gardner Fox, the main writers of the early Silver Age Flash comics, addressed many

---

* A freak electrochemical accident of this nature would never be seen again—until *Flash Comics # 110*, when another lightning bolt splashed young Wally West with identical chemicals, endowing him with superspeed as well. Wally then began his career as a junior crimefighter, under the imaginative name Kid Flash.

of them. Thanks to his ability to run very fast, the Flash was frequently depicted running up the sides of buildings or across the ocean's surface; he would catch bullets shot at him, and drag people behind him in his wake. Are any of these feats consistent with the laws of physics? It turns out that all of them are, granting, of course, the one-time "miracle exception" of the Flash's super-speed in the first place.

* * *

In his very first Silver Age appearance, "The Mystery of the Human Thunderbolt" in *Showcase # 4*, the Flash ran up the side of an office building, because with his "great speed he is able to overcome gravity." Earlier we explored the simple relationship between Superman's initial vertical velocity and the final height he can leap. As the Man of Steel rises, he slows down due to gravity, until at a height h his final speed is zero. We calculated in Chapter 1 that for Superman to leap to a height of 660 feet, equivalent to a thirty- to forty-story building, his initial liftoff velocity needs to be at least 140 mph. But the Flash can run much, much faster than this, and he should therefore be able to reach the top of a forty-story building with velocity to spare. So, as he approaches the side of a building, as long as he has a speed greater than the minimum $v^2 = (2gh)$, he should be able to leap up its side without violating any laws of physics (aside from the fact that he is running several hundred miles per hour, that is). In contrast, the fastest that a non-superpowered human can run is on the order of 15 mph (though faster sprints are possible)—which would enable the runner to scale the side of a small tool shed.

The question, however, is not whether the Flash is able to move fast enough to leap a vertical height h, but whether he can maintain traction to actually run up the vertical side of the building. Some interesting physics underlies the simple act of walking, related to Newton's third law, which states that forces come in pairs. When you run or walk, a force must be applied horizontal to the ground by your feet, opposite to the direction you wish to move. The ground exerts an equal and opposite force back on your feet, parallel to the ground's surface, that counters the back-directed force exerted by your shoes. The origin of this parallel force is friction. Imagine trying to walk across a floor covered with a uniform layer of motor oil, and you will realize how crucial friction is to a process as simple as walking. Without friction between his boots and the ground, the Flash would never be able to run anywhere. Captain Cold, one of the first and most persistent supervillains that the Flash would regularly combat, possessed a "freeze ray" gun that could ice up any surface. Time and again, Captain Cold (who, incidentally, isn't really a captain) would simply create a layer of ice directly in front of the Scarlet Speedster, denying him traction and rendering his superspeed useless.

No doubt due to its ubiquity and fundamental role in everyday life, the phenomenon of friction is generally taken for granted, despite its complexity. Exactly why does an object resist being dragged across another surface? While friction's basic properties were first scientifically addressed by Leonardo da Vinci in the early 1500s and Amontons in the mid-1600s, a true understanding of the root cause of this phenomenon would not arrive until the atomic nature of matter was properly resolved in the 1920s.

There are primarily two ways in which atoms can be arranged to form a macroscopic object: (1) in a uniform, periodic, crystalline structure, or (2) in a random, amorphous agglomeration. Of course, most solids lie somewhere between these two extremes, and typically there will be regions of crystalline order randomly connected, sometimes separated by amorphous sections. The net result will be that even the smoothest macroscopic surface will not be truly flat when viewed on an atomic scale. In fact, one doesn't have to go to such extremes: Even on length scales of a thousandth of a millimeter—much, much bigger than an individual atom—an object's surface will more likely resemble a jagged mountain range than the stillness of a quiet lake. Consequently, when two objects are dragged past each other, regardless of the apparent smoothness of their finishes, on the atomic scale it is not unlike taking the Rocky Mountain range, turning it upside down, shoving it atop the Himalayas, and then dragging the upside-down Rockies at a steady speed across the Himalayan mountaintops. One would naturally expect enormous geological upheavals and large-scale distortions in this extreme form of plate tectonics, and the results are no less catastrophic at the atomic level. With every footstep, bonds between atoms are broken, new bonds are formed, and atom-size avalanches and atom-quakes are produced. All of this requires a great deal of force in order to keep these atomic-scale mountain ranges sliding past and through each other. The resistance to such atomic rearrangements is called "friction," and without it, the Flash would only be running in place.

The amount of friction opposing the motion of an object along a horizontal surface is proportional to the weight of the object pressing down on the surface. The greater the weight of an object, the deeper the atomic "mountain ranges" interpenetrate, and the greater the frictional force that must be overcome to move the object. It is harder to get a big, heavy block to start moving than a smaller, lighter one. Engineering solutions for lifting heavy objects have been known since the time of the ancient Egyptians, who developed various ingenious schemes for moving giant limestone blocks during the pyramids' construction.

One obvious trick is to use a ramp. On a horizontal flat surface, all of a block's weight presses down perpendicular to the surface. On a sloped surface, on the other hand, the weight is still straight down, directed toward the center of the Earth (think of a plumb line held on the ramp). Only some of the weight is perpendicular to the surface of the tilted ramp, and the rest is directed down the ramp. The smaller the force pressing the atomic mountain ranges against each other, the less they will interpenetrate, and the easier it will be to move them past each other. So the frictional force, which is proportional only to the component of the weight perpendicular to the surface, is less for a block on a tilted surface compared with one on a horizontal surface. No matter how rough the surface, if the ramp is tilted at too steep an angle, the friction force holding the block in place will be insufficient to counteract the downward pull of the weight down the ramp, and the block will slide down the ramp. However, as the Flash runs up the vertical side of a building, there is *no* component of his weight perpendicular to the surface upon which he is running, that is, the building's face. In principle, therefore, there should be no friction between his boots and the building's wall, and without friction he cannot run at all.

So can he in fact run up the side of a building? Technically, no. At least, not "run" as we understand the term. He can, as he leaps up the side of the building, move his feet back and forth against the building's side, which would make it appear as if he were running. In essence he is traveling a distance equivalent to the height of the building in the time between steps. Typically, as the Flash runs, his foot pushes down on the ground at an angle with the road's surface so that the force the road exerts back on him (thanks to Newton's third law) is also at an angle with the surface. The net effect is that he accelerates in both the vertical and horizontal direction. The vertical velocity gives him a bounce up off the ground, and the horizontal component propels him in the direction he is running. The greater the vertical velocity, the higher the bounce, while the larger the horizontal velocity, the farther he advances before gravity overcomes the small vertical velocity and brings his feet back to the ground, ready for another step. Very fast runners, which would certainly include the Flash, can have both feet up off the ground between steps. The faster they run, the longer their time "airborne" between steps. If the Flash bounces about 2 cm vertically with every step, then he is in the air for about one eighth of a second before gravity pulls him down for another step. But one eighth of a second is a long time for the Crimson Comet. If his horizontal velocity is 5,250 feet/sec or 3,600 mph, then the horizontal distance he travels between steps is more than 660 feet. This is approximately one eighth of a mile, which we used as the benchmark for the tall building that Superman leapt in Chapter 1. As long as the Flash maintains at least this minimum speed, he needn't worry about losing his footing along the way, simply because he will scale the height of the building between steps.

Before he can scale a skyscraper, the Flash has to radically alter his direction from the horizontal to the vertical. As will be discussed in a later chapter, any change in the direction of motion, whether it is Spider-Man swinging on his webbing or the Flash changing his path at the side of a building, is characterized by an acceleration that requires a corresponding force. Rotating his trajectory by ninety degrees up the building's face entails a large force, provided by the friction between the Sultan of Speed's boots and the ground. In addition to superspeed, the Flash's "miracle exception" must therefore also extend to his being able to generate and tolerate accelerations that few superheroes not born on Krypton could withstand.

Newton's laws of motion can also explain how the Flash is able to run along the surface of the ocean, or any body of water, for that matter. Just as Gwen Stacy had to be concerned as she was about to strike the water while moving at her large, final velocity, the great speed of the Flash's strides enables him to run across its surface. As one moves through any fluid, be it air, water, or motor oil, the fluid has to move out of your way. The denser the medium, the harder this is to accomplish. It requires more effort to walk through a swimming pool, pushing the water out of your way, than to walk through an empty pool (that is, one filled only with air), and it is harder still if the swimming pool is filled with molasses. The resistance of a fluid to flow is termed "viscosity," which typically increases the denser the medium and the faster one tries to move through the fluid.

The density of water is much greater than that of air—water molecules are in contact with one another, while there are large, open spaces between air molecules. It is even more difficult to move through water when traveling at high speeds. But for the Flash, when running on top of the water's surface, this is a good thing. Just as someone is able to water-ski if he or she is towed at a large velocity, the Flash is able to run faster than the response time of the water molecules. As his foot strikes the water's surface at speeds greater than 100 mph, the water acts more like a solid than a liquid beneath the Flash's fleet feet (to test this, try rapidly slapping a pool of water), and therefore his oft-shown ability to run across bodies of water is indeed consistent with the laws of physics. In fact, at the speeds at which he typically runs, it is practically impossible for him to not run across the water's surface. However, in order to acquire forward momentum, the Flash must push back against the water. That is, even if the water does behave like a solid under the rapid compression under his feet, would the Flash be able to obtain traction in order to run? One way he could accomplish this is by generating backward propagating vortices under his feet, thereby gaining a forward thrust under Newton's third law. This mechanism was proposed as the means by which water-strider insects propel themselves along the water's surface. Here again comics were ahead of the curve. The Flash's ability to run across a body of water was likened to a rapidly skipping shell skimming over the water in *Flash # 117*, more than thirty years before scientists understood the water strider's method of locomotion.*

When it comes to air, there is a lot of space between neighbor-

---

* There are important exceptions to this general principle that viscosity increases with speed, such as with tomato ketchup. When rapidly stressed, the viscosity of ketchup decreases, while as just argued, a sharp shock to

ing molecules. At room temperature and pressure, for example, the distance between adjacent air molecules is about ten times larger than the diameter of an oxygen or nitrogen molecule. Moreover, each air molecule at room temperature is zipping around with an average speed of approximately 1,100 feet/sec or 750 mph (which is the speed of sound in air). When we run through air, we don't build up a high-density region in front of us, because our speed is much less than the average air molecule's velocity. Think about herding cattle: If the cows are running when you try to push one into the herd, the others will just run away. If they are walking very slowly, and you push at the same rate, the others don't have time to get out of the way, and they pile up into a herd. One can, of course, move faster than the speed of sound (a feat first performed by Col. Chuck Yeager in 1947), but the expended effort is large. When trying to displace a volume of air faster than the air molecules are moving, a high-density region (that is, a shock front) would pile up in front of you.

In fact, in "The Challenge of the Weather Wizard," the Flash uses just such a shock front to knock out the Weather Wizard. Mark Mardon, a small-time crook, stole his deceased scientist brother's "weather stick," a device that enabled him to control the weather. Much like any other self-respecting comic-book villain, once in possession of a weapon giving him mastery over the fundamental forces of nature, he immediately adopts a colorful costume, calls himself the "Weather Wizard," and sets upon robbing banks and vandalizing police stations. The finale of the story, as shown in fig. 9, comes when, "with a tremendous surge of speed, the Flash slams toward his foe so fast that the air in front of him piles up into a wave-front and a split instant later strikes Mardon like a solid sheet of glass." This is indeed a physically accurate consequence of the Flash's supersonic velocity, and the variation in the compressibility of air at high velocities bedeviled fighter pilots in the 1940s attempting to break the "sound barrier." Whenever the Flash runs at or faster than the speed of sound, the pressure waves

---

water increases its resistance to flow. This is why fast, hard raps to the bottom of a ketchup bottle momentarily reduce its viscosity and speed up its egress from the bottle.

Fig. 9. *Panels from "The Challenge of the Weather Wizard" (Flash # 110) demonstrating that the faster one moves, the harder it is to get the air out of the way.*

he generates create a "sonic boom." Just such a loud crash heralded the first appearance of the Flash in *Showcase # 4.*

Once the Flash has moved the air before him out of his path, he leaves a region of lower-density air in his wake. Compared with the surrounding air at normal density, this lower-density trail behind the Flash can be considered a partial vacuum. Air rushes in to fill any vacuum, and anything standing in the way of this rushing air behind the Flash will be pushed into the wake region. The faster he runs, the greater the pressure difference between the air behind him and the surrounding air, and the larger the force as this pressure imbalance is corrected. This effect is noticeable even for slower-moving objects, such as when a subway train enters a tunnel. The enclosed geometry of the tunnel accentuates the updraft behind the departing train, pulling loose newspapers and litter in its wake. Lacking a confined space, the Flash can generate a low-pressure region that can slow the descent of falling people, cars, or giant bombs, or as shown in fig. 10, or help detain and levitate a crook by using a vortex created by running in a circle.

Returning to the topic of the speed of sound in air, whenever the Flash runs faster than a velocity of 1,100 feet/sec (or 750 mph), his communication with others must become visual only. The Flash would not be able to hear anyone standing behind him or even at his sides, as he would outrace the sound waves trying to

Fig. 10. When the Flash runs at high velocity in a circle, he leaves a low-pressure region in his wake, which makes it easy to bring Toughy Boraz (yes, that's actually his name) and his stolen loot to police headquarters. From Flash # 117.

reach him. Of course, for anyone standing in front of him, the Flash outracing the sound waves would not be a problem, but there would still be a barrier to communication. Even when he can hear someone talking to him, the speech will have a high and tinny quality as heard by the Flash.

What we term "sound waves" are variations in density consisting of alternating regions of expansion and compression. The "wavelength" of a sound wave is the distance between adjacent compressed (or expanded) regions in the medium (whether it be air, water, or a solid), which is related to the pitch that we hear. The pitch, or frequency, measures the number of complete wave cycles that pass a given point per second. Long wavelengths have low pitches (think of the deep tones from a bass violin, where the length of the strings is related to the wavelength of the sounds they can produce) while shorter wavelengths are heard as higher pitches. As the Flash runs, even if he does not outrace the sound wave, his high-speed motion affects the pitch that he hears. Let's say he runs toward someone who is yelling a warning to him. The sound waves have some wavelength, which marks the average distance between adjacent compressed or expanded regions. If the Flash were standing still when these alternating density regions reach him, the tone he would hear would be determined by the wavelength originating from the speaker. But as the Flash runs, one region of compressed air reaches him, and as he is running toward the speaker, the next region of compressed air reaches his eardrum sooner than it would if he were standing still. The Flash thus hears a smaller wavelength and hence a higher frequency due to the fact that he is running toward the source of the sound. The faster he runs, the greater this shift in the wavelength and frequency of the detected sound.

This phenomenon is known as the Doppler effect, and if one knows the wavelength of a stationary source of waves, and measures the wavelength of the detected waves with a moving detector, one can determine the speed of the detector. Alternatively, if one sends out a wave of a known wavelength and it bounces off a stationary target, it should return with the exact same wavelength. If the target is moving toward the source, the reflected wave will have a shorter wavelength, while if the target is moving away from the source, the detected wavelength will be longer. Doppler radar, as often used in predicting weather, involves detection of this wavelength shift, which enables meteorologists to calculate the wind velocity in an approaching storm front.

This is also the basic premise underlying radar guns, which use radio waves of a known wavelength. From the shift in wavelength of the reflected wave to the incident wave, they can determine the velocity of the object (such as a thrown baseball or a speeding automobile) that reflected the waves. The faster the target is moving, the greater the wavelength shift, and the higher the pitch of the detected wave. If the Flash were to run at 500 mph toward someone who was using a normal speaking voice with a pitch of about 100 cycles per second, the sound waves reaching the Viceroy of Velocity's ears would be shifted to 166 cycles per second. It would sound strange, but the Flash could hear the person. Could he run so fast that the frequency shift would make the sound undetectable? In order for the pitch of the sound waves reaching the Flash to be greater than 20,000 cycles/sec, the upper range of human hearing, the Flash would have to run toward the speaker with a speed greater than 150,000 mph (that is, 0.02 percent of the speed of light). There would be other problems, aside from Doppler shifts, were the Flash to run toward someone at that speed.

The Flash's preferred technique for stopping bullets is also consistent with Newton's laws of motion. You don't need to be bulletproof when you can outrun a bullet. But what about the innocent bystanders caught in the line of fire? Fig. 11 shows a physically accurate use of superspeed in this situation. As narrated by one such potential victim in *Flash # 124*, "the amazing speedster merely made his hand travel at the same speed as the bullets whizzing at him, and with a sweeping motion plucked them right out of the air before they could harm him." That is, the Flash would first match his velocity to the bullets so that the relative speed between him and the bullet is zero. Just as one can easily pick up a book or a cup on an airplane in flight if it is not moving relative to you, the Flash is then able to pluck the bullet out of the air, since he is also moving at approximately 1,500 feet/sec or more than 1,000 mph in the same direction as the bullet. An "editor's note" in *Flash # 124* correctly points out that "Flash's action in stopping the bullets is similar to that of a baseball fielder who stops a hot grounder by letting his glove travel momentarily in the same direction as the ball."

As discussed Chapter 3, the problem with high velocities is not the speed but the deceleration. For Gwen Stacy, the braking time was very short, so the stopping force was large. The boxer rolling with a punch, as noted earlier, deliberately increases the contact

Fig. 11. *The Flash demonstrates that Impulse and Momentum principles are still important, even when you can run as fast as a speeding bullet. From* Flash # 124.

time in order to minimize the stopping force. The Flash, as the editor's note correctly points out, is applying the same principle in this situation. In addition to being able to run at amazing speeds, Barry Allen also apparently gained the ability to withstand crushing accelerations every time he sped up or slowed down. Thus, when the Flash stops running, the bullet he is holding stops as well, and he can then drop the slug at the gunman's feet for dramatic effect.

IF THIS BE MY DENSITY—

PROPERTIES OF MATTER

BEFORE DR. HENRY PYM began moonlighting, fighting crooks and capturing Communist spies as Ant-Man, he was a fairly typical biochemist. In his first appearance in "The Man in the Ant Hill," Pym was shown struggling with the bane of a modern scientist's life: funding! As difficult as it is nowadays to secure funding for scientific research, Pym's travails as relayed in *Tales to Astonish # 27* indicate that we have it pretty good compared with the early 1960s. As we learn in a flashback, at a recent scientific convention, a panel of scientists went beyond merely rejecting Pym's request for financial support of his search for a shrinking potion and took the additionally cruel step of personally taunting him. "Bah! You're wasting your time with your ridiculous theories," jeers one professor, "but they never work!" Another counsels: "You should stick to practical projects!" To which Pym replies, "No! I'll work only on things that appeal to my imagination...like my latest invention!" I should point out that two aspects of this exchange ring particularly true: namely (1) there continues to this day a tension at universities and research laboratories between research that is motivated by the search for practical applications and those investigations that are "curiosity-driven," and (2) unlike the public at large, scientists routinely use the expression "Bah!" in everyday conversation.

Pym's first accidental exposure (and you would not be far wrong to conclude that nearly all superheroes are fairly accident-prone, at least when it comes to gaining their powers) to his shrinking potion led to a harrowing adventure inside an ant hill,

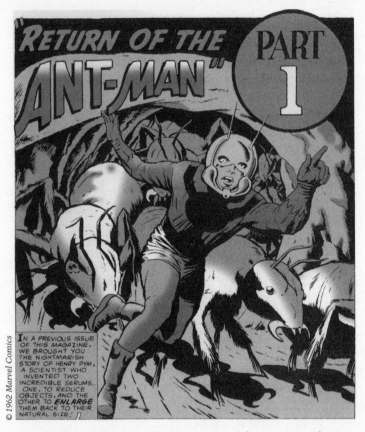

Fig. 12. *The opening page of "The Return of the Ant-Man" from* Tales to Astonish # 35, *in which we meet the costumed superhero alter-ego of Dr. Henry Pym for the first time.*

reminiscent of the 1954 science-fiction story *The Incredible Shrinking Man*. At the end of this story, Pym regains his original height through the application of the growth potion, and, once normal size, pours them both down the sink. Realizing that the potions are "far too dangerous to ever be used by any human again!" he vows, "From now on I'll stick to practical projects!" What Pym considers more practical than developing a reversible miniaturization process is left to the reader's imagination.

Dr. Pym's vow remained unbroken until the sales figures came in for *Tales to Astonish # 27*. As shown in fig. 12, by *Tales to Astonish # 35*, the good doctor was back in "The Return of the Ant-Man" (though in the prior story in *Tales to Astonish # 27*, he had never referred to himself by that title), having replicated his shrinking potion and designed a snazzy red-and-black jumpsuit and a "cybernetic" helmet that enabled him to electronically communicate with ants. Given that ants actually communicate with one another by excreting pheromone chemicals, we won't look too closely into how Pym's helmet might actually function. So arrayed, scientist Henry Pym fought common criminals, Communist spies (it was the early 1960s, after all), invading aliens, and bizarre supervillains such as the Porcupine and Egghead as the astonishing Ant-Man. It seemed as if no evildoer could win in a face-off (though not a literal one given the size differential) with a crimefighter whose superpower was that he was only a quarter-inch tall.

Adventures involving characters reduced to the size of insects have been a staple of science-fiction movies and comic books for at least fifty years. Yet here we are in the twenty-first century and we've still not achieved this radical form of weight reduction. What's the holdup?

After all, it truly seems that every other day brings a news report confirming the equation: Science fact equals science fiction plus time. Robots assemble automobiles or vacuum your apartment; a computer has beaten the world champion in a chess tournament; and therapeutic cloning promises to alleviate many devastating diseases and medical conditions. Man has gone to the moon, walked upon its surface, and returned safely to Earth—not just once but several times—and travel to other planets, at least within our own solar system, has been accomplished, if only by unmanned craft so far. Scientific papers have even been published in prestigious physics journals discussing the construction of "time machines," whose operation involves the concept of "negative energy." (The "negative energy" is necessary to prevent the collapse of wormholes—a concept developed in the General Theory of Relativity—which has been postulated as providing a theoretical mechanism for warp speed; that is, faster-than-light travel.)*

---

* This "negative energy" is associated with squeezed quantum states, and is beyond the scope of this book. And no, your brother-in-law cannot be considered a vast, untapped source of negative energy.

On the technology end of futuristic projections, the handheld communicators from *Star Trek* are now an everyday item—current cell phones, with digital cameras, image storage and transmission, and Internet access, exceed the imaginations of *Star Trek* writers from the 1960s. *Star Trek*'s "tricorders"—handheld devices the size of a hardcover book that enable chemical and biological analysis—may soon be available for purchase: PDAs such as the iPhone are already common, and the technology to perform "DNA analysis on a chip" and other functions is in development. From flat-panel television screens to microwave ovens and magnetic resonance imaging that provides three-dimensional views of the human interior, we are indeed living in the World of Tomorrow, even if we still lack portable jet packs and android butlers.

Yet despite all the fantastic marvels and concepts that are either already here or seem within our grasp, we still can't shrink or enlarge people at will. Compared with miniaturization, warp drives and time travel are right around the corner. However, back in the 1960s, it was shrink-rays that were promised by our comic books and movies, coming soon to a top-secret underground military scientific laboratory near you.

The 1966 science-fiction film *Fantastic Voyage* described the adventures of a surgical team and a mini-sub that is miniaturized to the size of a bacterium and injected into a scientist's bloodstream to remove a blood clot in his brain that is inoperable from the outside. Before the movie begins, a title card appears, reading: THIS FILM WILL TAKE YOU WHERE NO ONE HAS EVER BEEN BEFORE. NO EYE-WITNESS HAS ACTUALLY SEEN WHAT YOU ARE ABOUT TO SEE. BUT IN THIS WORLD OF OURS WHERE GOING TO THE MOON WILL SOON BE UPON US AND WHERE THE MOST INCREDIBLE THINGS ARE HAPPENING ALL AROUND US, SOMEDAY, PERHAPS TOMORROW, THE FANTASTIC EVENTS YOU ARE ABOUT TO SEE CAN AND WILL TAKE PLACE. Three years later, man did indeed walk on the moon, and it is certainly true that compared with thirty years ago, the most incredible things are indeed happening all around us. Yet we have much longer to wait before a team of doctors can make this ultimate house call. What is the insurmountable barrier that prevents an aspiring Dr. Henry Pym from radically changing size?

The reason that miniaturization is physically impossible (as far as we know) is that matter is made of atoms, and the size of an atom is a fundamental length scale of nature, not open to continuous adjustment. As discussed in Isaac Asimov's novelization of *Fantastic Voyage,* to make something smaller requires either (1) making the atoms themselves smaller, (2) removing some (large) fraction of its atoms, or (3) pushing the atoms closer together.

* * *

First let's consider the size of the atoms. In cartoon representations of atoms, in DANGER! RADIOACTIVITY! warning signs for example, the orbits of electrons around the nucleus are represented as elliptical trajectories, like those the planets make about the sun in our solar system. We would mark the "size" of our solar system as the distance from the sun at its center to the outer boundaries of the orbits of the planets, and similarly the "diameter" of an atom would be determined by the range over which the electrons buzzed around the nucleus. The typical size of an atom is about a third of a nanometer, where one nanometer is one billionth of a meter (a meter is roughly 39 inches long). This seems small, and it is: Looking along the cross section of an average human hair, about 300,000 atoms lie end to end across its width.

Every atom has a nucleus consisting of a number of positively charged protons and a comparable number of uncharged neutrons. In addition to the positively charged protons, the atom contains an equal number of negatively charged electrons. If oppositely charged objects attract one another, then why don't the positively charged protons pull the negatively charged electrons toward them until the electrons sit on the nucleus? Well, they would, if the electrons were standing still. After all, as discussed in Chapter 2, the Earth and moon pull toward each other through their mutual gravitational attraction, and the moon's orbit is such that its distance from the Earth and its speed exactly balance the inward gravitational pull. Similarly, the electrons reside in "orbits" around the nucleus at the center of the atom. Interestingly enough, all atoms are roughly the same size, to within a factor of three. The number of protons in the nucleus is countered by an equal number of "orbiting" electrons. Heavier atoms have more protons that pull the electrons more strongly toward the nucleus, but more electrons mean there is more repulsion between the negatively charged electrons trying to get away from each other. This balancing act results in a "size" of the atom that is roughly twenty or thirty billionths of a centimeter.

I must note, for reasons that we will get into in Section 3, that this picture of electrons moving in precise elliptical orbits around the nucleus is not correct. Rather, quantum mechanics tells us not where the electrons are, but provides a mechanism for calculating the probability of finding an electron at a certain distance from the nucleus. It is the largest distance for which the probability of finding an electron is appreciable (the extent of the "probability cloud") that is referred to as the "radius" of the atom. The diameter of the atom (twice its radius) can also be thought of as the atom's size. The expression for the most probable radius of the atom depends only on such terms as the mass of the electron, its electric charge, the number of positive charges in the nucleus, and a fundamental constant of the universe represented by the letter h, known as Planck's constant (the value of which determines the magnitude of all quantum phenomena). We will discuss h in more detail in Section 3, but for now, all we need worry about is that h is a fixed number, just as the mass of an electron or the magnitude of its electric charge is in the expression for the atomic radius. Once the number of positive charges in the nucleus (the quantity that determines which element we are dealing with) is set, there's nothing to change. The size of an atom is determined by a collection of fundamental constants and is not open to adjustment.

In Isaac Asimov's follow-up to his novelization of *Fantastic Voyage*, titled *Fantastic Voyage II: Destination Brain*, the mechanism proposed to enable miniaturization involve the creation of a "local distortion field" that somehow manages to change the value of Planck's constant. If h becomes a tunable parameter, a factor-of-ten reduction in its value would shrink the size of an atom to one hundredth of its present size. Needless to say, we have no idea how to begin to accomplish this in the real world, which is, after all, why h is considered an invariable constant. Our lives would be profoundly different if we ever discovered a way to change the fundamental constants of nature so that the speed of light or the charge of an electron becomes open for our adjustment. Until that day comes, these constants are just that, and because the radius of an atom is described by said constants, the size of an atom cannot be changed. We therefore cannot make atoms themselves smaller—at least not without also changing the type of universe we live in.

What about the second suggestion for size reduction—that is, removing a fraction of the atoms in an object? Everything is made of atoms; consequently, removing some of them should make an object smaller. Certainly the reduction in the size of electronic devices suggests that some objects can be made with less material and still retain their functionality. Problems arise, however, with complex, living things, for which the removal of a significant number of atoms would have serious consequences. To go from a height of six feet to six inches is a reduction in height by a factor of twelve. Of course, people are three-dimensional, and a factor-of-twelve reduction in the width and breadth would be required as well. To accomplish this by removing atoms (assuming that one could do this, had a safe place to store them, and could replace them later when you wanted to regain your original height) means that you would get to keep only one atom for every 1,728 atoms subtracted. Even assuming that this removal is performed uniformly—such that the same fraction of atoms is removed from all of your cells—biological functionality would be lost or at least severely compromised.

Consider the neurons in your brain. It is a myth that man uses only 10 percent of present brain capacity; evolution theory argues against such a monumental waste of available resources. If neurons could be smaller and still fulfill their role in the brain, then there would be a strong competitive advantage to any mutation in this direction. Not only would it require fewer atoms to build a person (so the demands for raw materials through food would be greatly reduced), but one could have many more neurons and hence synaptic connections if our brains retained their current size. The typical neuron has a width of approximately one thousandth of a centimeter, and this is true whether one is considering an ant's neuron or a human's. People are smarter than ants (on average—I'm sure we can all think of a counterexample from our personal experiences) because we possess roughly four hundred thousand times more neurons, not because our neurons are a thousand times bigger. Remove 99 percent of the atoms, and you can make your body's cells 99 percent smaller, but they won't work as intended.

Finally, what about the third possibility—shrinking a person by making him denser by pushing the atoms closer together? Unfortunately, this is not a successful strategy for miniaturization either, for the same reasons that the planet Krypton could not simply be fifteen times denser than Earth. Atoms in most solids are already tightly packed, with each atom essentially in direct physical contact with its neighbors. Moreover, thanks to the repulsion of negatively charged electron probability clouds around each atom, they are fairly rigid objects and resist being squeezed together. When a box is filled to the top with a child's collection of marbles, nearly all of the container's available space is taken up by the marbles. With few exceptions, every marble is in physical contact with several of its neighbors. Certainly there are open gaps between the marbles, but these spaces are not big enough for us to add more than possibly a few percent more marbles. If the marbles are hard spheres and not compressible, then squeezing on the walls of the container will not lead to a significant reduction in its volume. Reducing the size of the container by a factor of ten would require pressures that would significantly deform or crush the marbles. Trying to shrink a person by applying similar pressures would result in comic-book stories that would be both brief and messy and almost certainly would not garner approval by the Comics Code Authority.

* * *

Well, if miniaturization is so difficult, how does biochemist Dr. Henry Pym, also known as the Ant-Man, accomplish it? In *Tales to Astonish* # 27, Pym had devoted years to discovering a potion that would shrink any object, until treated with an antidote growth serum. Later, he would convert his shrinking potion into an easy-to-swallow pill form. When it eventually became necessary to explain how he was able to shrink other objects, such as his costume, helmet, and weapons that he carried, it was revealed that he had developed a small generator of "Pym particles" that were able to increase or decrease an object's size. No explanation has ever been put forth for how these potions or Pym particles actually work, and their physical basis must fall under the "miracle exemption" we frequently invoke when considering the source of a hero's superpowers.

SO HE TALKS TO FISHES. WANT TO MAKE SOMETHING OF IT?—

FLUID MECHANICS

WHILE SUPERMAN, Spider-Man, Batman, and the Hulk are considered A-list superheroes, there are those who fight the never-ending battle, yet seem to garner little respect. Despite facing world-shaking threats and being present at the creation of a variety of superhero teams, these heroes are, unfairly, I would say, relegated to the ranks of second-stringers or worse. I am here referring to such champions as Ant-Man (a founding member of the Avengers), Bouncing Boy and Matter-Eater Lad (both early members of the Legion of Super-Heroes), as well as an original participant of the Justice League of America: the King of the Seven Seas, the Aquatic Ace, Aquaman.

In Chapter 8 we will see how Henry Pym, despite being only a quarter of an inch tall when fighting for justice as the astonishing Ant-Man, is still able to punch his way out of a paper bag (a minimum prerequisite in the superhero game). Bouncing Boy will make an (albeit brief) appearance later on, when we discuss electrostatics and grounding. Matter-Eater Lad, whose real name in Tenzil Kim, from the planet Bismoll, whose inhabitants have evolved the ability to chew and digest any and all types of solids, from stone to metal to plastic, is not really relevant to explain any important physics principles. Nevertheless, villainous foes of the Legion have discovered to their regret that no jail can hold Matter-Eater Lad (who always takes a bite out of crime). But we will see in this chapter that the most undeservedly underrated superhero is Aquaman.

Fig. 13. *Aquaman in his first comic adventure, in* More Fun Comics # 73, *punching a hole through the plates of a submarine, which is designed to withstand the pressures at the ocean's floor.*

Making his debut in *More Fun Comics* # 73 back in November of 1941, Aquaman was created by artist Paul Norris and writer Mort Weisinger; the latter would go on to edit the Superman comic line in the 1950s. In this first story, a German U-boat mistakenly torpedoes a refugee ship, which had been granted immunity from harm. In order to cover up their mistake, the submarine crew prepares to sink the lifeboat carrying the passengers and crew from the refugee ship. However, the refugees are saved by the sudden appearance from beneath the ocean's surface of a strange hero, wearing an orange top and green slacks, who then goes on to mete out justice to the German submariners. While he never achieved the popularity of Superman, Batman, or Wonder Woman, he did share with them the distinction of maintaining a monthly publication schedule even during the dark period of the mid 1950s, when nearly all of the other superhero titles went out of print. Not bad for a guy who talks to fish.

From this very first appearance in 1941, it was clear that Aquaman could breathe underwater, possessed great strength, could swim at remarkable velocities, enjoyed keen vision that enabled him to see even in the murky ocean depth,* and could communicate with fish (though in his first adventure he used plants as an intermediary to send a message to some porpoises—it would be four issues later when he gave direct orders to a sawfish). All of these skills are consistent with known properties of fluid mechanics (well, "fish telepathy" requires a little electromagnetic theory—we will hold off discussing this superpower until later on in Chapter 20).

FRESH AIR UNDERWATER

The most striking ability of Aquaman, as well as that of Marvel Comics' Prince Namor, the Sub-Mariner,† and all the other denizens of comic books' many distinct underwater cities of Atlantis, is the ability to extract oxygen directly underwater. Without this superpower, there doesn't seem to be much point to being a water-based superhero. It turns out that this is the one special power that requires the smallest miracle exception from the laws of nature. Why shouldn't Aquaman breath through water—after all, we do!

Everyone knows that drowning results when the lungs fill up with water. What is less commonly recognized is that normal breathing would be impossible without a small amount of water in the lungs. Fresh air comes in through the nose, and travels down the bronchial tube, where it is warmed to the body's temperature and premoistened. In fact, the air has to be at 100 percent relative humidity as it moves down the ever more finely branching tubes on its way to the alveoli—small little spherical buds

---

* He rarely made use of this ability. Noteworthy is *More Fun Comics # 106*, where a darkly inked page invites the reader to "look at things not through Aquaman's dark-adapted eyes, but through our own. On the ocean floor, all is blackness. But here and there, quivering, gleaming streaks of phosphorescence knife through the gloom...."

† It's pronounced Sub-Mariner, not Submarine–er, by the way. Prince Namor's creator, Bill Everett, joined the merchant marines at age fifteen (and left two years after) and presumably was very familiar with this synonym for seamen. His half-human/half-Atlantean hero, capable of breathing underwater, is best described as a *sub*-merged *mariner*. Imperius Rex!

where the exchange of oxygen and carbon dioxide occurs. These pockets are roughly 0.1 to 0.3 mm in diameter, smaller than the period at the end of this sentence. On the other side of the walls of the alveolar bud are the capillaries—very narrow blood vessels in which plasma and red blood cells flow to drop off carbon dioxide molecules and pick up oxygen molecules on their way to the heart. The capillaries are narrow for the same reason that the alveolar spheres are so small—to maximize the ratio of surface area to volume. Since the gas exchange takes place only through the walls of the alveoli and the capillaries, the more surface area there is, the more regions there are for possible gas diffusion to occur.

There has to be some transition for these gas molecules between the interior of the alveoli—which are connected through the bronchial tubes to the outside world—and the capillaries that carry the blood. This is provided by a thin coating of water on the interior of the alveolar surface. This water layer facilitates the transfer of gases by ensuring that the inner cell walls of the alveoli do not become dried out by direct contact with air, which would cause them to lose their functionality. Only after it is has dissolved from the gas phase to the liquid phase can an oxygen molecule diffuse through the two cell walls and get picked up by speeding red blood cells. The alveoli can be considered air bubbles in water, and we could not breathe without (a little) water in our lungs, though, just as so often in life, too much of something turns a necessity lethal. Aquaman, who lacks the gills of a fish that facilitate our finny friends' oxygen extraction directly from the surrounding water, must have some sort of superpower adaptation that enables him to continue breathing even when completely underwater.

But even this very thin water layer in the alveoli should be physically capable of causing asphyxiation. The same physics responsible for glistening dewdrops should produce acute shortness of breath, or worse. The magnitude of surface tension in the water layer is sufficient to cause the small alveolar buds to close up entirely, so that even deep breaths would not be enough to provide the necessary pressure to drive the oxygen molecules into the bloodstream. What saves us from choking on an amount of water that could not fully fill a thimble? Soap!

Surface tension is the name given to the pulling force that results from the attraction of molecules in the fluid (let's say water) to each other. Such an attractive force must of course exist—or else the atoms or molecules in the liquid would fly away from each other as they return to the vapor state. For most liquids, this force is a relatively weak electrostatic cling (called the van der Waals attraction) that arises from fluctuating charge distributions in the molecule. The force can't be too strong, for the water molecules must be able to move past each other and flow through hoses or fill up the volume of a container in exactly the manner that a solid doesn't. We'll discuss van der Waals later on, when we consider the physics that enables gecko lizards and Spider-Man to climb up walls and across ceilings.

This attractive force tends to pull the water molecules equally in all directions—it is not stronger in the up-down direction than it is in the left-right direction. For water molecules in the middle of a liquid, the pull is balanced on all sides. A molecule on the surface of the liquid only feels an attractive pull from the water molecules beneath it, as the air above does not exert an upward attractive pull. These surface molecules therefore experience a net downward pull that curls the water into a perfectly spherical drop in the absence of gravity. For water on a blade of grass at dawn, condensing from the atmosphere owing to the lower temperatures in the absence of sunlight, the water adheres to the surface of the grass, and surface tension curves the top layer of the morning dew into a hemisphere. This curved surface of water acts as a lens, concentrating the early-morning sun's rays and accounting for the glistening light of dawn before the sun rises higher in the sky and the more intense sunlight evaporates the water droplets.

This tendency of water to curve is less charming when it forces the walls of our alveoli to constrict, requiring extreme pressures to keep the air buds open. When faced with the problem of decreasing the surface tension in alveolian water in the development of our physiology, natural selection chose the same solution we employ when washing our clothing. The cells in the alveolar walls generate a substance known as "pulmonary surfactant." The first term just refers to the lungs, while a "surfactant" is a long, skinny molecule with different chemical groups at either end. Electrostatic interactions result in one end of this molecule being attracted to the charge distributions in water molecules, while the other end is repelled by those same charges. If the long skinny molecule* is fairly rigid, like a spine, then a large collection of such molecules will orient themselves so that all of the regions that are repelled by water are pointing in one direction (typically where there is a low concentration of water), while those ends that are attracted to water will extend into the fluid. The region where the surfactant molecules can satisfy both ends at the same time is at the water-air interface, with the water-attracting end inserted into the water and the water-avoiding end protruding out into the air. In such a configuration, the surfactant interferes with the water-water bonding at the surface of the water layer. This reduces the cohesive force between water molecules that was the source of the surface tension. Without pulmonary surfactants, the alveoli—essentially air bubbles in water—are unable to effectively facilitate gas exchange with the bloodstream. These crucial surfactants do not develop in the fetus until late in its gestation, which is why premature babies may suffer from respiratory distress syndrome, an often-fatal condition prior to the development of effective artificial surfactants.

A moment ago I referred to the reason why surface tension arising from even a thin layer of water in the lungs does not kill us as "soap." While not technically correct, in that pulmonary surfactants are not soaps, the converse is true, in that soaps are surfactants, with water-attracting and water-repelling chemical groups at either end of long skinny, chain-like molecules. Soap helps one clean up by reducing the surface tension of water, so that it can make direct contact with the dirt. That is, surfactants make water wetter, and help us breathe easy as well.

THE PRESSURE OF THE JOB NEVER GETS TO HIM

In his first adventure in 1941's *More Fun Comics # 73*, Aquaman is shown in the first oversized panel (which is commonly referred to in comic books, appropriately enough, in this case, as the "splash panel") swimming in the ocean, rescuing a woman from

---

* A long, typically carbon-based molecule is called a "polymer" because it contains many (poly) similar chemical structures (called "mers" from the Greek "meros" meaning "part") along its chain.

the lifeboat in his right arm while he deflects a shell fired from a German U-boat with his left. We can calculate how tough Aquaman's skin must be by considering the force necessary to change the momentum of this missile (as discussed in Chapter 3). A one-pound shell traveling at 500 feet per second would require Aquaman to generate a force of 30,000 pounds to deflect it in a time of one millisecond (one thousandth of a second). This force is applied over a very small area, let's say three square inches. The pressure (force per area) on Aquaman's forearm is consequently 10,000 pounds per square inch, which is more than 660 times atmospheric pressure and comparable to the strength of cast iron. A year later, in *More Fun Comics # 85*, at the North Pole the Aquatic Ace tosses a full-grown polar bear at a group of poachers from more than 100 yards away as if the ursine projectile were a football (a seemingly odd thing to do with one of nature's creatures when battling poachers, but the polar bear was unharmed in the incident). It is surprising that the King of the Seven Seas relies on fish to save him from tight spots or to help capture malefactors as often as he does, for he clearly has the stuff that's tough enough!

The standard explanation for Aquaman's superstrength is that it is a natural consequence of his being able to withstand the pressures at the ocean's floor. As in the case of breathing underwater, Aquaman's great strength in being able to support the weight of an ocean is merely different in degree, rather than in kind, from your own great strength, Fearless Reader. After all, every time you stand up from a chair, you are lifting up an ocean as well—only in your case it is an ocean of air. But if Aquaman has to contend with a half mile of water pressing on his shoulders, we must deal with approximately fifteen miles of air above us. The source of this pressure, for both Aquaman and us, is basically Newton's laws of motion.

Air molecules move at the speed of sound, roughly 700 miles per hour. As we discussed earlier, any object in motion, whether a Mack truck or an oxygen molecule, will remain in motion until acted upon by an outside force. That outside force can come into play when the oxygen molecule collides with the surface of your shoulders; the force the shoulders exert on the molecule to deflect its trajectory is balanced by an equal and opposite force (Newton's third law) that the molecule exerts on the shoulder. The greater the number of molecules that strike your shoulder per second

(that is, the denser the air), the larger the net force that these collisions exert on your shoulders. Every single square inch is subjected to a force of nearly fifteen pounds from these atmospheric
impacts. We never really notice this pressure, for basically two
reasons. The first is that we have lived our entire lives under the
weight of our atmosphere, and have evolved a skeleton structure,
muscles, and skin capable of withstanding this pressure. The second is that the force exerted by collisions of air molecules is fairly
uniform in all directions. Air is pretty dilute, with the spacing
between molecules roughly ten times the diameter of any given
molecule. Thus, while it is technically true that the entire weight
of miles of air presses down upon us, the number of collisions
coming up from below is essentially identical to the number
pressing down from above, and we never experience an unbalanced force from atmospheric pressure under normal circumstances.

While a pressure of fifteen pounds per square inch seems pretty
high, it pales compared with the pressures experienced by Aquaman beneath the ocean's surface. Oxygen molecules consist of
two oxygen atoms and are heavier than water molecules (which
are comprised of only one oxygen atom and two hydrogen atoms).
The momentum of a typical air molecule is 60 percent greater
than for a water molecule in liquid (the water is moving roughly as
fast, on average, as the air molecules, but has a smaller mass).
Nevertheless, the pressure exerted by collisions from liquid water
is much greater than from the atmosphere, as the liquid state is
over eight hundred times denser than the atmosphere. The water
molecules are essentially in direct physical contact thanks to the
weakly attractive van der Waals force (which is also responsible
for surface tension) described above. Unlike gases, liquids are
essentially incompressible, and the more fluid above you, the
greater the weight of water pressing down on you.

The force exerted when you are underwater is larger than that
on dry land, and the further below the water's surface you are, the
heavier the water's weight presses down on you. In fact, the pressure you experience is directly proportional to how deep under the
surface you are:

**PRESSURE = ATMOSPHERIC PRESSURE +
(DENSITY × ACCELERATION DUE TO GRAVITY × DEPTH)**

The depth is measured from the water's surface (which is why, if the depth is zero, the pressure you experience is just that of the atmosphere). The density of water (one gram per cubic centimeter) multiplied by the acceleration due to gravity is just another way of stating the weight of the water per volume. When multiplied by the depth, the units of the second term on the right-hand side of the equation above represent weight per area, which is the same as the pressure.

One eighth of a mile beneath the surface of the ocean, the pressure is equivalent to twenty atmospheres, and across Aquaman's broad shoulders, with an area of approximately ten square inches, lies a weight of three thousand pounds. To stand under this crushing pressure requires an equivalent force supplied by the Aquatic Ace's legs—just as Superman's muscles and skeletal structure had to be strong enough to withstand the higher gravity of Krypton. We proposed in Chapter 1 that Superman would weigh roughly three thousand pounds on the planet of his birth. Aquaman's legs have to be at least this strong, as he can swim much deeper in the ocean. The King of the Seven Seas must be roughly at least as strong as the Golden Age Man of Steel!

Or perhaps even stronger. In *Justice League of America # 200*, Aquaman must battle an alien creature that has assumed an artificial body composed of pure glass. Now, glass is actually pretty tough, and in its pure form is stronger than many metals. However, scratches and cracks, inevitable when fighting the Justice League, will severely weaken glass's ability to withstand compression. Even so, it would take a pressure of more than a ton per square inch to shatter such glass. Using the previous equation, we conclude that roughly one mile beneath the ocean's surface, the pressure is indeed 2,600 pounds per square inch. Aquaman is able to withstand such forces—the alien invader, not so much.

At the water's surface, there is obviously no weight of water overhead, and the pressure is just that of the atmosphere. Since the ocean of air is the same over the surface of the planet, any body of water will have the same height, regardless of shape or size. In fact, two bodies of water, if they are connected by an underground channel, will by necessity have the same height, which we refer to as "sea level." What if there was no atmosphere—what would be sea level—the height of the ocean's surface—then? This is not such an academic question, and in fact we

© 1982 National Periodical Publications Inc. (DC)

Fig. 14. The Aquatic Ace dispatches an alien creature that has assumed a body composed of glass by dragging him to the ocean's depths in Justice League of America # 200.

make practical use of this equation, which describes the height of a column of water in the absence of, or at least reduction in, atmospheric pressure, whenever we use a drinking straw. We will now show that physics places severe limits on the long-distance thirst-quenching ability of Superman!

The name of the inventor of the drinking straw is lost to the mists of time, but there is evidence that its use dates back to ancient Mesopotamia, the region that is now part of modern-day Iraq. The reeds that grow on the shores of the Tigris and Euphrates rivers (an area known as the Fertile Crescent, one of the founding sites of agriculture) were useful not just for sipping a tasty beverage, but when pressed edge first onto wet clay (also found in abundance at these rivers' shores) could be used to form wedge-shaped characters that left permanent markings when the clay had dried and hardened. The Sumerians of about 4000 B.C.E. are credited with being the first to use writing to record and transmit information, and their distinctive "letters" are referred to as "cuneiform," which is Latin for "wedge-shaped." Drinking straws have therefore been around nearly as long as writing, though the twisty "crazy straw" had to await innovations in injection plastic forming.

When a straw is inserted into a glass of water, the two "bodies of water"—that is, the water inside the straw and outside the straw—are connected by the fluid in the rest of the glass. We'll assume that the straw is not firmly pressed against the bottom of the glass, and the water can move between these two regions freely. The atmospheric pressure on the water outside the straw is the same as that inside the straw, when first placed in the glass. The force on the liquid inside the straw is thus the same as for the liquid outside the straw, and the liquid inside and outside the straw will be at the same level. When we place our lips around the end of the straw and suck, we form an airtight seal that enables us to reduce the pressure inside the straw. The atmospheric pressure outside the straw is unchanged, and there is consequently a pressure difference between the "two bodies of water." This pressure difference results in an unbalanced force on the incompressible fluid, the result being that the normal air pressure on the drink outside of the straw pushes the water inside the straw to a greater height.

What is the limit on the height that the water can rise up such a straw? The lower the pressure inside the straw, the larger the pressure difference between the water outside the straw and inside the straw. Suppose Superman (in particular, the Silver Age Man of Tomorrow who can fly and has super-breath) had an extremely large straw that he dipped into a large body of water. Is there a length of straw that would be too long for even Superman to be able to sip from? Yes! The greater the pressure difference between the outside and the inside of the straw, the larger the unbalanced force that pushes the water up the straw. But the pressure on the outside of the straw cannot be larger than the atmospheric pressure of fifteen pounds per square inch. When we suck on a soda straw, we can possibly decrease the interior pressure to ten or five pounds per square inch. Superman can obviously do much better, but even he can't improve on a perfect vacuum. The lowest possible pressure inside the straw is zero pounds per square inch, achieved when there is not even one single air molecule remaining inside the interior of the straw. To find out how high the water could rise under this extreme suction, we use the equation relating pressure and depth on page 84, only now the pressure on the left-hand side of the equation is zero, and we'll calculate a negative depth, which is equivalent to a positive height above the water's surface. The pressure difference, between fifteen pounds per square inch and zero, can push the water inside the straw to a height of thirty-four feet, regardless of the width of the straw. Which is quite high, but also quite finite. If Superman tries to drink from a straw a "mere" thirty-five feet long, he will come up—literally—high and dry!

Of course, in the spirit of the somewhat goofy Silver Age adventures of Superman, it's always possible to concoct a mechanism by which the Man of Steel can drink from any length straw he needs. As stipulated, he cannot decrease the pressure inside the straw to below zero pounds per square inch. But perhaps he can increase the pressure on the water outside the straw to a value greater than normal atmospheric pressure. If he were to give a quick puff of super-breath directed down toward the body of water he was attempting to drink from immediately before suctioning on the straw, he could induce a much larger pressure difference, and consequently bring the water up the straw to any height that the story might require.

A large pressure imbalance can lead to more dramatic effects than simply enabling one to sip a cool drink. In the above discussion, we have assumed that the straw is constructed from some sort of high-strength metal that will not collapse due to this pressure differential. A lower pressure inside the straw means that there are fewer molecules striking the inner wall of the straw per second than there are outside the straw. That is, there is an unbalanced force on the walls of the straw, just as the unbalanced force on the liquid led to the water rising up the straw to our lips. If this force is large enough, it can cause the straw to collapse along its length—as anyone who has drawn too forcefully on a paper straw can confirm.

The impact of air molecules too small to see can crush objects more substantial than a paper straw—fifteen pounds per square inch is a pretty significant pressure if it acts in only one direction. Boil about three quarts of water for ten minutes or so in a fifty-five-gallon steel drum by placing it over a flame, and then seal off the drum and remove the source of heat. The steam that has displaced the air and filled the drum will now condense back into the liquid state, taking up much less volume and exerting a much smaller pressure on the interior of the drum. With the can sealed from the outside world, no air can replace the condensing steam, and the pressure differential between the inside and outside of the steel drum will grow. In a few minutes time, the steel barrel will violently implode, as if struck by the fist of an angry, invisible giant. Such dramatic lecture demonstrations have practical consequences when railroad tanker cars are steam cleaned. When the tanks are filled with hot water and then closed up, the resulting pressure difference when the steam condenses back into liquid water can cause even a tanker car to suffer a violent implosion. If atmospheric pressure can do such damage, it puts into context how tough Aquaman, who can withstand the pressures of the ocean's depth, must be.

FASTER THAN FLIPPER

Aquaman is not only known for his ability to breathe underwater, for his superstrength, or for his communication with his finny friends, but also for his great swimming speed. He has been known to swim at speeds of up to 100 mph, much faster than any mammal or fish. How does he do it? How do any of us swim?

The first rule of swimming is to avoid sinking. The best way to do this is to float, and the best way to do that is to have a density less than that of water. Imagine a box held underwater. If the box weighs exactly the same as an equivalent volume of water, then aside from possibly a different color, there is no difference between the box and an equal volume of water. If we drew imaginary lines around a volume of water, we would not expect that this mental exercise would cause the water enclosed to move either up to the surface or down to the bottom of the container. In such a situation, we say that the box is "buoyancy matched" to the surrounding fluid, and in still water it should remain at its original position indefinitely. If the mass of the box were greater than the mass of water occupying the same volume, then the box would sink, and if the box had less mass, it would rise to the surface and float. While humans are mostly water, we have about two quarts of air in our lungs, and some of our cellular material, notably fat, has a lower density than water. Consequently, we have, on average, a lower density (mass per volume) than water, and can float. Salt water is denser than fresh water, which is why it is easier to float in the ocean than in a freshwater lake or pool. A solid block of steel has a much greater density than water, and it thus sinks when placed in open water. We can lower the density of the metal by changing the volume it occupies. The same mass of steel, when fashioned into the shape of a hollow, open-topped boat, takes up a volume that has an outer wall of steel and an interior of air, so that its average density is less than water's, and it will consequently float.

Aquaman can change his depth underwater presumably the same way fish do: by varying the air content in a special organ called, variously, a swim bladder, an air bladder, or a gas bladder. This organ holds a reservoir of air. When a fish swims to too great a depth, the larger water pressure squeezes the fish, and the fish responds through a complex chain of chemical reactions by, in essence, taking a deep breath, transferring air into its bloodstream, and injecting air into its swim bladder, thereby expanding it. The fish's average density thus lowers, and it is pushed toward the surface. Similarly, it can extract air from the bladder, increasing its density, when it wishes to descend. Aquaman, being of unique half-human, half-Atlantean parentage, has obviously developed an analogous organ that enables him to swim at any depth in the seven seas he chooses.

Being able to float in water is necessary in order to swim, but not sufficient. One must also be able to provide a backward force against the surrounding water, so that, by Newton's third law, the water can provide an equal and opposite thrust that propels one forward. The hand motion of the breaststroke, (technically referred to as "scooping") pushes the water backward and the swimmer forward. A question that has long puzzled those interested in the physics of swimming, and has recently been resolved experimentally, is whether one could swim faster if the viscosity, that is, the resistance to flow, of the fluid were increased. Clearly, thicker fluid, such as molasses, would provide more resistance to the swimmer's stroke, and the harder the swimmer pushes against the fluid, the more a counterforce is provided that moves the athlete forward. However, it is harder to move through molasses than water, just as there is more resistive drag when one tries to run through a swimming pool compared with running through the air on dry land. Which effect plays a greater role, positive or negative, in determining a swimmer's progress?

At my home institution, the University of Minnesota, just such an experiment was conducted in 2004 by chemical-engineering professor Ed Cussler. The swimming pool at the university was filled with nine hundred pounds of guar-gum powder so that the pool fluid had twice the viscosity of normal water. Athletes from the university swim team completed laps in the thicker soup, and their times were compared to normal conditions. The net result: zero. The swimmers were indeed able to increase the efficiency of their strokes, but this turned out to be exactly compensated by the increase in drag in the thicker fluid. The only effect was that the swimmers were significantly slimier when leaving the pool—but at least it was in the interest of science!

In order to achieve a velocitiy in the water of 100 miles per hour, Aquaman was making use of a miracle exception, along the lines of that invoked for the Scarlet Speedster, the Flash. But perhaps he also makes use of some of the adaptations of one of nature's fastest swimmers, the dolphin. While not being in the Aquatic Ace's league, the dolphin is able to achieve speeds of twenty to twenty-five miles per hour, so fast, in fact, that until recently, scientists thought that the dolphin was somehow violating the principle of conservation of energy. It was argued that dolphins would require more muscles than they actually possess in order to swim at such speeds. Of course, a dolphin's speed is not in conflict with the laws of physics. Dolphins are nearly all muscle, which provides great strength for each of their strokes. Muscle is denser than water, but these mammals avoid sinking to the ocean's bottom not through reservoirs of fat, but instead with blubber. Blubber has a lower density than fat and the spring-like resiliency of muscle, and it thus provides crucial assistance to the dolphin as it swims. It is as true in swimming as it is in comedy: Timing is everything. Dolphins change the shape of their flukes into arches at precise points in the stroke cycle in order to maximize their thrust. Furthermore, dolphins are extremely streamlined, with a bare minimum of appendages extending from their body that might contribute to water drag.

Interestingly, dolphins are continuously shedding their skin, at such a rate that their epidermis is completely replaced every two hours. What benefit could this provide that would justify the excess energy the dolphin must expend in order to be essentially always rebuilding its skin? Physical experiments and computer simulations suggest that the little flakes of skin that slough from the dolphin's body as it swims perturb and break up turbulent vortices generated in the dolphin's wake. By inhibiting the formation of these vortices the transition to turbulence is damped (in a situation not dissimilar to that of a dimpled golf ball, which will be discussed in Chapter 12), and lower turbulence translates to greater speed. Perhaps the key to the Aquatic Ace's super swimming speed is not the speckled orange shirt we associate him with, but a severe case of underwater dandruff!

CAN HE SWING FROM A THREAD?–

CENTRIPETAL ACCELERATION

PRACTICALLY EVERY ISSUE OF *The Amazing Spider-Man* features scenes of him using his webbing to swing from building to building through the canyons of New York City. But is this realistic? Specifically, is Spider-Man's webbing strong enough to support his own weight, as well as the weight of any falling criminal, victim, or innocent bystander whom he happens to catch mid-flight as he swings in his parabolic trajectory? As Spider-Man swings in an arc, there is an extra force in addition to his weight that the webbing must supply. Let's now consider why.

Remember that Newton's second law of motion, **F = ma**, told us that a force is needed to change an object's motion. A change in motion, or acceleration, refers to either a change in magnitude (either speeding up or slowing down) or to a change in direction. If no force acts on the object, then it persists in "uniform motion," which means constant motion in a straight line. Any change in motion, whether in magnitude or direction, can only come about if a force acts on the object. When an automobile negotiates a hairpin turn, an external force (friction between the tires and the road) changes the auto's direction, even if the speed remains unchanged.

In order to change the direction of motion, an external force is needed—and the corollary of this is that a force can only produce an acceleration in the direction that the force acts. For example, gravity pulls an object toward the ground, regardless of its initial motion. More importantly, gravity can only pull an object toward the ground because that is the only direction it acts. If the Golden Age, pre-flying Superman runs off the edge of a cliff with a steady horizontal speed, he will start falling due to gravity. Since gravity does not act in the horizontal direction, his horizontal speed will not change as he falls! No force, no change, after all. His vertical velocity does increase, becoming greater the longer he falls, just as in the case of Gwen Stacy considered earlier, because there is a force acting on him in the vertical direction. The net effect of his constant horizontal speed plus an ever-increasing vertical speed is a parabolic trajectory that becomes steeper the longer he plummets.

Put another way, a 90-mph fastball, thrown without spin perfectly parallel to the ground, falls to the ground at the exact same rate as a ball simply dropped out of the pitcher's hand at the same moment. Both balls would strike the ground at the same instant (if released from the same height), because the only force changing their motion is gravity, in the vertical direction. Any change in either the direction or magnitude of an object's motion can only arise from an external force acting in the direction of the change.

As Spider-Man swings from building to building, his trajectory is a semicircular arc rather than a straight line. Therefore, even if the magnitude of his speed does not change during his swing, his direction of motion is continually being altered, which can only be accomplished by an external force. It should be obvious that this force comes from the tension in the webbing. The webbing, therefore, has to do double duty and supply two forces: (1) a force to support Spider-Man's weight, which it would have to maintain even if he were simply hanging from the vertical line, and (2) a second force to divert him in a circular trajectory. If the webbing line were to snap in mid-swing, then the only external force acting on Spider-Man is gravity, and his motion at this point would not differ at all from that of a ball tossed with the same velocity that Spider-Man possessed at the instant the webbing broke.

The acceleration that this additional force in the webbing provides as Spidey swings in a circular arc is identical to the acceleration experienced by the moon as it makes its circular orbit about the Earth. In one case, the force arises from the tension in the webbing, while in the other it is Newton's force of gravitational attraction, but for both, it changes straight-line motion into circular motion. Gravity is the moon's "webbing," causing its direction to change. If the tension in the webbing or gravity were suddenly to disappear, both Spidey and the moon would depart from their circular trajectories, and continue moving with the velocity they had at the moment the external forces were removed. With a little bit of geometry or calculus one can show that the acceleration a of an object being constantly deflected onto a circular orbit with a velocity v is $\mathbf{a} = (\mathbf{v} \times \mathbf{v})/\mathbf{R} = v^2/\mathbf{R}$, where R is the radius of the circle.

Spider-Man's webbing has to supply a force mg, in order to support his weight, and an additional force mv^2/R in order to change his direction as he swings. The faster he swings (the larger his velocity v) or the tighter his arc (that is, the smaller the radius R), the greater will be this centripetal acceleration v^2/R. When Spidey swings from a web strand 200 feet long at a speed of 50 mph, the centripetal acceleration is 27 feet/sec^2, in addition to the acceleration due to gravity of 32 feet/sec^2. If Spider-Man's mass in the metric system is approximately 73 kilograms, then his weight mg is 160 pounds, and the additional force the webbing must supply just to change his trajectory from straight-line motion into a circular arc is roughly 135 pounds. The total tension in the webbing is nearly three hundred pounds, and will be more if Spidey is carrying someone as he swings.

Three hundred pounds or greater of tension may seem to be more than a thin strand of fiber can withstand, but if Spider-Man's webbing is anything like real spider silk, he has nothing to worry about. Dragline silk webbing, which spiders use for their webs and while rapidly fleeing predatory birds, is actually five times stronger per pound than steel cable and more elastic than nylon. The webbing's properties result from thousands of rigid filaments only a few billionths of a meter wide (providing great redundancy so that no one filament is crucial for the integrity of the webbing), interspersed with fluid-filled channels that distribute the tensile force along the length of the webbing. Spider-Man is able to alter the material properties of his webbing by adjusting its chemical composition as it jets from his web-shooters. Similarly, real spiders can control the tensile strength of their webs by varying the relative concentration of crystallizing and noncrystallizing proteins.

There is considerable interest in commercial applications of webbing, which would require large quantities of spider silk. As it is not practical to harvest spiders for their silk (they are too territorial to farm in a conventional manner), genetic-engineering experiments have inserted spiders' web-making genes into goats, so that the goats' milk will contain webbing that can be more easily sieved and acquired. While the development of web-producing goats has hit some snags,* other scientists have reported preliminary success with infecting spider cells in the laboratory with a genetically engineered virus that induces the cell to directly manufacture the proteins found in spiderwebs. The silk-producing gene from spiders has also been successfully introduced into E. coli and plant cells. Such research could have far-reaching practical applications. As Jim Robbins discussed in his article "Second Nature" in the July 2002 issue of *Smithsonian*: "In theory, a braided spider-silk rope the diameter of a pencil could stop a fighter jet landing on an aircraft carrier. The combination of strength and elasticity allows it to withstand an impact five times more powerful than Kevlar, the synthetic fiber used in bulletproof vests."

The high tensile strength of real spider silk enables it to support a weight of more than ten tons per square centimeter. Even a webbing strand with a diameter of only a quarter inch could support more than six thousand pounds safely, well below the three hundred pounds of weight and centripetal force we estimated earlier. Unless Spider-Man is trying to carry both the Hulk and the Blob simultaneously, his webbing should be more than able to do the job.

Therefore, according to Newton's laws of motion, it is entirely plausible that Spider-Man can swing from building to building, stop a runaway elevated train (as in the 2004 film *Spider-Man 2*), or weave a bulletproof shield out of very narrow lines of webbing. So, to answer the question posed in the title of this chapter: Simply take a look overhead!

---

* Sorry.

CAN ANT-MAN PUNCH HIS WAY OUT OF A PAPER BAG?-

TORQUE AND ROTATION

EVERY COMIC-BOOK HERO has some Achilles' heel, and Ant-Man's was only a millimeter big. There are certain obvious drawbacks to being the size of an ant. For example, just as Superman is susceptible to Kryptonite, Ant-Man must be ever vigilant against the more common hazard of being stepped on. In addition, his stride being only a few millimeters, he requires thousands of steps to cover the same distance he could walk in one step while normal height. The time required for him to walk a few feet would therefore increase correspondingly.* No doubt this was his motivation for frequently hitching rides atop carpenter ants. The fact that he could ride on top of an ant without crushing it suggests that Ant-Man's mass decreased along with his size, implying that his density remained constant as he shrank. (Remember that density is the mass of an object divided by its volume; if the volume decreases by a factor of one thousand, and the mass is reduced by an identical factor, their ratio and, hence, the object's density, is unchanged.) Pym made good use of his reduced mass and constructed a spring-loaded catapult that could shoot him across town. Of course, as we've discussed at length in Chapter 3, it's not the journey but the stopping that is problematic. In

---

* In a 2008 issue of the *JLA Classified*, the Atom, a size-reducing superhero in the DC comic universe, is shown deliberately increasing his size when he must walk across a room, in order to increase his stride and reduce the number of steps he must take. Ant-Man had only one size he could shrink to, and consequently employed ants as transportation.

order to avoid a messy finale to his trajectory, Pym called upon his special rapport with ants and used his cybernetic helmet to instruct hundreds of them to form a living air bag to cushion his landing. Ant-Man's kinetic energy would be distributed among the many, many ants so that no one insect would suffer too much for its participation in breaking his fall.

If Ant-Man is such a lightweight that he could be propelled across several city blocks by a coiled spring and not harm the ants that stopped his motion, then how is he able to disable such foes as the Protector or the Hijacker, or meet "The Challenge of Comrade X"? In particular, how is Ant-Man able to punch his way out of a vacuum-cleaner bag (shown in fig. 15), as thrillingly rendered in *Tales to Astonish # 37*, or swing a crook overhead using a nylon lariat in the very next issue? As explained in *Tales to Astonish # 38*, Henry Pym retained "all the strength of a normal human," even when ant-size. Not to nitpick, but the average normal-size human, not to mention the average biochemist, would be hard-pressed to swing a full-grown man overhead, even using a "practically unbreakable" nylon lasso. But leaving that issue aside, what does it mean to say that Ant-Man has the strength of a normal-size person, such that he can break a vacuum bag, but only the mass of an ant, whereby he is easily sucked up by the vacuum cleaner in the first place? Perhaps the more basic question is: Why do you have the strength that you do, in that you can easily lift a twenty-pound object, but struggle with one weighing two hundred pounds and cannot possibly lift two thousand pounds? Our strength comes from our muscles and skeletal structure that comprise a series of interconnected levers. It turns out that these levers are not that well suited to lifting things.

Let us stipulate that by "strength" (of its many definitions) we mean the ability to lift an object. Mankind's ingenuity has led to the development of a wide range of machines to perform tasks such as heavy lifting. One of our earliest inventions employed to lift objects is the simple mechanical device of a lever. Many of us first encountered a lever as children in the form of a playground's teeter-totter or seesaw, consisting of a horizontal board supported by a fulcrum point placed beneath the midpoint of the board. When seated at one end of the seesaw, you are able to lift a play-mate high in the air, through the mechanical advantage of the lever. With the fulcrum placed at the exact middle of the board,

Fig. 15. *A scene from "Trapped by the Protector" from* Tales to Astonish # 37, *in which it is demonstrated that Ant-Man is both as light as an ant (and hence easily captured by a vacuum cleaner) and as strong as a normal-size human (and consequently able to punch his way out of the vacuum bag).*

you can lift only a mass roughly equal to your own. If, however, the fulcrum point is placed much closer to one end, then a small child can lift a full-grown adult, provided the adult sits at the end of the seesaw nearer to the fulcrum point. This is because seesaws, and levers in general, do not balance forces but rather "torques."

If a force is defined as the ability to push or pull an object in a straight line, then a torque is a measure of the ability to rotate an object. A torque is mathematically defined as the product of the applied force and the distance between the force and the point where the object is to be rotated. While both "torque" and "work" are defined mathematically as the product of force and distance, in the case of work, the distance is the displacement of the object— that is, the distance over which the force pushes or pulls the object (more on work, in Chapter 12). The force must be acting in the same direction as this distance in order to change the object's energy. In contrast, for a torque that causes a twist, the force should be at a right angle to the separation between the applied force and the point about which the object is to be rotated. This distance is referred to as the "moment arm" of the torque. For a given applied force, the larger the force's distance from the point where the object is to be rotated, the greater the torque.

This is why doorknobs are placed at the end of the door as far away from the hinges as possible. Try closing a door by pushing it at the end that's immediately adjacent to the hinges, and then apply the same force to the other end, where the doorknob is located. The same force is used, but increasing the moment arm by increasing the distance between the push and the hinges mag- nifies the torque, and makes closing the door much easier. A wrench is another simple machine that amplifies a force applied at one end to produce a rotation at the other. When trying to loosen a particularly stubborn nut, one sometimes makes use of a "cheater," basically an extension arm for the wrench by which the moment arm, and hence the applied torque, can be increased when the available force that can be applied is already at a maxi- mum. Returning to the example of the seesaw, a small child is able to lift a full-grown adult only when the fulcrum of the seesaw is placed near the adult's end (in a playground seesaw, the adult usually sits closer to the fixed fulcrum in the center). In this case, the moment arm for the child is increased, and the torque she applies is large enough to lift the adult up into the air, which the child could not accomplish without the mechanical advantage provided by the lever.

Levers also play a role in determining the strength of Ant-Man's tiny punch. Our arms are able to lift and throw by making use of the principle of levers. An object, let's say a rock, is placed on one end of the lever, which we'll call a "hand." A force is exerted by the compression of the bicep muscle, causing the other end of the lever (the forearm) to move down, which in turn raises the far end of the lever—that is, the hand holding the rock. The biceps pulls the hand upward—when we need to lower the rock, the triceps contracts and in so doing pushes the hand back downward. Muscles can only contract and pull; they cannot push. Consequently, an ingenious series of levers, consisting of muscles attached to various points of our skeletal structure, have evolved in order to enable a wide range of movement. The fulcrum of the lever that is your forearm is located at the elbow. It may seem odd to have both forces applied on the same side of the fulcrum, but this type of lever is essentially the same as a fishing rod, where the force applied to one end—very near the fulcrum located near the reel—causes a rotation and consequent lifting of a fish at the other end of the rod. Your bicep applies a pulling force approximately two inches in front of your elbow, and most people's forearms are fourteen inches long. The ratio of moment arms is thus 1:7, which means that the force applied by your biceps is reduced by a factor of seven at the location of your hand. That's right, reduced—in order to lift a rock weighing 20 pounds, your biceps has to provide a lifting force of 140 pounds.

A reasonable response to this news would be: What's the point in that? Why have a lever built into your arm that increases the force needed to lift an object? There wouldn't seem to be any point at all if the primary function of our arms were to lift rocks. Because the bicep is attached much closer to the fulcrum point (the elbow) than your hand, the biceps contracts two inches and the hand rises fourteen inches, due to the same ratio of 1:7 in moment arms. This ratio also holds when we want to get rid of this rock we are holding. In this case a muscle contraction of less than two inches produces a displacement of the hand of roughly twelve inches. This requires only 0.1 seconds to occur, and the hand holding the rock can get rid of it with a velocity of 12 inches in 0.1 second, or 10 feet per second (that is, 7 mph). This is a low estimate, and the average person can provide a much larger release velocity using other levers connecting her upper arm to her shoulder. A very, very small subset of the general population can throw baseball-size objects at speeds of up to 100 mph. And that's the point of the inverse lever in our arms—it's not intended to lift up large rocks; it's there to enable us to throw smaller rocks at high velocities. Those of our ancestors who were better rock- or spear-throwers were, on average, better hunters. Being a better hunter increased one's chance of securing dinner, and that in turn increased the odds of getting a date. In this way, certain hunters were able to pass these "good throwing-arm" genes down to their progeny.

Meanwhile...I haven't forgotten about Ant-Man trapped in a vacuum-cleaner bag. For the tiny crime-fighter, all length scales are obviously reduced, but the ratio of moment arms of 1:7 in his arms still holds for Henry Pym, regardless of whether he is ant-size or normal height. Punching involves the same muscles and similar motions as throwing, only instead of a rock, one is throwing a fist. The force provided by your muscles does not depend on their length, but on their cross-sectional area (that is, the area one would measure in a magnetic resonance imaging slice though the thickness of the bicep, not the surface area around the outside of the arm). If Ant-Man is 0.01 times his normal height, then the force his muscles can provide is reduced by a factor of $(0.01)^2 = 0.0001$. If Pym can punch with a force of two hundred pounds when full size, his scaled-down punch delivers a wallop of two hundredths of a pound. At his miniature size, his fist is much smaller and has a cross-sectional area of 0.0005 square inches (assuming his hand is just a millimeter wide). The pressure that his punch provides is defined as the "force per unit area," which is 0.02 pounds divided by 0.0005 square inches—or 40 pounds per square inch.* This is to be compared with his normal-height punching force of 200 pounds divided by his normal-size fist's cross-sectional area of 5 square inches, for a pressure of 40 pounds per square inch. That is, Henry Pym's punches exert the same pressure when he is ant-size as they do when he's at his normal height. If he can punch through the vacuum bag while normal sized, then he can do so at his reduced height. It appears that Ant-Man can indeed punch his way out of paper bag. In this way, he serves as a role model and inspiration to all comic-book fans.

---

* When concerned about fracturing or tearing a surface, we must consider the force and the area over which it is applied. When attempting to cross a barely frozen lake that has only a very thin layer of ice on its surface, you are more likely to successfully get across wearing snowshoes than stiletto high heels. Your weight is the same regardless of footwear, but the higher pressure in the high heels will lead to an icy dunking.

WHY BEING BITTEN BY A RADIOACTIVE SPIDER ISN'T ALL IT'S CRACKED UP TO BE

While we're on the subject of one's strength while the size of an insect, I would like to take a moment to dispel a myth concerning Spider-Man. As we have just argued, if Henry Pym shrinks at a constant density, then while the force of his punch is not as great as when he is at normal height, the pressure his fist is able to supply to an unsuspecting vacuum-cleaner bag is unchanged. A common misconception is that this scaling works in both directions, so that if one were to be bitten by a radioactive spider, just to take a random example, then one would gain the proportionate leaping ability of a spider. That is, if a spider or flea can jump one meter high—which is roughly 500 times higher than its body height—then a human with a comparable leaping ability would be able to leap a distance roughly 500 times his body height. For someone six feet tall, this implies a jumping range of 3,000 feet! If this were indeed the case, then Spider-Man would have the Golden Age (pre-flight, pre-yellow-sun-derived superpowers) Superman beat by—and here the expression could be taken nearly literally—a country mile. However, this is nowhere near the case. If Peter Parker did indeed gain the leaping ability of a spider, then he would be able to jump the same distance as a spider—namely, one meter. For the sake of exciting and engaging comic-book stories, it is a good thing that Stan Lee and Steve Ditko did not understand this scaling problem. Let's see where they went wrong.

What determines how high you can leap? Two things only: your mass and the force your leg muscles can supply to the ground. These two factors determine how much acceleration you can achieve as you lift off the ground. Once you are no longer in contact with the pavement, the only force acting upon you is gravity, which slows you down as you ascend. So there are two accelerations we have to concern ourselves with: the initial liftoff boost that gets you airborne, and the ever-present deceleration of gravity that eventually halts your rise. Once you are moving with some large velocity v, the height h you will climb is given by the familiar formula from before $v^2 = 2gh$, where once again g represents the deceleration due to gravity.

There is a surprising aspect of this equation that we have not yet remarked upon—namely, that nowhere does the final height that the leaper reaches depend on the mass of the person jumping! Big or small, if you start off with a velocity v and the only thing pulling you back to Earth is gravity, then your eventual height depends only on the deceleration due to gravity g and your initial velocity v. Of course, there is another acceleration that enters into the leap—that provided by your leg muscles at the start of the jump. And this acceleration does depend on the mass of the leaper. Using Newton's second law of motion, that force equals mass times acceleration, or $\mathbf{F = ma}$, it is clear that for a given force F, the larger the person (that is, the bigger his mass m), the smaller will be his acceleration a, and the less of an initial liftoff velocity he will achieve. A smaller starting velocity means a lower height h you will be able to jump.

It's not that spiders are such great leapers that they can jump many times their body length. Rather, it's that small insects have tiny muscles (providing a small force), but they only have to lift an equally tiny mass to leap one meter, which just turns out to be many times larger than their size. Humans have much bigger muscles than spiders and can achieve much greater forces, but they have to lift much greater masses, so the net effect is that the height they can jump is also about one meter. Of course, some humans such as Olympic high jumpers can leap much higher than this, while most of us slugs can jump barely more than a third of a meter (that is, one foot). In fact, for a flea to leap two hundred times its body length requires a great deal of cheating on nature's part: In addition to being particularly streamlined to minimize air drag, the flea pushes off with its two longest legs to maximize the lever arm. These are its hind legs, so in fact fleas always jump backward when they alight.

It is a natural mistake when scaling up the abilities of the insect and animal kingdoms to human dimensions to assume that it is the proportions that are important, rather than the absolute magnitudes. In the nineteenth century, many distinguished entomologists made the same error. As succinctly put in a footnote in the classic *On Growth and Form* by D'Arcy Thompson: "It is an easy consequence of anthropomorphism, and hence a common characteristic of fairy-tales, to neglect the dynamical and dwell on the geometrical aspect of similarity." But such misconceptions make for much more interesting fairy tales, and comic-book stories.

THE HUMAN TOP GOES
OUT FOR A SPIN—

ANGULAR MOMENTUM

WHEN ROSCOE DILLON was entering adolescence, he developed an obsession that would consume his life as an adult. What makes Roscoe unique is that unlike most young teenagers, his fixation involved spinning tops. Whether it was a simple children's toy or a sophisticated gyroscope, Roscoe was fascinated by tops. But he gave up playing with them as an adult, when he drifted into an unsuccessful life of crime. During his second stint in prison, he concluded that he needed a gimmick in order to be a successful thief, and saw that his youthful love of tops could be just the novel angle that his criminal career lacked. Upon release from the penitentiary, he studied all aspects of top rotation, and constructed special tops that emitted gas bombs, bolos, or entangling streamers, with which he would embark on what initially appeared to be a successful crime spree. Naturally he committed these thefts wearing a bright green unitard accented with narrow horizontal yellow rings running along the length of his body—the better to strike fear into the hearts of men. He also trained himself to spin rapidly about an axis passing through the length of his body, whereby he discovered that "the spinning action increases my brain power!"*

Of course, having set up his base of operations in Central City, he soon attracted the attention of the Crimson Comet, as relayed in "Beware the Atomic Grenade" (always sound advice) in *Flash # 122*

---

* What?

from August 1961. Initially the Top was able to escape from the Scarlet Speedster, but he eventually overplayed his hand. The Rotating Rogue managed to trap the Flash inside a giant "atomic grenade" that spun about its central axis like a top. Dillon threatened that unless all of the world's governments made him, the Top, the supreme ruler of the world (which is quite a jump from the beginning of this comic-book story, where we meet Dillon stealing the payroll from Wimbel's Department Store (note to lawyers—*not* to be confused with the major 1960s retailer Gimbels department store), his atomic grenade would explode and destroy half of the planet. In case the world's governments did not accede to his demand, Dillon had a backup plan. The Top planned on relocating to the other side of the globe, in North Africa, where he would be "safe from harm!" The devastation that would visit his "safe" side of the Earth if the other half were obliterated by atomic devastation seems to not have occurred to this "mentally enhanced" master criminal. Not that you should be worried, Fearless Reader, as the Flash eventually managed to run at such a great speed around the atomic grenade that he built up a mass of compressed air that launched the bomb at escape velocity,* never to return to Earth. He then searched the other hemisphere until he located the Top, and delivered him in to the Central City police. In fact, Flash stories in the 1960's often ended with the perpetrators of attempted global terrorism being handled as a local, municipal matter. It should be noted that all of the Top's battles with the Flash ended this same way—with the Top behind bars. In appears that Dillon's fixation with tops was not exactly the key missing ingredient that would propel him to a successful crime career.

* * *

DC Comics was not the exclusive home of rotating supervillains—Marvel's Giant-Man was bedeviled by his own whirling dervish: Dave Cannon, also known as the Human Top. Cannon was a mutant whose superpower involved the ability to spin at high speeds. In his first appearance in 1963's *Tales to Astonish # 50*, the Human Top battled Giant-Man and the Wasp when they attempted

---

* Flash Fact: An object will escape the Earth's gravitational field if launched with a speed of seven miles a second!

to stop his theft of the payroll of Danly's Department Store (the elucidation of the basis for the attraction between spinning supervillains and department-store payrolls remains a great open question in science). One difference between DC Comics' Top and Marvel's Human Top, aside from the fact that Marvel's villain's code name proudly proclaims his species (and while the Top was originally just a guy who liked to spin, the Human Top was in fact a mutant), is that Cannon never claimed that his spinning invoked any improvement in his mental faculties. In fact, he was considered so irresponsible that he was dropped from the Masters of Evil by Egghead. When other villains do not consider you up to the same high standards exhibited by the Beetle and the Shocker, you really should take a long look at your life. Cannon modified his costume, adopting a helmet that looked like the top half of a bright green missile with two broken hockey sticks protruding from it, going shirtless, and changing his moniker to Whirlwind. Shockingly, the addition of a green pointy helmet did not earn Whirlwind any new respect.

The difficulties that the Top and Whirlwind had breaking into the upper echelons of villainy is surprising, for their superpower involves one of the most important concepts in physics—Angular Momentum. We have discussed "linear momentum" in Chapter 3 as part of our description of the physics underlying the death of Gwen Stacy. There we did not need the modifier "linear," because only one type of momentum was discussed—the straight-line kind. Now we'll consider a more general situation where an object rotates or orbits about an axis. Linear momentum is simple, as there is only one way that an object can move in a straight line, but there are in principle an infinite number of axes that a mass can revolve about. The axis of rotation could pass through the spinning object, as in the case of the Top and Whirlwind, who twirl about an imaginary line passing through their body from their feet to the top of their head. But the axis of rotation need not pass through the volume of the object, such as when the mighty Thor twirls his mystical hammer Mjolnir. In this case, the axis of rotation is a line at a right angle to the plane formed by the spinning hammer, along the length of Thor's outstretched arm. Alternatively, the axis of rotation could be a great distance from the object in question, as in the Earth orbiting the sun, where the trajectory of our planet defines a disc (technically an ellipse), with the axis of rotation again being perpendicular to this plane and passing through the sun.

The Principle of Conservation of Linear Momentum states that a mass moving in a straight line will continue to do so if no outside forces act upon it—and that when such forces are present the change in momentum can be related to the product of the force and the time it acts, which we called the Impulse in Chapter 3. For Gwen Stacy, knocked from the top of the George Washington Bridge by the Green Goblin, her momentum steadily increased as she fell due to the external force acting upon her—namely, gravity. In order to change this momentum when Spider-Man stopped her descent, he needed to apply another force, through his webbing.

Physics has identified a corresponding Principle of Conservation of Angular Momentum that has important similarities to the Conservation of Linear Momentum. Linear momentum is mathematically defined as the product of an object's mass and velocity. The inertial mass reflects how difficult it is to change the motion of the object—it is easier to deflect a mosquito than a Mack truck. For a given mass, the larger its velocity, the greater its momentum and the more force is needed to alter its motion.

Similarly, an object's angular momentum is mathematically defined as the product of its "moment of inertia"—which reflects how difficult it is to rotate the object about a particular axis—and its rotation speed. An object rotating about an axis passing through itself or orbiting about some external axis will continue to do so, unless some outside force acts to change its rotation.

It is easier for an ice skater to twirl about an axis passing along the length of her body, the way the Top or Whirlwind does, if her arms are at her sides. Extending her arms away from her body places more mass further from the axis of rotation, and increases her moment of inertia. The more weight distributed away from the axis of rotation, the harder it is to spin. In this case we say that the moment of inertia is larger. For a given object, the faster the rotation speed, the greater the angular momentum. The harder it is to rotate an object at a given rotation speed, the harder it is to stop its spinning, and the larger the angular momentum.

An object moving in a straight line has a constant momentum, if no outside force acts on it. Thus, if it were to lose some of its weight, its speed would have to increase in order to keep the product of mass and velocity constant (that is, the linear momentum unchanged). This is the principle by which rockets and jets fly (hey—it *is* rocket science after all!). Similarly, if a spinning object changes its distribution of mass, then it alters how easy or difficult it is to rotate about an axis, that is, its moment of inertia. If there is no outside torque, the angular momentum cannot change, so an increase in the moment of inertia leads to a decrease in the rotation speed. When a figure skater wants to twirl at a faster rate, he or she pulls their arms in toward their body. Having more mass closer to the axis of rotation makes it easier to spin, and the decrease in the moment of inertia leads to a faster spinning rate. Whirlwind's motivations in designing a helmet with partial hockey sticks protruding from the sides, moving mass away from the rotation axis and increasing his moment of inertia and thus making it harder for him to spin further demonstrates that he is not the sharpest knife in the supervillain kitchen drawer.

Recall from our discussion of Ant-Man punching his way out of a paper bag (in Chapter 8) that torque is defined as the ability to rotate an object. Mathematically, it is described as the product of a force and the distance between where this force is applied and the axis of rotation. When the force acts right at the axis of rotation, it will push the object, but not change how it is spinning. Any change in the angular momentum of an object is determined by the product of the external torque and the time for which it acts, similar to the corresponding equation for Impulse (a force multiplied by time) for linear momentum. What we have seen in the preceding chapters is that there can be no change in the motion of an object, whether it is moving in a line or rotating, if there is not an outside force acting on it. No force, no change. A twirling object, isolated in outer space, will never stop spinning, for if there is no outside force to change its rotation, why would it? This is the basis by which gyroscopes, as part of inertial guidance systems, keep airplanes and missiles flying at an even keel.

Roscoe Dillon may have been wrong that developing a persona revolving around tops* would lead to a successful criminal career, but he was right about at least one thing in *Flash # 122*—a gyroscope is at its heart a spinning top. The crucial element of the gyroscope that makes it useful for determining orientation is that

---

* Sorry.

it is a spinning top isolated from the rest of the world. In this way, its Angular Momentum does not change. Imagine a top, rotating in a clockwise direction when looked down upon from above. Curl the fingers of your right hand in the direction of the top's rotation and stick out your thumb; your thumb will point down toward the surface on which the top is spinning, The direction of your thumb can be considered to be the "direction of the top's angular momentum" (for consistency's sake you should always use the same hand—either right or left—when comparing the rotation of different objects. The convention is to use the right hand, but which hand is not as important as always using the same hand). A gyroscope consists of a spinning disc, held in "gimbals" that act to isolate the disc from outside torques. Recent innovations in gyroscope design employ electrostatically levitated micro-discs, which avoid mechanical couplings altogether.

For a disc spinning counterclockwise, the angular momentum of the disc "points" up—let's call this direction "north." This is the "direction" of the angular momentum, and in the absence of an outside torque, nothing can change this. So if I want to turn the disc such that the direction of its angular momentum is east, I can only do so if I apply a twisting force (a torque) to the disc. Try to turn a spinning bicycle wheel, and you'll note that it resists moving away from the direction of its rotation. For a gyroscope, the spinning disc is housed in the gimbals, so that if the gimbals are turned, the rotating disc will keep spinning in the same orientation.

This is very useful as a guidance system. If the spinning disc of the gyroscope is rotating in such a way that the direction of its angular momentum is pointing due north (for example), then it will keep pointing due north, regardless of the orientation of the surrounding jet plane or missile. If your rocket starts to veer from a heading relative to north, adjustment thrusters can return the projectile to its proper course. With another gyroscope pointing due east and a third pointing perpendicular to the north-east plane, one can accurately position oneself in three-dimensional space, without visual information or any signals from the ground. Not bad for a child's toy.

Whirlwind's mutant ability to twirl about at great rotational velocities proved to be too much for the police force when he initially embarked on a life of crime. Eventually he acquired a cos-

tumed superhero nemesis, as do all costumed supervillains:
Giant-Man. Just two issues previously, before he shifted gears and
became Giant-Man, Henry Pym was fighting crime and foiling
Communist saboteurs as the astonishing Ant-Man. David Cannon
was therefore particularly unlucky in that he had to deal with
Pym's twelve-foot-tall alter ego, rather than his more diminutive
quarter-of-an-inch persona. However, his command over angular
momentum enabled Whirlwind to be able to look his supersized
adversary in the eye—so to speak.

As shown in fig. 16 from *Avengers # 139*, by increasing his rotational speed, Whirlwind is able to create a cushion of air that carries him aloft so that he can fight Giant-Man face-to-face.
(Technically, at this stage, Henry Pym was using the superhero
identity of Yellowjacket, a character who, like Ant-Man, employed
the power of miniaturization to fight evildoers. However, when
Whirlwind threatened his wife, Janet Van Dyne (who also is a
shrinking superhero, known as the Wasp), who was lying ill in a
hospital, Pym thought it best to battle Cannon as a twenty-foot-
tall protective husband rather than an insect. Whirlwind boasts
that by increasing his rotation speed he can "fly like a helicopter,"
and thus match Pym's height advantage.

Helicopters hover in the air over a stationary point through the
same mechanism that enables planes to fly—Newton's third law
of motion, which states that forces come in pairs. The blades of a
helicopter are angled so that they deflect and push air molecules
downward as they rotate, providing an upward thrust on the
blades themselves. It is doubtful that this is what Whirlwind was
doing as he used his mutant power of "super-rotation," for no
other reason than it is not obvious in fig. 16 what part of Cannon's
anatomy is filling the role played by the helicopter's blades (the
blades on his helmet being too small for the job).

More likely he is using the same trick that the Flash employs, as
described in Chapter 4, when he was able to levitate the crook
Toughy Boraz* by creating a partial vacuum in the center of a column of rotating air. In this way David Cannon is making use of the
physics of tornadoes, where the fast moving air in the twister cre-

_____

* Only in Silver Age comic books would a serious criminal go by the name
"Toughy."

Fig. 16. Whirlwind, utilizing his mutant ability to spin very rapidly about a vertical axis passing through his body, demonstrates the decrease in pressure in the center of a twister.

ates a low-pressure region in the center. The faster the rotation, the lower the pressure and the greater the lifting force. Conservation of angular momentum is also the reason that helicopters have a tail rotor that is positioned at a right angle to the main lifting blades. Without this tail stabilizer, there can be no change in the total angular momentum of the helicopter when it's off the ground and isolated from any external torques. As the blades spin faster, their angular momentum increases. However, as an isolated system, the total angular momentum of the helicopter is a constant. Consequently the increase in the blades' angular momentum can only be achieved if it is compensated by the body of the helicopter itself rotating in the opposite direction. This ensures that the sum of the angular momentum of the blades and the helicopter body does not change. The body of the copter is fairly massive and has a large moment of inertia, so it won't (thankfully) spin as fast as the blades, but it will twirl around, making navigation a challenge. This is prevented by the tail rotor, whose rotation in a direction perpendicular to the spinning top blades provides a balancing torque that keeps the body of the helicopter pointing in one direction.

Providing a counter-rotation to balance and negate the angular momentum of a rotating object, such as a tornado, is an old superhero trick. Whenever the Flash needed to counteract a destructive meteorological phenomenon such as a cyclone, he would do so by running at superspeed around the cyclone in the opposite direction of the twister's rotation, creating what is technically known as an anti-vortex. The angular momentum of the anti-vortex is equal and opposite to that of the original vortex, and the two annihilate, as if the twister did not exist.

Speaking of rotation, surprisingly enough, a signature characteristic of a magic-based hero turns out to be physically accurate, provided we allow the standard miracle exception, of course. When the Norse god Thor needed to travel quickly from one location to another, he used his great strength to twirl his hammer at high speed. Throwing it in the direction he wanted to go, he would momentarily let go of the hammer's handle strap and then grab on to it again, flinging himself through the air as an unguided missile. This would appear to be a perfect violation of the Principle of Conservation of Momentum. In fact, in *Bartman Comics # 3* (featuring the adventures of Bart Simpson's superhero alter ego), Radioactive Man is so angered when he spies a Thor-like character taking flight in this manner that he socks the mock-Thor, intoning, "This is for breaking the laws of physics!" And yet, such a means of transportation is physically plausible.

When Thor twirls his hammer, the mighty Mjolnir, he plants his feet firmly on the ground. This is presumably what makes the X-Men villain the Blob so difficult to move—his mutant ability enables him to plant his feet so strongly that he effectively couples his center of mass to the Earth's, such that dislodging the Blob requires moving the entire Earth unless the connection is broken. When Thor is ready to let fly, all he must do is jump slightly (breaking his connection with the Earth) at the moment he throws his hammer in the desired direction. He doesn't even need to go through that business of releasing and re-grabbing the handle strap. If one is as strong as a thunder god, one can use this technique to fly through the air with the greatest of ease. No wonder they named a day of the week after this guy!

* * *

The universality of angular momentum is reflected in the fact that nearly everything undergoes some form of rotation, from the electrons, protons, and neutrons within all atoms to massive clusters of galaxies. We will consider subatomic rotations in Chapter 19, when we discuss how they can account for magnetism in iron and certain other metals. Studies of galactic rotations provide evidence for phenomenon even more mysterious than magnetism— Dark Matter.

Back in 1933, when Siegel and Shuster first conceived of Superman, astronomer Fritz Zwicky proposed that most of the mass of the universe was invisible, demonstrating once again that comic-book writers have always had to struggle to keep up with scientists' seemingly fantastic proposals. (This is the same Fritz Zwicky who, along with Walter Baade, also in 1933, reported the first observation of neutron stars.) Observations of the Coma cluster of galaxies indicated to Zwicky that they were rotating too fast to be stable. In Chapter 7 we discussed centripetal acceleration, which is present whenever any object moves in a circle or parabolic arc, such as when Spider-Man swings on his webbing through New York. As Newton's second law of motion insists, any acceleration must result from an external force. For Spider-Man the force causing his centripetal acceleration is provided by the tension in his webbing. For the rotating Coma cluster, it is the gravitational attraction of all of the stars in the galaxies that holds it together as it spins about an axis of rotation extending from the center of the cluster.

Consider the spiral galaxy NGC 3198, in the rather nearby Ursa Major constellation. As fortune would have it, this disc is oriented so that from Earth it appears nearly edge on, so we see little more than a thin strip. It is as if this spiral galaxy were an old-style music record on a turntable, and we are able to look along the edge of the record. We know that this massive collection of stars is rotating about an axis perpendicular through the plane formed by the galactic disc, from the Doppler effect. Recall our discussion in Chapter 4 of the shift in wavelength of sound or light waves when either wave source moves relative to the observer. If the source is moving away from the observer, the wavelength of light is increased compared with a stationary source, and is decreased when the source moves toward the observer. The faster the source is moving, the greater the wavelength shift, which is the basis for radar guns that measure a moving object's speed.

The Doppler shift for light emitted by stars at the edges of rotating spiral galaxies indicates the speed at which the disc of stars is rotating. The intensity of light provides a determination of how many stars are present in the galaxy and how massive they are. But there must be more mass in this galaxy than we can see (which basically means more stars—for everything else is too small to contribute significant mass) in order to hold the galaxy together at the measured rotational speed. A lot more mass—as in nearly ten times more mass. If Spider-Man tries to swing too fast at the end of his line, so that the tension required is more than the webbing can support, the line will snap and he will be flung away from the smooth trajectory he started. Similarly, galaxy NGC 3198 should not be a stable object in the night sky, but should have flown apart based upon its rotational speed and the gravitational pull that can be accounted for by the stars it contains. Searching for dust clouds in the infrared portion of the spectrum, or for any other astronomical objects that might be hidden, has proven insufficient to account for the "missing mass."

In fact, this problem is found throughout the universe. There are regions of space that contain gases that emit X-rays that are very hot—one million degrees hot. These gases collect around galaxies, yet once again there is not enough mass in the observable stars to provide sufficient gravity to hold these hot gases in place. Astronomers have concluded that there is not enough mass in stars, interstellar dust, or planets to account for the gravitational stability of these hot gases. There must be another, unseen source of gravity that we cannot yet detect.

There are only two options—either we don't really understand gravity (which is possible, although the current theory of gravity, expressed in Einstein's General Theory of Relativity, has been experimentally verified enough times for us to have some measure of confidence in it), or there is a lot of mass in the universe that we just can't see. Scientists have concluded that the second option is the more palatable, and there is general agreement that nearly all of the universe's mass is actually contained in what has been termed Dark Matter. In fact, as will be discussed in Section 2, there is in fact a precedent, dating back to the 1930s, for inventing a "miracle mass" to avoid having to throw out all previously accepted and verified physics principles. As of this writing, scientists do not know anything, really, about the composition of this ubiquitous material. Everything we see in a clear night sky, the billions of galaxies, each cluster containing billions of stars—is nothing more than a few percent of all of the mass and energy in the universe. The Top and Whirlwind aspired to be master criminals, but they never dreamed that the source of their super-power—angular momentum—could lead us to conclude that 96 percent of the universe is missing. The ultimate crime, indeed.

IS ANT-MAN DEAF, DUMB, AND BLIND?—

SIMPLE HARMONIC MOTION

THERE ARE SEVERAL SUPERHEROES whose special power is the ability to shrink. In addition to Ant-Man and the Wasp in the Marvel universe, the Atom, Elasti-Girl of the Doom Patrol, and Shrinking Violet of the Legion of Super-Heroes in DC Comics all share the ability to miniaturize themselves. The "explanation" for their shrinking powers varies, but the laws of physics dictate one certainty about these tiny superheroes: They will find it very difficult to communicate. I'm not referring to Ant-Man's and the Wasp's relationship problems that led to their divorce, but rather the physical limitations involving conversations with anyone in the non-shrunken world.

If you're only a few millimeters tall, no one can hear you, nor will your own hearing be too keen, so you will have to depend on nonverbal means to communicate with normal-size people. As Ant-Man shrinks, his voice becomes higher pitched, until when he has reached the size of an ant, his normal speaking voice will be near the upper range of normal-size-human hearing. At the same time, his hearing threshold also shifts to the higher end of the register, such that he will miss most of what people are saying to him. To make matters worse, everything our tiny hero sees will be an out-of-focus blur. Let's see why being an "inch-high private eye," while not cutting you off completely from the outside world, will nevertheless lead to a host of problems.

First we address the basic question: What determines the range at which we are able to speak and hear? To answer this, let's look at a pendulum. A pendulum is just a mass connected to a thin

string (we'll ignore the weight of the string) with the other end
connected to a frictionless pivot point on the ceiling. The mass is
usually assumed to be some dense sphere, such as a billiard ball or
a bowling ball, but it could be a rock or Spider-Man. This mass is
lifted to some height, such that the string makes a small angle
with the vertical direction and, once released, the forces acting on
it are (1) gravity, always pulling straight down in the vertical direction, and (2) the upward tension in the string. The tension's direction makes an angle relative to the vertical and is continually
changing as the mass swings back and forth. Part of the bob's
weight can be viewed as directed along the line of the string, while
the part of the weight that is not compensated by string's tension
is the force that changes the mass's speed and is responsible for the
acceleration of the mass as it moves to and fro.

The time that a pendulum takes to move back and forth—from
some high initial point, through a complete arc, and back to its
original position—is called the "period," and its motion is called
"periodic." Whether it's Spider-Man swinging back and forth on
his webbing, or a billiard ball attached to fishing line, the time to
complete a full oscillation depends on only two factors: the acceleration due to gravity (what we have called g) and the length of the
string. Surprisingly, one thing that the period does not depend on
(at least for small angles of oscillation) is the initial height at
which the mass starts its motion.

Galileo was perhaps not the first person to notice that the
period of a pendulum is an intrinsic property and independent of
how high or low the swinging mass's starting point is, but he is
properly credited for determining what controls the oscillation
rate. As surprising and counterintuitive as it seems, the time necessary for a playground swing to go back and forth depends neither
on how heavy or light the person sitting on the swing is, nor on
how far back he starts his motion, but only on the length of chain
between the seat of the swing and the top pivot point of the swing
set. We are assuming that the swing is not being pushed by a stationary helper nor is the person in the swing pumping his legs as
he moves back and forth. It is certainly true that the higher the
starting position of the swing, the faster you will be moving at the
bottom point of your swing, since the component of the tension in
the string that deviates from the vertical (responsible for accelerating along the swing) is greater the larger the starting angle.

Shouldn't this mean that it should take less time to complete an oscillation? No, because while you may be moving faster owing to your higher initial position, you have farther to travel until you reach that bottom point of the arc. The combination of the greater speed but longer distance to be traversed balances out, such that the time needed to complete the arc remains the same, regardless of the starting point. This is why a pendulum or any other device undergoing simple harmonic motion makes a great timekeeper. Two identical clocks, using either a swinging bob like that in a grandfather clock, or a coiling and uncoiling spring like that found in old-fashioned pocket watches or metronomes, will keep identical time, independent of the starting push that begins the oscillation. A metronome is an upside-down pendulum, and its frequency is independent of how it is set in motion, but it can be altered by changing the location of the mass on the swinging arm (which has the effect of changing the length of the pendulum).

If the period of a pendulum does not depend on the starting point of the swing, why does it depend on gravity and the length of the string? It's not hard to see that the weaker the acceleration due to gravity, the less force there will be pulling on the mass and the slower will be its motion. On the moon a pendulum will take longer to complete a cycle than it does on Earth, and in deep space, where we may take the acceleration due to gravity g to be zero, the bob will never move at all. Why does the length of the string enter into the pendulum's period? Because of torque. There is a torque rotating the bob along its trajectory—the force is the component of the bob's weight along the arc and the moment arm is the length of the pendulum's string. The mass shows up on both sides of the rotational form of $F = ma$, and cancels out. The two remaining factors, the acceleration due to gravity and the moment arm l determine the time necessary to complete a cycle.

The frequency of the pendulum—the number of back-and-forth oscillations it completes in one second—is just the inverse of the period, defined as the time needed to finish one cycle. If the period is one tenth (0.1) of a second, then it will have a frequency of ten cycles per second. The shorter the period, the higher the frequency. The square of the period, in turn, is proportional to the ratio of the length of string in the pendulum l and the acceleration due to gravity g. That is, the $(\text{period})^2 = (2\pi)^2 \times (l/g)$. To explain why it is the square of the period, and not just the period, that depends on the ratio of l over g, and why the factor of $(2\pi)^2$ enters, would require us abandoning our "algebra-only" pledge. For our purposes, the important point is that one has to increase the length of the pendulum's string by a factor of four to double the period. Conversely, shrinking the length of the string (say, by using Pym particles) decreases the period, and the smaller the period, the higher the frequency.

Speaking involves the generation of sound waves, which are created by forcing air passed vibrating vocal cords. A human vocal cord is not a mass swinging back and forth on a string, but the beauty of the pendulum as a description of simple harmonic motion is that it captures the important physics of any oscillating system.* When Henry Pym shrinks down to ant size, he reduces his dimensions by roughly three hundred times. The fundamental frequency of the oscillator is, correspondingly, seventeen times bigger (that is, by the square root of three hundred). Normal human speech occurs at a pitch of roughly two hundred cycles per second, but for an ant-size person the frequency is shifted up by this factor of seventeen, to 3,400 cycles per second. Our hearing range extends from twenty cycles per second at the low end up to twenty thousand cycles per second, so we should still be able to hear Ant-Man, but he will have a high-pitched voice, as his chest cavity similarly shrinks. When they hear a quarter-of-an-inch-tall superhero order them to surrender in such a squeaky voice, it is surprising that Ant-Man's foes do not succumb to fits of laughter rather than his tiny right hook.

Not only will his voice change as he shrinks, but Ant-Man's hearing will also be affected by his reduced height. The resonant frequency of a drum also rises as its diameter shrinks. A large bass drum has a deep, low tone, while a smaller snare drum emits a higher pitch when struck. When Dr. Pym's eardrums shrink upon exposure to Pym particles, the frequencies that he is able to detect shift accordingly. (The physics underlying the range of human

---

* The periodic harmonic motion of a swinging pendulum is one of the cornerstones of much of the theoretical modeling one does in physics. When attempting to describe some complicated natural phenomenon, we so often begin by invoking a simple pendulum that one is tempted to paraphrase Yogi Berra and state that 90 percent of physics is "simple harmonic motion" and the other half is the "random walk."

hearing is actually pretty complicated, but for our purposes we'll assume that it is determined by the eardrum.) The lowest frequency he can hear when at his normal six-foot size is roughly twenty cycles per second, which upon shrinking becomes seventeen times greater at nearly 340 cycles per second. A normal person's speech, at a pitch of two hundred cycles per second, will therefore be below the range of detection for our tiny titan. For this reason, Ant-Man and his miniaturized colleagues will need to be astute students of body language as they interact with the normal world.

In addition to the change in the threshold frequency of his eardrums, Henry Pym's hearing sensitivity will also be affected as he shrinks to the size of an ant. As the vocal cords vibrate back and forth, they cause alternating compressions and expansions on the air rushing past them, forced through the throat by contractions of the diaphragm. This variation in density is slight—only one part in ten thousand on average distinguishes the adjacent regions of compressed and rarefied air. The larger the density variation, the greater the volume or loudness of the sound wave. You only have control over the initial density variation as the sound leaves your mouth. The compressed region of air expands, compressing the region in front of it, which in turn expands and squeezes the next region of air. What you hear is the instruction set for the sound wave, generated from your vocal cords and transmitted to your ears. The air from your mouth does not physically travel from you to the listener. If I tell you that I had garlic for lunch, you hear this information before you come close enough to receive independent confirmation of this fact. As the information is spreading out in all directions, the farther the hearer is away from the speaker the more the variation in air density—the sound wave—will be attenuated. Beyond a certain point the information will not be detectable at all.

Alternatively, if one is too close to the source, the eardrum is unable to linearly respond to the density variations, and the ability to distinguish differing sounds is degraded. This can be handy, as the DC Comics miniaturized hero the Atom discovers in "The Case of the Innocent Thief" in *Atom # 4.** In this story, a crook named Elkins discovers a hypnotic ray that compels people to obey his oral commands. While the Atom is only a few inches tall, the crook exposes him to this ray and yells a command forbidding the Atom to try to capture him. Yet almost immediately the Atom knocks out Elkins, using a pink rubber eraser as a trampoline to come within punching distance. The Atom is able to resist the mesmerizing command because, as he explains at the end of the story, "in his excitement, Elkins shouted his orders at me—words which sounded like thunder to me! And since I couldn't understand a word he said to me, I didn't have to obey him!" (Interestingly enough, my kids put forth this same argument nearly daily, though they have not yet mastered miniaturization technology, nor am I shouting.)

If the hearing and speaking issues weren't bad enough, Ant-Man's vision should also be blurred. The average spacing between adjacent peaks or valleys (that is, the wavelength) in the alternating electric and magnetic fields that comprise a light wave determines the color of the light. Let us say that on average, white light—which consists of light of all wavelengths from red (650 nanometers) to violet (400 nanometers) added together in equal magnitudes—has a wavelength of 500 nanometers (one nanometer is one billionth of a meter). In order for light to be detected, it must hit the rods and cones in the back of your eye, and in order to get to these photoreceptors, it must first pass through your pupil. This opening in the front of your eye, depending on the brightness in the room in which you are reading this, is roughly five millimeters in diameter. One millimeter is equal to a million nanometers, so the opening in your pupil is roughly ten thousand times larger than the wavelength of visible white light. From the point of view of the light waves, the pupil is a very large tunnel through which they can easily pass. When Ant-Man shrinks to the size of an insect, however, the opening in his pupil will be three hundred times smaller than it was at his normal height. The orifice in his eye is now roughly a factor of thirty times larger than the wavelength of visible light, which is still 500 nanometers. The light waves can still fit into this "tunnel," but just barely.

To understand what the consequences are when the size of the opening is only a few times bigger than the wavelength of light, consider water waves on the surface of a large lake. A channel is

---

* We'll have more to say about the Atom and his shrinking ability in Chapter 13.

formed between two docks that ride low on the surface of the water. When the separation of the docks is very large, say half a mile apart, compared with the spacing between peaks of the water waves, the waves pass through this region with no noticeable perturbation. Right near the dock, as the waves break, there is a change in the wave front, but in the middle between the docks the waves are hardly affected by the docks. This is the situation for Henry Pym at his normal height, when his pupil is ten thousand times larger than the wavelength of light. For the miniaturized Ant-Man, it is as if the two docks narrow to a bottleneck, so that the separation of the docks is only a few times more than the separation between adjacent wave peaks. The waves still move through the constriction, but as they scatter off the edges of each dock they set up a complicated interference pattern on the other side of the obstruction. This effect is termed "diffraction" and is most noticeable when the dimensions of the object scattering a wave are comparable to the wavelength. If you were hoping to gain information about the cause of the water ripples by examining the wave fronts, you would have obtained a clean, sharp image when the docks were thousands of feet apart and a distorted and confusing picture when the docks were only separated by a few feet.

The effect for Ant-Man is that the image he observes through his shrunken pupil will be blurry and out of focus. This is why an insect's eye, and in particular its lens, is radically different from the lens in a human's or larger animal's eye. Insects use compound lenses that adjust for the diffraction effects. Even so, it would be hard for a fly to read a newspaper, even if he did care about current events. An insect's eye is very good at detecting changes in light sources (such as the moving shadow created by the rolled-up newspaper of doom), but poor at discerning the contrast between sharp edges. Consequently they rely on other senses, such as smell and touch (hair filaments detect subtle variations in air currents) to navigate their way through the wide world. Unfortunately for Ant-Man, the one sense that is least affected by miniaturization, smell, is the one that is the least sensitive in humans.

LIKE A FLASH OF LIGHTNING—

SPECIAL RELATIVITY

IN AN EARLIER CHAPTER I MENTIONED the sonic boom that the Flash creates whenever he runs faster than the speed of sound. Why is there a "boom" when an object moves at or faster than the speed of sound? And how can this help us understand Einstein's Special Theory of Relativity?

First, let's consider the boom, and then get to Einstein. Imagine yourself standing out in the countryside, and the Flash is running toward you at the speed of sound—that is, at one fifth of a mile per second. If he starts ten miles away from you, he'll reach you in fifty seconds. When he is ten miles away, he says, "Flash" and when he is only five miles in front of you, he says, "Rules." What do you hear? If the Flash was running slower than the speed of sound, then the "Flash" he spoke would reach the five-mile mark before he did, and then he would utter "Rules" while the "Flash" was about to reach your ears. You would clearly hear "Flash Rules," followed a short while later by the sound of the Scarlet Speedster running past you.

If the Flash were instead running faster than the speed of sound, then he would arrive at the five-mile point before the sound he emitted at the ten-mile mark. He would then say "Rules" and continue on toward you. Since the "Rules" has less distance to travel, it would reach you before the "Flash," so you would hear the words in the reverse order that they were spoken— that is, to you it would sound like he'd said, "Rules Flash." (Actually it would be more like "Selur Hsalf") This backward speech would not reach you until after he had passed you. Running faster than the speed of sound, he can cover the distance from five miles away to you in less time than the sound waves.

If he were running at exactly the speed of sound, then when he yelled "Flash" at the ten-mile point it would reach the five-mile mark at the exact same instant as the Tornado Titan himself did. When he then says "Rules," it takes off from the five-mile point toward you at the same moment that the "Flash" does, so that they both reach your eardrums at the same moment, twenty-five seconds later. You don't hear "Flash Rules" or "Rules Flash" or even "Suler Hsalf" but the two words superimposed at the same instant. Sound is a pressure wave, so the waves from the two words add up and create a larger vibration than if heard separately. The Flash would not even have to be talking or making noise as he advanced toward you—the disturbance created as he pushed the air out of his way would form a pressure wave that you would hear as a thunderous roar (or a "sonic boom") at the exact instant the Sultan of Speed cruised by. If the Flash ran faster than the speed of sound, this disturbance would still be created. In this case he would race past you in relative silence, and then later on the sonic boom, traveling at the speed of sound, would eventually reach you, with explosive consequences. (The "Rules Flash" he spoke in the previous paragraph would be lost in the sonic crash.) The "crack" of a gunshot, or of Catwoman's whip, are mini sonic booms created by the bullet or the tip of the whip moving faster than the speed of sound in air.

The danger posed by the indiscriminate creation of sonic shock waves by the Flash is recognized by current comic-book writers. In *DC: The New Frontier*, a 2004 revisiting of DC Comics' Silver Age heroes, set in the late 1950s when they historically made their first appearances, the writer Darwyn Cooke describes a scene where the Flash races from Central City (hazily located in the American Midwest) to Las Vegas, Nevada. In caption boxes that describe the Flash's thoughts as he runs cross-country, he tells us, "I wait until I'm clear of the city limits before I hit the sound barrier. That's one I figured out the hard way a few times. Flying glass and pedestrians don't mix." Indeed they don't, as graphically demonstrated in *Flash # 202 (Vol. 2)*. In this story, our hero has lost his memory and, in civilian clothes, does not realize that he possesses superspeed. Acting instinctually when mugged by a street gang, his high-velocity movements blow out every window on the block and cause massive structural damage to the surrounding buildings.*

Regardless of the order in which you hear what the Flash says, whether it's "Flash Rules" or "Rules Flash," if you have keen eyesight and can read lips you can be sure of the order in which the Flash actually said these words. This agreement is due to the fact that the light reflecting from the Flash travels much faster than sound (186,000 miles in one second, compared with one fifth of a mile per second). This is how we are able to determine the distance a thunderstorm is from us by comparing the timing between the lightning and the sound of thunder.†

But what if the Flash were running close to the speed of light? All sorts of strange things happen concerning length, time, and mass for objects moving near light speed, as described by Albert Einstein in 1905 in his Special Theory of Relativity (it's called "special" because it ignores gravity, which is accounted for in the "General" Theory of Relativity, developed in 1915). This is neither the time nor place to go into a full discussion of relativity. A fair treatment of the topic would overwhelm the present book. But I do want to mention a basic point about traveling near the speed of light that won't take too long, and that we'll build on when we consider the connection between electricity and magnetism in Chapter 19.

The Special Theory of Relativity can be boiled down to two statements that appear simple, but contain a wealth of physical insight. They are (1) nothing can travel faster than the speed of light (sorry, Superman and Flash), which is the same speed for everyone, no matter how fast they are moving, and (2) the laws of

---

* The intensity of the pressure wave from a sonic boom decreases with distance from the source. Consequently, as long as a supersonic jet stays well above street level, a similar concern over building damage can be avoided.
† The flash of lightning originating from the thunderhead covers a distance of one mile in roughly five millionths of a second (we are physiologically unable to detect events happening that quickly; for us it is instantaneous), while the sound of thunder that is created simultaneously takes nearly five seconds to reach us. Counting the number of seconds between the two events, and making use of the fact that sound takes five seconds to cover one mile, allows us to easily calculate the distance of the thunderstorm from us.

physics are the same for everyone, regardless of whether you are moving or not. The first point is the really weird one. If the Flash is running as fast as a speeding bullet, to us the bullet is traveling at 1,000 miles per hour, while to the Flash, running in the same direction at the same speed, the bullet appears stationary (which is why he is able to, "with a sweeping motion," pluck it from the air so easily). But the speed of light is 186,000 miles per second to both you, standing still, and to the Flash, regardless of how fast he runs. Even if he is racing at half the speed of light—at 93,000 miles per second—the speed of light relative to him is not 93,000 miles per second, but still 186,000 miles per second, the same as for you standing on the street corner. How can this be?

When the Flash runs toward you, from the Flash's point of view it is as if he is stationary but you are racing toward him. The Special Theory of Relativity states that for both you and the Flash, you must agree that the speed of light is 186,000 miles per second. In order for this to be true, Einstein argued that from your point of view, the Flash will appear thinner (that is, his length in the direction he is running will appear compressed) and time will seem to pass slower for him than for you. From the Flash's point of view, a yardstick he holds would still be one yard long, and his watch keeps time just as it always has, but to him it is you who are moving rather than him, and he will make similar determinations about you (your length will be shortened and time will move slower for you, as it appears to the Flash). This is because in order to measure the length of a yardstick that the running Flash is holding, for example, you have to consider the front and back ends of the stick, and record the times when they pass a given point. For two people moving relative to each other (say, a running Flash and a stationary observer) it becomes impossible for them to agree as to whether or not two things happen at the same time when the events are separated in space and time. The shortening of lengths as a function of velocity will play a key role in helping us understand how an electrical current generates a magnetic field in Chapter 18.

* * *

The concept that two observers can disagree as to the ordering of events was certainly not invented by Einstein, nor is it unique to the Special Theory of Relativity. As pointed out by Lev Landau and G. B. Romer in their excellent short book, *What is Relativity?*, even when both observers are stationary they can disagree depending on where they're situated relative to an object (Go to the house on the left. Which house? Which left? My left or your left?) Just such an ambiguity played a central role in the climax of "The Race to the End of the Universe," in the December 1967 issue of *Flash # 175.* Space aliens force Superman and the Flash to run a race across the Milky Way galaxy, ostensibly as a competition on which they intend to gamble. However, the aliens are actually disguised supervillain foes of the Flash, and their true intent is to destroy the Scarlet Speedster and the Man of Steel. At the end of the race, the Justice League of America watches on monitors, yet no clear winner is declared. Those members of the Justice League watching the finish line from monitors placed on the left-hand side of the line claimed Superman had won the race, while those watching monitors on the right-hand side of the finish line are sure that the Flash is the victor. While this comic provided a graphic illustration of one of the fundamental principles of the Special Theory of Relativity, namely that differing observers could disagree about the simultaneity of events when objects are moving at high velocities, at least one reader in 1967 found the lack of a clear winner at the conclusion of this story to be a titanic rip-off.

Information cannot travel faster than light, so there will always be a discrepancy as to the order in which events occur. In order to balance out so that the one thing everyone agrees on is the value of the speed of light, it will appear that lengths are shortened in motion and that time passes more slowly. Comic books frequently get this wrong when dealing with characters who can travel at the speed of light (such as Negative Man of the Doom Patrol and Captain Marvel—not the "Shazam" guy, or the late 1960s–early 1970s officer of the Kree military, but an African-American heroine in the Avengers in the late 1980s who could transform her body into coherent photons of light). The other characters in the story should not be able to see these heroes, as there's no way for light to be scattered from them if they are moving as fast as light. At best, they might appear as a spark of lightning from a great distance away, but would be invisible when nearby. The highest velocity the universe allows is the speed of light. As the Flash runs faster and faster, one would think that he should be able to pass this limit, but this can't happen (in the real, physical universe—not the comic-book universe where it happens all the time). To explain this phenomenon from the point of view of a stationary observer, it must be that the faster he goes, the harder it becomes for him to further accelerate. From Newton's second law (force is equal to mass times acceleration), we know that if the force that his running shoes apply stays constant but there is no corresponding acceleration, it must be because his mass has increased. So in addition to time seeming to be slower and lengths appearing to shrink, the mass of the running Flash will seem (to us stationary slowpokes) to increase the faster he runs. This is occasionally referred to in the comic books. In *Flash # 132*, the Monarch of Motion is attempting to run at light-speed, but finds that he is in fact "slowing down. I'm going slower and slower, despite all my efforts..." As shown in fig. 17, the "grim truth dawns on the harried speedster," namely that he can't run any faster because he is getting heavier. Of course, those of us with an understanding of Einstein's Special Theory of Relativity understand perfectly well what is happening to the Flash—an alien is shooting him with a "gravity increasing ray," as shown in fig. 18! But taken out of context, fig. 17 provides an excellent illustration of Einstein's Special Theory of Relativity.

In the more recent *JLA # 89*, the Flash has to move the entire population of 512,000 men, women, and children in Chongjin, North Korea, away from the imminent blast of an atomic bomb in a fraction of a second. In order to accomplish this feat, he must move at speeds very close to the speed of light. The relativistic

Fig. 17. *The Flash, attempting to run at light-speed, discovers that he becomes heavier the faster he runs in* Flash # 132. *Is this a textbook illustration of Einstein's Special Theory of Relativity?*

Fig. 18. *No! The Flash is the victim of an alien's "gravity increasing ray!"*

consequences of his high speeds are alluded to once he has saved the town and collapses to his knees on a hilltop. As described in a caption box, "as his body sloughs off the screaming aftereffects of near light travel, eyes of almost infinite mass turn towards the blaze engulfing Chongjin." Of course, to the Flash, he is stationary and it's the rest of the world that moves and gains mass. By the way, it was the realization that an increase in the kinetic energy of an object is directly connected to a rise in its mass that led Einstein to derive $E = mc^2$.

If nothing can travel faster than the speed of light, then how does the Flash's Cosmic Treadmill allow him to move forward or backward through time? I'm sad to say that time travel doesn't work in a reversible manner. The closer an object's velocity comes to the speed of light, the slower time appears to pass, as regarded by a stationary observer. As will be discussed in a moment below, there is a barrier (I suppose one could call it a "time barrier") associated with the change in mass with velocity that prevents any object from accelerating up to and beyond the speed of light. So time will appear to slow, but will never move backward. However, it is true that time travel into the future, in great leaps rather than a second at a time, is possible, if one can move fast enough.

In the *Legion of Superheroes # 16 (Vol. 5)*, published in 2006, Supergirl from the twenty-first century suddenly flies a thousand years into the future. She had been pursuing an alien missile, sent at great speed by a race of would-be world conquerors to destroy the Earth. This missile was traveling at near light-speed, and the Maid of Steel had been chasing it for three solid days. After finally catching and destroying the missile, she meets a collection of teenage superheroes in the year 3006, the aforementioned Legion, and winds up joining them on a series of adventures. Supergirl and the Legion periodically express puzzlement over how she wound up a thousand years later from when she began chasing the missile. The comic eventually explained her time displacement as arising from her having crossed the path of a "zeta beam," which apparently can move you dramatically forward in time. But Einstein's Special Theory of Relativity provides a more natural explanation.

In a reference frame moving along with Supergirl, three days elapse as she chases after the missile. But from her point of view, she and the missile are essentially stationary, and the rest of the universe is moving past her at great speed. If the universe's velocity, relative to Supergirl, is close to the speed of light, she will observe a dramatic time-dilation effect. Using Einstein's equations, if Supergirl's velocity were 99.9999999963 percent of the speed of light, then a journey that she experiences as lasting three days would, for stationary observers on the Earth, take one thousand years. When she eventually slowed down and stopped, she would find herself a thousand years ahead of when she began. There is thus a physically valid way to travel into the future, with the caveats that (1) it would require extreme amounts of energy for an object of any real size to be accelerated to such a velocity and (2) it is a one-way trip, and flying backward won't send you back in time. (Eventually the "zeta radiation" wore off, sending Supergirl back to the twenty-first century, enabling her to continue having adventures in her own comic book.)

But what about Superman, who frequently travels back in time by flying at great speed? The Warner Bros. 1978 film *Superman: The Movie* included a scene where, in order to undo the effects of a devastating earthquake (particularly Lois Lane's death), the Man of Steel flies around the Earth so fast that he reverses the direction of time. Is this at all possible? Close examination of the scene in question, estimating the height above the Earth that Superman is flying, calculating the distance of an orbit and counting the number of circuits completed and the time he takes doing so, we conclude that Superman is indeed traveling at or faster than the speed of light! But that alone won't save Lois Lane. Traveling that fast, he would jump forward into the far future, as his cousin did in the

Legion of Super Heroes story, and might be in time to help some distant descendant of Ms. Lane, but not be able to save Lois herself.

While the Flash did not have to worry about length shortening (technically referred to as "Lorentz contraction") too often, the slowing of time, or "time dilation" nearly cost him his secret identity. While the Scarlet Speedster is the fastest man alive, one of the quirks of his secret identity is that his alter ego, Barry Allen, is perennially late for his appointments. Often his fiancée, Iris West, would complain about Barry's tardiness and wish he were more like the Flash, not realizing that they were one and the same. As fate would have it, Iris's father would also notice Barry's habitual lateness, and as described in "Slowdown in Time" in *Flash # 141*, was very intrigued by the fact that Barry's watch always seemed to run slow. For Iris's father was physics professor T. H. West, and he suspected that Barry's watch was running slow because of time dilation effects arising from his superspeed crimefighting exploits as the Flash. Spotting a robbery in progress, he called Barry, reported the crime underway, and asked Barry to synchronize his watch. Knowing that the Flash would have to run at great velocities to catch these criminals, he intended to meet up with Barry later on and compare the times on their watches. And it would have worked, too—had Barry not reset his watch before meeting with Professor West. Tipped off by the strange request to synchronize watches, as the story concludes, Barry thinks: "As a scientist myself I'm also familiar with Einstein's Theory of Relativity—and the effect of ultra-speed on clocks!" Sometimes even physics professors fall short when they try to outthink the Viceroy of Velocity!

In theoretical physics, there is one thing that can travel faster than the speed of light: particles termed "tachyons" that in fact can never travel slower than the speed of light. Under certain circumstances, they can appear to travel backward in time (which proved useful for Adrian Veidt in *Watchmen*, who employs tachyons to obscure the precognition abilities of Dr. Manhattan). Tachyons were proposed as a test of certain consequences of the Special Theory of Relativity. As far as we know, they do not exist, and more importantly, even if they were as common as crabgrass it doesn't appear that they can interact with our physical world, in which no object can move faster than the speed of light. The Flash may travel backward and forward in time using his Cosmic Treadmill, but its only real value is in providing the Scarlet Speedster with a cardio workout.

The fact that Superman and the Flash would move through time thanks to their superspeed powers has not escaped the attention of physicists. Following the printing of the first edition of this book, a story was relayed to me concerning Superman's meddling with the space-time continuum. Back in the 1950s a group of physics majors at M.I.T. wrote to Mort Weisinger, the editor of Superman comics at the time. They complained that a recent issue had shown Superman flying faster than the speed of light, in direct contradiction to Einstein's theory. What did he have to say about that? To which Weisinger allegedly replied: Einstein's is only a theory—Superman is fact!

SECTION 2

ENERGY— HEAT AND LIGHT

THE CENTRAL CITY DIET PLAN—

CONSERVATION OF ENERGY

THE FLASH MAY BE ABLE to run across the ocean and pluck bullets from the air, but a more pressing question is: How frequently does he need to eat?

The short answer is, a lot! Before we determine exactly what his caloric intake must be as he rounds up his rogue's gallery of supervillains in Central City, we address the more basic question: Why does he need to eat? What exactly does food contain that is essential for any activity, whether running, walking, or just breathing while sitting still? And why do we only obtain these qualities from organic matter, and not from rocks or metal or plastic?

The Flash eats for the same reason we all do: to provide raw materials for cell growth and regeneration, and to provide energy for metabolic functioning. At birth, your body contains a certain quantity of atoms that was insufficient to accommodate all of the growth that will occur during your lifetime. As you grew and matured, you needed more atoms, typically provided in the form of complex molecules that your body would break down and convert into the building blocks necessary for cell replacement and growth. As noted in our discussion of the explosion of Krypton, all of the atoms in the universe—including in the food we ingest—were synthesized via nuclear reactions in a now long-dead star where hydrogen atoms were squeezed together to form helium atoms, helium fused to form carbon, and so on. An additional by-product of these fusion reactions in our sun provides the second essential component of the food we eat. Matter-Eater Lad of the

Legion of Super-Heroes may be able to subsist by consuming inert objects such as metal or stone, and the cosmic menace Galactus must consume the life-energy of planets, but for us, for the most part, the food we eat must have been previously alive. Only such foodstuffs provide the necessary component for life, as mysterious as its name is mundane: energy.

The use of the word "energy" is so common that it is unnerving to realize how difficult it is to define without using the word "energy," "heat," or "work" in the explanation. The simplest non-mathematical definition is that "energy" is a measure of the ability to cause motion. If an object is already moving, we say it possesses "kinetic energy," and it can cause motion if it collides with something else. Even if it is not moving, an object can possess energy, such as when it is pulled by an external force (for example, gravity) but is restrained from accelerating (say, by being physically held above the ground). Since the object will move once it is let go, it is said to possess "potential energy."

All energy is either kinetic or potential, though depending on the circumstances, a mass can have both kinetic and potential energy, such as when Gwen Stacy fell from the top of the bridge in Chapter 3. When she was on the top of the bridge, she had a large potential energy, as gravity could act on her over a long distance, from the top of the tower to the river below. But her motion was constrained, as the bridge was holding her up. When she was knocked off the tower, this constraint was removed, and the force acting on her (gravity) then began her acceleration. As she plummeted, she had a shorter and shorter distance over which to continue falling, so her potential energy decreased. This energy didn't disappear, but rather her large potential energy at the top of the bridge was converted to ever-increasing kinetic energy as she fell faster and faster. At any given point in her fall, the amount of kinetic energy she gained was exactly equal to the amount of potential energy she lost (ignoring the energy expended in overcoming air resistance). If she had struck the water at the base of the bridge, her potential energy relative to the ground would have been zero (once at the base of the tower she has no more potential to fall), while her speed and, hence, kinetic energy would have been at its greatest. In fact (again ignoring air drag) her kinetic energy at the base would have been exactly equal to her large potential energy at the top of the bridge when she started to fall.

This kinetic energy is then transferred to the water, which supplies a large force that changes her high velocity to zero, with a similar dire result as when she was caught in Spider-Man's webbing, as described in Chapter 3.

Or imagine Spider-Man swinging back and forth on his webbing, like a pendulum. At the top of his arc, he is not moving (which, again, is why it's the highest point of his swing), but he is high off the ground, and has a large potential energy. His potential energy at the bottom of his swing is a minimum, and if he had started at this point, he would not move. Starting at a higher point, his earlier potential energy changes to kinetic energy, and at the lowest point in his arc, his loss of potential energy is exactly equal to his gain in kinetic energy. The only force acting upon him at this lowest point is gravity (straight down) and the tension in the webbing (straight up). Neither of them acts in the horizontal direction of his swing at this point. But he is already moving, and an object in motion will remain in motion unless acted upon by an external force. As he overshoots this lowest point and starts to rise again, his kinetic energy changes back to potential energy. If no one pushes him, he can never have more total energy than what he started off with (where would it come from?), and so the final point of his swing cannot be higher than his initial starting height. In fact, some of his kinetic energy is used up in pushing the air out of his way (air drag), so on each swing he will rise to a lower height than his previous swing, never returning to the high point.

This accounting of how much energy is potential and how much is kinetic leads to one of the most profound ideas in all of physics: Energy can neither be created nor destroyed—it can only be transformed from one form to another. This concept goes by the fancy title of the Principle of Conservation of Energy. We have never been able to catch nature in a slipup where the energy at the start of a process does not exactly equal the energy at the end. Never.

Even when physicists thought that energy was not being conserved, it eventually turned out that it was. Studies of the decay of radioactive nuclei in the 1920s and 1930s found that the final energies of the emitted electrons and resultant nuclei was less than the starting energy of the initial nuclei. Faced with the possibility that energy was not being conserved in the decay reaction, Wolfgang Pauli instead suggested that the missing energy was being carried away by a mysterious "ghost particle" that was invisible to their detectors. Eventually, devices were constructed to observe these "ghost particles." They turned out to be real, and did in fact account for the energy imbalance in the radioactive decay. In fact, neutrinos (as these mystery particles were named, somewhat whimsically, by Enrico Fermi, describing them in Italian as "little neutral ones") are some of the most prevalent forms of matter in the universe.

In more mundane circumstances, one must still be careful in accounting for all of the energy in any process. Consider driving a nail into a wooden board. The potential energy of the hammer, held over a carpenter's head, is converted to kinetic energy as she swings. When the hammer strikes the nail, the hammer's kinetic energy causes the motion of the nail (hopefully deep into the board of wood) and, as a side effect, also causes the atoms in the head of the nail to shake more violently, warming up the nail. The partitioning of the incident kinetic energy of the hammer into additional vibrations of the atoms in the nail head, the forward motion of the nail itself, and the breaking of molecular bonds in the wood (necessary if the nail is to occupy space previously claimed by the board) can be summarized by describing the "efficiency" of the hammering process. If one carefully adds up all of the small and large bits of kinetic energy in the nail, wood, and even the air (the "bang" one hears upon striking the nail results from a pressure wave—sound—induced in the surrounding atmosphere), the net result will exactly equal the initial kinetic energy of the hammer right before it strikes the nail head. However, warming the nail and creating a sound effect are "waste energies" from the point of view of the carpenter, and count against the efficiency of the hammering process.

Sometimes this waste energy is not so insignificant. An automobile traveling along a level road has a certain amount of kinetic energy. This energy arises from a chemical reaction during the combustion of gasoline vapor with oxygen, ignited by an electrical jolt from the spark plug. The gases resulting from this small explosive reaction are moving at great speeds, so that they may displace a piston. The piston's up/down motion is translated via an ingenious system into the rotation of the car's tires. Of course, not all the energy of this chemical reaction goes into the displacement of the auto's pistons—much of it heats up the engine, which is useless from the point of view of locomotion. In addition, as the car travels down the highway, energy is also needed to push the air out of the way of the car. An automobile's efficiency is limited in large part by the effort it must expend displacing air from the immediate volume it intends to occupy—over six tons of air for every mile traveled by an average-size car! The larger the profile of a car or truck, the greater the volume of air that must be displaced, and the more energy must be devoted to this task, in addition to propelling the car forward. This same principle also explains why it is easier to run underwater through a swimming pool with your hands flat at your sides than if you hold them out away from your body. The smaller the surface area, the greater the fuel efficiency for comparable-mass vehicles (assuming the vehicles have the same engine).

The Green Goblin expended energy carrying Gwen Stacy to the top of the George Washington Bridge. This increase in her potential energy was stored as she lay atop one of the towers. The increase in her potential energy came from the chemical energy in the fuel in the Goblin's glider, and so on. Taken to its logical conclusion, the Principle of Conservation of Energy states that, if one can never create new energy or destroy current energy, but simply convert it from one form to another, then all of the energy and matter currently in the universe was present at the moment of the Big Bang that heralded the universe's creation. At this primordial instant, the entire universe was compressed within an inconceivably small volume. As the universe expanded, the total energy and mass content remained unchanged but was now spread over an ever-increasing volume.

"Energy density" is the energy per volume. Consequently, if the amount of energy remains the same but the volume increases, the energy density will decrease. Energy and matter are able to interconvert through a process represented by Einstein's famous equation $E = mc^2$. Ordinarily, when photons collide and form matter, an equal amount of matter and antimatter is created. Within a fraction of a second after the Big Bang, as the universe expanded and cooled down, protons and neutrons started forming from a "quark-gluon plasma." Through a process that remains poorly understood, slightly more protons and neutrons formed in the early universe than anti-protons and anti-neutrons. This formation of matter occurred only once—early in the universe's history the energy density was high enough to allow matter to condense into existence. At later times (such as now) when the energy density is below the $E = mc^2$ threshold, there is not enough background energy in outer space to spontaneously form matter.* The protons, neutrons, and electrons created in the early moments of the universe came together to form hydrogen and helium nuclei. Gravity pulled some of these atoms together to form large clumps that became stars. In the centers of these stars, held together by gravitational potential energy, a nuclear reaction transforms these hydrogen atoms into heavier elements and kinetic energy.

Now, it is all well and good to say that all of the energy (and consequently all of the matter) found in the universe today was present at the moment of the Big Bang. But this only leads to two deeper questions about energy: What is it really? And where did it originally come from? Science provides the same answer to both questions: Nobody knows.

FAST FOOD

In order to figure out how much the Flash must eat to be able to run at superspeed, we need to calculate his kinetic energy. Physicists are always looking to conserve energy—consequently, we'll recycle the math from Chapter 1 so we won't have to do any more work. Speaking of work, in order to change the kinetic energy of an object by either speeding it up or slowing it down, one must do Work. "Work" is capitalized here, because in physics the term has a very specific meaning that is slightly different from its common usage.

When a force acts on an object over a given distance, we say that the force does Work on the object and, depending on the force's direction, will either increase or decrease the object's kinetic energy. In this way Work is just another term for energy,

---

* In 1997 scientists were able to directly verify that a sufficiently large energy density could cause matter to spontaneously come into existence. By colliding high-energy gamma-ray photons of light together, they were able to create electron/anti-electron particle pairs in the laboratory, in essence re-creating the mechanisms operating in the first seconds of the universe.

and they will have the same units. For a falling mass m, the force acting on it is its weight due to gravity $F = mg$, and the distance the force acts upon the object is just the height h that it falls. So **Work = (Force) × (distance) = (mg) × (h) = mgh.** This turns out to be the potential energy that the object had at a height h, so in this example, Work can be viewed as the energy needed to increase an object's potential energy.

Consider the falling Gwen Stacy from Chapter 3 or the leaping Superman from Chapter 1. In either case the Work that gravity does is given by Work = mgh. For Gwen, the force pulls her down in the direction of her motion, so the Work increases her kinetic energy, and for the Man of Steel, the force is still downward, but is opposite to the direction of his leap, so the Work decreases his kinetic energy. Gwen starts with no kinetic energy, but the gravitational force acting over a distance (the height of the bridge tower) provides her with quite a large final velocity right before striking the water. The connection between her final speed v and the distance she fell h was given by $v^2 = 2gh$, where g is the acceleration due to gravity. This is a true statement, and according to our Rule of Algebra (see Preface), we can multiply and divide both sides of a true statement by the same quantity, and it remains a true statement. So if both the left- and right-hand sides of $v^2 = 2gh$ are divided by 2, the result is $v^2/2 = gh$. If we now multiply both the left- and right-hand sides by Gwen's mass m, we obtain $\frac{1}{2} mv^2 = mgh$. The right-hand side is the Work that gravity does on Gwen. The left-hand side must therefore describe her change in kinetic energy—that is, her final kinetic energy minus her initial kinetic energy. Since she started with no kinetic energy (no motion, no kinetic energy, though she had plenty of potential energy), her final kinetic energy is stated as Kinetic Energy = $\frac{1}{2} mv^2$. Congratulations—you've just done another physics calculation.*

---

* It turns out that we could have started our discussion in Chapter 1 with a definition of energy and, using the Rule of Algebra, worked backward to "derive" the expression **F = ma**, instead of starting with **F = ma** and determining the expression for energy as we did. Where one starts in the calculation is a matter of personal taste. In the end we will always find the expression for Gwen Stacy of $v^2 = 2gh$. This connection between her final velocity and the height from which she falls is the important thing, and Gwen does not really care which equations one uses to obtain it.

* * *

When the Flash stops running, the Work of changing the Scarlet Speedster's kinetic energy is done by friction between his boots and the ground. From time to time, the acceleration or deceleration that the Flash must experience is (more or less) realistically addressed, and we see the consequences of these decelerations. In *Flash # 106*, the Flash needed to stop suddenly while chasing an object that was traveling at 500 mph. The comic shows him gouging giant ruts in the ground with his feet as he attempts to quickly bring himself to rest. Here, the forces that would accompany his rapid deceleration—friction in particular—are represented accurately. In bringing himself to rest from a speed of 500 mph, the large change in kinetic energy requires a correspondingly large Work. The comic panel (fig. 19) shows the Flash stopping in approximately fifteen feet, so the distance is relatively short, and since Work = (Force) × (distance), the force that his feet exert on the ground must be correspondingly very large. In fact, to change his velocity of 500 mph in a distance of fifteen feet requires a force of more than 80,000 pounds!

Similarly, in *Flash Comics Vol. 2, Issue # 25* (April 1989), Wally West* runs so fast that, in his attempt to stop suddenly, he leaves mile-long gashes across North America. From the length of the skid marks, the scientists who are tracking Wally are able to determine both how fast he was going and his probable stopping point, using the same techniques that the police employ when reconstructing an automobile accident from the length of the tire skid marks. Realistically, one should always know where the Flash has been from the deep gouges his feet excavate every time he suddenly starts or stops running. Fortunately for the Central City Department of Roads and Transportation, this physically accurate portrayal of the Flash's powers only occurred occasionally.

Returning now to the Flash's eating habits, if kinetic energy KE is written mathematically as $KE = (\frac{1}{2}) mv^2$, then the Flash's caloric intake requirements increase quadratically the faster he runs. If he

---

* Wally was originally Kid Flash, but by 1985 he had dropped the "kid" portion of his title, taking over the mantle of the Flash after Barry Allen, the Silver Age Flash, had died saving our universe from the Anti-Monitor. It's a long story.

Fig. 19. *A rare example from* Flash # 106 *of the realistic effects of the Flash's sudden deceleration. The shorter the stopping distance, the greater the force his boots must exert on the ground when braking.*

runs twice as fast, his kinetic energy increases by a factor of four, and thus he needs to eat four times more in order to achieve this higher speed. Back in the Silver Age (late 1950s to 1960s), artist Carmine Infantino would draw Barry Allen as fairly slender and not as a hulking mass of muscle, since he was, after all, a runner (the Flash, that is, not Carmine). If the Flash weighed 155 pounds on Earth, then his mass would be 70 kilograms. When running at 1 percent of the speed of light (nowhere near the Flash's top speed), his speed would be $v = 1,860$ miles/sec, or 3 million meters/sec. In this case his Kinetic Energy KE is $(\frac{1}{2}) \times (70 \text{ kg}) \times (3,000,000 \text{ meters/sec})^2 = 315$ trillion kg-meter^2/sec$^2 = 75$ trillion calories. Energy is so frequently used in physics that it has several units of measurement, one of which is termed the "calorie," and is defined as 0.24 calories $= 1$ kg-meter^2/sec^2. That is, roughly one quarter of a calorie is equal to the Work resulting from applying a force of 1 kg-meter/sec^2 over a distance of one meter.

The reason that 1kg-meter^2/sec^2 is equal to 0.24 calories is that in the mid-nineteenth century, physicists were confused about energy, a situation that has not greatly improved in the intervening years. A calorie was originally defined as a unit of heat, as heat was thought to be a separate quantity distinct from work and energy. Hence, one system of measurement for heat was developed, while a different unit was employed to measure kinetic and potential energy. The physicist who recognized that heat was simply another form of energy, and that mechanical work could be directly transformed into heat, was James Prescott Joule, in whose honor a standard unit of energy (1 Joule = 1 kg-meter^2/sec^2 = 0.24 calories) is named. While physicists use Joules when quantifying kinetic or potential energy, we'll stick with the more cumbersome kg-meter^2/sec^2 in order to emphasize the different factors that enter into any determination of energy.*

It should be noted that a physicist's calorie is not the same as a nutritionist's calorie. To a physicist, one calorie is defined as the amount of energy needed to raise the temperature of one gram of water by one degree Celsius. This is a perfectly valid, if arbitrary, way of defining energy in a laboratory setting. But this definition leads to the observation that a single soda cracker contains enough energy to raise the temperature of 24,000 grams of water by one degree. In other words, to a physicist, the energy content of one plain cracker is 24,000 calories! In order to avoid always dealing with these very large numbers, a food calorie is defined to be equal to 1,000 of these "physics calories." Therefore the twenty-four food calories in a single cracker are actually equivalent to 24,000 calories using the physics laboratory definition of the term. Just as well—it's bad enough to think of the roughly 500 food calories in a cheeseburger, but if we considered that it actually contained 500,000 physics calories, we might never eat anything ever again!

To convert the Flash's kinetic energy of 75 trillion calories into food calories, we should divide his energy by 1,000. This helps, but he still expends 75 billion food calories running at 1 percent the speed of light. Put another way, he would need to eat 150 million cheeseburgers in order to run this fast (assuming 100 percent of the food's energy is converted into kinetic energy).* If he stops, his kinetic energy goes to zero, and in order to run this fast again, he

---

* Note that kg-meter^2/sec^2 is also the unit for gravitational potential energy PE = mgh, when kg, meter/sec^2, and meters are used for mass m, acceleration due to gravity g, and height h, respectively. This is reassuring, for if kinetic energy is equivalent to potential energy, they should be measured with the same units. Something would be very wrong with our analysis if kinetic energy had units of kg-meter^2/sec^2 and potential energy or Work had units of sec^3/kg or something equally strange.

needs to eat another 150 million burgers. At one point in Flash comics (during the mid-1980s) it was briefly acknowledged that he would need to eat nearly constantly (even chewing at superspeed) in order to sustain his high velocities. In the Golden Age, the Silver Age, and now in the Modern Age, conservation of energy is conveniently ignored. Nowadays the Flash's kinetic energy is ascribed to his being able to tap into and extract velocity from the "Speed Force," which is a fancy way of saying: Relax, it's only a comic book.

CHEESEBURGERS AND H-BOMBS

The next in a seemingly never-ending series of questions we ask is: Why does a cheeseburger, or any food, provide energy for the Flash? It's easy to identify kinetic energy when something is moving, and the potential energy due to gravity is also pretty straightforward, but there are many other forms of energy that require some thought as to which category—potential or kinetic—they belong. The energy the Flash gains by eating is not due to the kinetic energy of atoms shaking in his food (a hot meal has the same number of calories as a cold one) but from the potential energy locked in the chemical bonds in his food. As energy can never be created or destroyed, but only transformed from one state to another, let us follow the chain backward, to see where the stored potential energy in a cheeseburger comes from.

In order to understand the potential energy stored in food, we have to consider some basic chemistry. When two atoms are brought close to each other, and the conditions are right, they will form a chemical bond, and a new unit, termed a "molecule" will be created. A molecule can be as small as two oxygen atoms linked together, becoming an oxygen molecule (O_2), or it can be as long and complex as the DNA that lies within every cell of your body. The question of whether and when two or more atoms will form a chemical bond, and the elucidation of these conditions, is

---

* Since roughly half of our caloric intake goes toward maintaining metabolic functions, the Flash would likely need to eat twice the amount we have estimated.

the basis of all chemistry. All atoms have a positively charged nucleus around which a swarm of electrons hover. The chemical properties of an element are determined by the number of electrons it possesses, and how they manage to balance their mutual repulsion (as they are all negatively charged) with their attraction toward the positively charged nucleus. When an atom is brought very close to another atom, the most likely locations of the electrons from the two atoms overlap and, depending on their detailed nature, there will be either an attractive or repulsive force between the two atoms. If the force is attractive, the electrons create a chemical bond and the atoms form a molecule. If the force is repulsive, then we say that the two atoms do not chemically react. In order to determine whether the force is attractive or not involves sophisticated quantum mechanical calculations. (We'll have much more to say about quantum mechanics in Section 3.) If the force is attractive, and one restrains the two atoms to keep them physically apart, then there is a potential energy between the atoms, since once this restraint is removed, the two atoms form a molecule. In this way, we say that the two atoms, once chemically joined, are in a lower-energy state, just as a brick's gravitational potential energy is lower when placed on the ground. Work has to be done to lift the brick to a height h, just as energy has to be supplied to the molecule to break it apart into its constituent atoms.

We are finally (and I can almost hear you saying: Thank goodness!) in a position to answer the question of why the Flash needs to eat, or rather, why food provides the energy he needs to maintain his kinetic energy. When the Flash runs, he expends energy at the cellular level in order to flex his leg muscles. This cellular energy, in turn, came from the breakfast that Barry Allen ate. From where did the energy in the food arise? From plants, either directly consumed or through an intermediate processing step (such as when plants are eaten by animals and those animals are eaten by humans). This stored energy in food is simply potential energy on a molecular scale. Plants take several smaller molecular "building blocks" and process them, stacking them up into a subcellular "tower of blocks." This molecular tower of complex sugars, once constructed, is fairly stable. The process of lifting and arranging a group of blocks into a tall tower raises all of the blocks' potential energy (except for the bottom block).

Similarly, plants do Work when constructing sugars from simpler molecules, raising the potential energy of the final, synthesized molecule. The potential energy remains locked within the sugars until the mitochondria within our cells construct adenosine triphosphate (called ATP for short) and release the saved energy, just as the Work in building a tower of blocks is stored as the potential energy of the top blocks until the tower is knocked down, converting their potential energy into kinetic energy. The amount of energy released by the ATP in the Flash's leg-muscle cells is greater than the energy needed to "knock the complex-sugar tower down," though the gain to the Flash is much less than the plant cell's effort in raising the tower in the first place.

Where did the plant cell obtain this energy? Through photosynthesis, whereby the energy in sunlight is absorbed by the plant cell and employed in complex-sugar construction. The light comes from the sun (don't worry, we're nearly at the end of the line), where it is generated as a by-product of the nuclear-fusion process, in which hydrogen nuclei are fused together through gravitational pressure to create helium nuclei. Ultimately, all of the chemical energy in food is transformed sunlight that was generated by the same nuclear-fusion process in the center of the sun that occurs during the detonation of a hydrogen bomb. In this way, the majority of energy on Earth is solar energy at its source, just as all the atoms on Earth, from the ATP molecules throughout the Flash's body to the ring in which he stores his costume, were created in a solar crucible (in the massive star that preceded our own sun).

Ultimately, all of life is possible because the mass of a helium nucleus (containing two protons and two neutrons) is slightly less than that of two deuterium nuclei (deuterium is an isotope of hydrogen in which the single nuclear proton is joined by a neutron) combined in the center of a star. And by slightly less, I mean that the mass of a helium nucleus is 99.3 percent of two deuterium nuclei. This small mass difference leads to a large outpouring of energy, since from $E = mc^2$ the change in mass is multiplied by the speed of light squared.

And 99.3 percent really is the magic number. If the mass difference were 99.4 percent, then deuterium nuclei would not form, and hence fusion of helium could not proceed. In this case, stars would shine too dimly to synthesize elements, and no violent supernova explosions would occur to both generate heavier ele-

ments and expel them into the void, where they might form planets and people. On the other hand, if the mass difference were 99.2
percent, then too much energy would be given off from the fusion
reaction. In this case, too many protons would combine directly to
form helium nuclei in the early universe, and no nuclear fuel
would remain when stars formed. The source of this amazing fine-
tuning of the basic properties of nature is the subject of current
investigation.

POWER IS WORK OVER TIME!

Frequently, the constraint in performing a task is not how much
energy is required, but how fast that energy can be supplied and
used. Consider the Flash in fig. 19, stopping in roughly fifteen feet
from a speed of 500 miles per hour. The Work needed to decrease
his large kinetic energy to zero was a little over a million pound-
feet, which called for a force of more than eighty thousand pounds.
If the Scarlet Speedster were to stop over a longer distance, say a
mile rather than fifteen feet, then the force needed would be under
230 pounds, and it is doubtful that he would leave such deep ruts
and scarring in the ground. Put another way, it requires less force
to change his kinetic energy if he has a longer time to do so.

* * *

Power is defined in physics as the rate at which the energy of a system is changed (technically, by "changed," I mean transformed
from one form to another, such as the kinetic energy of the Flash
being converted to mechanical work done by his boots on the
ground). If I do Work on my automobile, pushing it from rest until it
is moving at 60 mph, its final kinetic energy is the same if this process takes six seconds or six hours. The rate of change of the car's
kinetic energy is obviously much higher in the first case, and if you
have a motor capable of accelerating an auto from 0 to 60 mph in six
seconds, we say your car's engine has more power than the engine
that takes six hours. As Work, "kinetic energy" and "heat" are just
different expressions for energy, the rate of change of energy, measured in units of kg-meter^2/sec^2 per second, is defined as a Watt
(after James Watt, an early pioneer of the study of thermodynamics).

We often need a lot of energy in a short period of time, so one convenient unit is one thousand Watts, which is termed a kiloWatt. A Watt is a unit of power—that is, the rate at which energy is used. To keep track of how much energy one uses in, say an hour, we multiply the rate of usage by the time—the resulting expression is referred to as kiloWatt-hours and is how the electric company keeps track of how much electrical energy you have used. Large power plants have to supply electrical energy to a great many homes at any given moment. A plant that can provide a kiloWatt of power (which a typical residential house requires) for one million homes is rated as capable of producing a gigaWatt of power.

It's the power and not the energy that determines why we don't have flying cars, despite the fact that we are presently well within the "far future" as imagined by Silver Age comic books of the 1950s and 1960s. Consider Nite Owl's flying "Owlship" from the 1986 graphic novel *Watchmen*. Nite Owl (technically Nite Owl II), in the novel by Alan Moore and Dave Gibbons, is a knockoff of the Charleton Comics superhero Blue Beetle, who in turn was a knockoff of the DC Comics hero Batman. Dan Dreiberg in *Watchmen* used his inheritance from his banker father to outfit himself with a brownstone with a large basement in which he stored an array of gadgets, specialized superhero costumes (for fighting evil-doers in such extreme environments as underwater, in the Arctic, and in a radioactive hot zone), a collection of trophies and mementos from memorable cases, and a large flying airship called *Archimedes* (or *Archie* for short). The airship, a large, rounded transport about the size of a minivan, is modeled on the Blue Beetle's flying beetle airship—and uses the same mechanism to achieve flight: imagination.

Archie in *Watchmen* has no visible means of levitation or thrust, and can hover in the air during a daring rescue of stranded citizens trapped in a tenement fire, or fly from New York City to Antarctica. It takes a lot of power to accomplish this. Earlier we considered the potential energy of an object raised a distance h above the ground. We determined that the potential energy (say, of Gwen Stacy on the top of the George Washington Bridge) is described by the expression mgh, where m is the mass and g is the acceleration due to gravity. The larger the weight of an object, the more Work one must do to raise it against gravity to a height h.

The full-scale model of the Owlship constructed for the Warner Bros. 2009 film *Watchmen* weighed an impressive four tons (the imaginary airship has to be strong to withstand the guards' gunfire as they break Rorschach out of prison). Lifting such a heavy object one eighth of a mile, so that it could fly above the towers of New York, raises its potential energy by more than seven million kg-meter^2/sec^2. That's a lot of energy, but more importantly, since the Owlship is just suspended in thin air, we must expend this amount of energy every second to keep it levitated. The rate at which the Owlship uses energy when simply hovering in the air, 200 meters above the ground, is 7,200 kiloWatts, or 7.2 megaWatts (where a megaWatt is a million Watts). If we wish to fly the *Archimedes* to some other location, then we need to provide kinetic energy as well, and this will add to our power needs. At a uniform cruising speed of 700 miles per hour, for long distance travel at 30,000 feet, the total power needed to fly *Archie* is over 500 megaWatts. A trip from New York to Antarctica would there- fore require more than 180,000 gallons of gasoline!* The *Watch- men* graphic novel required a significant suspension of disbelief when it came to the superpowered Dr. Manhattan, but perhaps the biggest miracle exception from the laws of nature involved the energy supply for Nite Owl's flying ship.

The Owlship as drawn by Dave Gibbons in the 1986 graphic novel displays no visible signs of a propulsion system. The airship in the 2009 film version has some jet thrusters added underneath and to the rear of the transport, to acknowledge some mechanism by which it flies. Personally, if I had designed and constructed a flying ship that could hover and fly for more than thirteen hours at the speed of sound using a novel lightweight and highly efficient energy source, I would not be sitting naked in my underground Owl Cave, moping about my former crime-fighting career. Rather I'd be too busy counting the kajillion dollars a day I would earn by licensing this process.

---

* That much gas, by the way, would weigh more than 560 tons—making *Archie's* gas tank the heaviest thing on the ship.

DEEP-BREATHING EXERCISES

In order to run, the Flash needs the energy stored in food, which is locked up in complex molecules. We have described this energy as similar to the potential energy of a tower of blocks that plants stack, expending Work. We transform this stored energy—after first consuming the plants—into kinetic energy when the tower is knocked over. But what is the trigger to topple this tower? How does the tower know when the cell needs the energy to be released? There's a lot of biochemistry that goes into the release of energy by the mitochondria in a body's cells, but the essential step involves a chemical reaction of oxygen going in and carbon dioxide coming out. Without oxygen intake, the stored energy in the cell cannot be unlocked, and there's no point in eating. The faster the Flash runs, the more kinetic energy he manifests, the more potential energy stored in his cells he needs to release, and the more oxygen he needs to breathe. We've already discussed the fact that he would need to eat a staggering amount of food in order to account for the kinetic energy he routinely displays. What about his oxygen intake? Would the Flash use up all of the Earth's atmosphere as he ran?

To answer this question, we first need to know how much O_2 the Flash uses when he runs a mile. The volume of oxygen used by a runner will depend on his or her mass, and has been measured to be about 70 cubic centimeters of O_2 per kilogram of runner per minute, for elite athletes at a pace of a six-minute mile. Taking the Flash's mass to be 70 kg, he then uses nearly 30 liters of O_2 for every mile he runs (a liter is one thousand cubic centimeters, a little under a quart). Let's assume that this rate of O_2 use remains the same even for much higher speeds. Thirty liters of O_2 contains under a trillion trillion oxygen molecules, and at a speed of ten miles per second, this means that the Flash inhales about a trillion trillion O_2 molecules every second. That sounds like a lot, but fortunately there are many more O_2 molecules in our atmosphere than that. A lot more. In fact, very roughly, the Earth's atmosphere contains more than ten million trillion trillion trillion O_2 molecules. So even at a rate of consumption of one trillion trillion molecules per second, he would have to run this fast (10 miles/sec) and breathe at this rate continuously for more than

100 billion years before he exhausted our oxygen supply. The faster he ran, the quicker he would use up our air, but even running at nearly the speed of light (which he is capable of, but doesn't do very often) it would take him more than two million years of continuous running and breathing at this rate to exhaust the atmosphere. So, at least regarding this aspect of his super-speed, we can breathe easier.

The Earth's atmosphere may be safe, but of course this assumes that the Flash is able to breathe at all while he runs—at several hundred miles per hour, would he be able to even draw a deep breath? Fortunately for the Scarlet Speedster, he carries a reservoir of air with him whenever he runs. In *Flash # 167*, this region of stationary air (relative to the Flash) is described as his "aura," while in fluid mechanics it is termed the "no-slip zone." Whatever you call it, it's the reason golf balls have dimples.

Try this simple physics experiment at home: Turn on the cold water in your bathroom sink, but only barely open the valve. For the best results, first remove the aerator in the faucet. When the water is just barely coming out, you may see it move very smoothly from the faucet, with the appearance of a polished cylinder, wider at the faucet outlet and tapering slightly due to surface tension. If you ignore the sound of the water splashing in the basin, it's difficult to tell that the water is moving at all, that it isn't actually a rigid structure. This type of water flow, where all the water molecules are moving smoothly in the same direction, is termed "laminar flow." At the opposite extreme, turn the valve all the way open. The water churns and swirls out, moving in different directions and with a wide range of speeds. This type of water flow is termed "turbulent." Naturally, if you want to get water through a pipe in the most efficient manner possible, you would like the flow to be laminar, where all water molecules are moving in the same direction down the pipe, rather than turbulent, where vortices and swirls by necessity mean that some water molecules are moving against the flow.

Even in laminar flow through a pipe, all the molecules may move in the same direction, but they won't all have the same speed. Those molecules at the outer edge will collide with the pipe's walls, transferring their kinetic energy to the pipe (which is rigid, so that the pipe warms up a bit but doesn't move) and coming to rest. Right next to the pipe's walls is a thin layer of water that is not moving. Water next to this non-moving layer loses some of its kinetic energy, but not all, because unlike the atoms in the pipe, the water molecules in this "no-slip zone" can move. In the next ring closer to the center of the pipe, the water is moving a bit faster still. So, even in uniform laminar flow, there is a continuous series of concentric rings, each ring moving progressively faster than the adjacent ring. The water dead center in the pipe moves the fastest of all. In laminar flow, all rings are uniform, while in turbulent flow there is chaotic motion across the width of the pipe.

The situation is mirror-reversed for a moving pipe being pulled through stationary water. The water closest to the pipe's walls is dragged along with the pipe, the water right next to this ring moves a little slower, and so on. But in either case, for either the water moving through the pipe or the pipe moving through the water, the water right next to the pipe is stationary, relative to the pipe. As long as the flow is laminar, then right next to a moving object, there will be a thin layer of air (the arguments for a fluid such as water apply for a fluid such as air, as well) that is not moving relative to the object. Just as in the example of the water faucet, this laminar no-slip zone is more robust the slower the motion is through the fluid. At too high a speed, the transfer of energy across concentric rings becomes disordered, and turbulence sets in. An object moving at a given speed must expend more energy generating turbulent flow than if the flow is laminar. Recall in Chapter 6 that dolphins shed skin cells in order to break up turbulent vortices in their wake, enabling them to swim at great speeds.

This is one reason why golf balls have dimples. The bumps on the golf ball decrease the cross-section of the turbulent wake behind the ball moving at higher speeds. In a crude sense, the dimples reduce the drag on the ball, as less energy is lost in the smaller turbulent wake. This effect was discovered accidentally. In the mid-1800s, golf balls were smooth, solid spheres of gutta-percha gum. It was empirically noted by golfers that old, scuffed-up balls with scratches and dings went farther for a given swing than brand-new, smooth balls. Experimental study and a theoretical understanding of fluid mechanics led to an optimal design of dimpled golf balls.

What's good for a golf ball is good for the Flash. As the Scarlet

Speedster runs, the layer of air right next to him remains stationary relative to his body, so he has a non-moving pocket of air that he carries around with him at all times. Even in a layer only a few centimeters thick, there are nearly a trillion trillion O_2 molecules. This "reservoir" of air must be continually refreshed with new air from outside the boundary layer in order for the Scarlet Speedster to run for more than a few seconds at a time.

In Flash comics, the no-slip zone "aura" that surrounds the Crimson Comet not only enables him to breathe as he runs, but also frees him from other untidy consequences of fluid drag. Consider: If a meteor burns up in the atmosphere when it turns into a meteorite due to the extreme frictional forces it experiences as it pushes the air out of its way entering the atmosphere at high velocity,* then why doesn't the Flash burn up when he runs at high speeds?

Flash Comics # 167 provided an answer to this question, but it was a solution that few comic fans found satisfying. According to this story, the "protective aura" the Flash gained along with his superspeed powers was provided by a "tenth-dimensional elf novice-order" named Mopee. In this story, using his magical abilities, Mopee (who bore more than a passing resemblance to Woody Allen) removed the Flash's aura but not his superspeed. Consequently, when the Flash ran at great speeds, he would burn up due to the tremendous air resistance he encountered.

That the Flash had an imp who bedeviled him was not as surprising as the fact that the Silver Age Flash went eight years in his own comic before encountering him. In the fifties and sixties, it seemed like nearly every superhero published by DC Comics had his or her own extradimensional pest. The first such character was Mr. Mxyzptlk, a fifth-dimensional derby-wearing sprite against whose magical powers Superman was powerless. Mxyzptlk could only be forcibly returned to the fifth dimension if he was tricked into saying his name backward, after which he was unable to return to our three-dimensional world for at least three months

---

* Not all meteors burn upon entering the atmosphere, however. In order to account for the large quantity of Kryptonite that managed to reach Earth intact, it was argued in *Superman # 130* that remnants of his planet's destruction possessed an invulnerability to air friction!

(presumably so that readers would not grow overly tired of him and his appearances would be noteworthy). Not to be outdone by the Man of Steel, Batman had his own magical imp, named Bat-Mite, whose attempts to honor his idol, the Caped Crusader, frequently backfired and created mayhem and difficulties for Batman and Robin. J'onn J'onzz, the Martian Manhunter, had an alien sidekick named Zook, while Aquaman had an imp named Quisp. Of the seven founding members of the Justice League of America, only Green Lantern and Wonder Woman have never had a supernatural or extradimensional spirit to call their own.

It wasn't the fact that the Flash finally acquired his own imp that upset comic-book fans, but rather that Mopee claimed to have used his magical powers to give Barry Allen his superspeed powers. The science-fiction aspects that writers John Broome, Gardner Fox, Robert Kanigher, and editor Julie Schwartz introduced into the Silver Age with the creation of the Barry Allen Flash seemed undermined by the claim that the Flash's powers were in fact "magically" derived. Mopee never had a return engagement in Flash comics and, as far as most fans of the Silver Age are concerned, *Flash # 167* never happened. Let us speak of it no more.

13

THE CASE OF THE MISSING WORK—

THE THREE LAWS OF THERMODYNAMICS

UNLIKE HENRY PYM, who found that he could reverse the polarity of "Pym particles" and enlarge himself into Giant-Man, and the actress Rita Farr (Elasti-Girl) of the Doom Patrol who, after breathing mysterious vapors from a previously undiscovered volcanic vent during an African movie shoot, was able to either grow to the size of a five-story building or shrink to the size of an insect, the DC Comics hero the Atom could only change his size in one direction: down.

The Atom is one of my favorite comic-book superheroes, for in his secret identity he is Ray Palmer, physics professor. *Showcase # 34* introduced us to Professor Palmer as he was fruitlessly trying to develop a shrinking ray, motivated by the economic benefits of such a device. As he records in his audio journal after Experiment # 145 fails (well, actually, Experiment # 145 succeeded in reducing the size of a kitchen chair to only a few inches, but the chair subsequently exploded, just as every other object that had been shrunken), "compression of matter [...] would enable farmers to grow a thousand times more than they do, on the same land! A single freight car could transport the goods of a hundred freight trains." Of course, the exploding part would make maintaining inventory difficult.

Palmer found the solution to his shrinking-ray problems one night when, on a nighttime drive, he spied a chunk of a white-dwarf star falling near him. This extraterrestrial material would turn out to be the one key missing ingredient* that enabled Palmer to safely reduce objects without subsequent explosion. Presumably because this stellar remnant has the word "dwarf" in its name, it possesses miniaturization properties previously unsuspected by physicists. We'll have more to say about the Atom's origin and the white-dwarf meteor at the end of the book. Suffice it to say that the mechanism for how Ray Palmer was able to reduce his height to six inches (his typical crime-busting size) and much further, all the way down to smaller than an electron, makes about as much physical sense as inhaling strange underground vapors or Pym particles.

What is significant about the Atom was that unlike Ant-Man or Elasti-Girl, he was not restricted to a constant-density reduction. That is, the Atom could independently control both his size and his mass. Apparently, white-dwarf matter provides two "miracle exceptions" for the price of one.

At normal height, Ray Palmer was six feet tall and weighed approximately 180 pounds, equivalent to a mass of 82 kilograms. When reduced to a height of six inches, he is twelve times smaller than normal. His width and breadth will decrease also by a factor of twelve—that is, if he wants to avoid looking like his reflection in a fun-house mirror. His volume therefore decreases by a multiple of $12 \times 12 \times 12$, or a factor of 1,728. If he were to shrink at a constant density, then his mass would also have to be reduced by this same factor of 1,728, leading to the Atom having a mass of only 47 grams, or 1.66 ounces. This is pretty light, and it is hard to see how such an ephemeral crime-fighter could tangle with the likes of Chronos the Time Thief or Doctor Light. Fortunately for those of us on the side of Good, the Atom could maintain his diminutive stature and increase his body weight up to his full-grown 180 pounds with just a click of his "size and weight controls" that he kept in his costume's belt buckle (later he added controls to his gloves as well). As shown in fig. 20, at his lighter weight, he could use a pink rubber eraser as a trampoline, propelling himself into the face of a crook, and at the moment of

---

* As John Carr (fig. 3) and Ray Palmer illustrate, in comic books, scientists tend to construct miraculous devices and leave an open space for the "key missing ingredient" that will make the mechanism operational.

Fig. 20. *Image from the cover of* Atom *# 4, where the Mighty Mite demonstrates his mastery of both his size and mass. Decreasing his density so that he is lightweight enough to use a pink rubber eraser as a trampoline brings him to within striking distance of a crook's chin. At this point he increases his mass dramatically, so that his punch kayos the "innocent thief."*

impact increase his weight to 180 pounds. The resulting blow landing on the crook's chin would have the same force as when the Tiny Titan was his full height.*

Another trick the Atom would employ was, while only a few inches tall, to grab hold of the necktie that a criminal wore and then increase his weight to 180 pounds. The crook's head would be violently jerked downward and would crash on a tabletop or

---

* Since Ant-Man shrinks at constant density, the force of his punch diminishes at the same rate as the cross-sectional area of his biceps. His punches pack an impressive pressure only because the surface area of his fist decreases along with the force supplied by his muscles.

some other hard surface, knocking him out cold. This is undoubtedly why criminals in the late 1960s were in the fashion avant-garde in adopting a more "casual Friday" approach to their work clothes when breaking the law.

Being able to control his mass (but only when his size was reduced from his normal height), the Atom could ride air currents generated by the wind or thermal gradients to get from place to place. (He would also make use of the fact that he could reduce himself to the size of an electron and telephone himself to distant locations—we will discuss the physics underlying this trick in Chapter 26.) The Atom would often be shown, as in fig. 21, gracefully gliding along air currents. In this panel, from *Atom # 2*, the Atom needs to get to the top of a barn that is on fire. He shrinks down to less than an inch and adjusts his density so that he is "lighter than a feather"—at which point he is then pushed aloft by the "hot air currents to the roof." In reality, he should suffer a much more disorienting and punishing trip, and would, via direct, firsthand experience, verify the statistical mechanical underpinnings of the branch of physics called thermodynamics.

THE FIRST LAW—YOU CAN NEVER WIN

We have considered Newton's three laws of motion; now we take up the three laws of thermodynamics. The field of thermodynamics,

Fig. 21. *Panel from* Atom # 2, *wherein the Tiny Titan utilizes his ability to reduce both his size and his weight to ride the thermal air currents from a burning building.*

the study of the flow of heat, was empirically developed by scientists in the nineteenth century, well before a proper understanding of the atomic nature of matter. As such, they came slowly to these laws, struggling with such problems as the connection between useful work and heat, a quantitative concept of temperature, the nature of phase transitions, and the intrinsic inefficiency of any mechanical process. Our consideration of this topic will hopefully be less painful, as we already know that matter is composed of atoms. We'll begin by elucidating the First Law of Thermodynamics with the simple question of why there is a net upward force on the Atom (fig. 21) as he rides the air currents created by the burning barn.

Back in the 1800s, scientists believed that matter contained a separate fluid termed "caloric" (previously named "phlogiston"). When mechanically deformed, objects would release their caloric fluid, and thus would feel warm to the touch. The caloric fluid was alleged to be "self-repelling" (which accounted for why objects expanded when heated). It may sound silly today, but if you did not know that matter is composed of atoms, the caloric model seemed a reasonable way to account for these and other observations. The mystery of the true nature of heat was solved by Joule, Benjamin Thompson, and others, who demonstrated that mechanical work could be directly converted to heat without the release of any special fluid. The term "heat" describes any exchange of energy without the system performing work (defined in the previous chapter as a force multiplied by a distance). That any change in the energy of an object can only come about by either a transfer of heat or the performance of work is the essence of the First Law of Thermodynamics.

When an object is "hot," the kinetic energy of its constituent atoms is large, while when an object is "cold," the kinetic energy of the atoms is lower. The "temperature" of an object is therefore just a useful bookkeeping device to keep track of the average energy per atom of the object. We say something has a high temperature when the atoms contain a larger amount of kinetic energy compared with another object with a lower kinetic energy per atom, which we call "cold."

In our earlier discussion of friction and the Flash in Chapter 4, we noted that when two objects are rubbed or dragged against each other, this process can be viewed at the atomic level as the scraping of one atomic mountain range across the other. The kinetic energy of the dragged objects is converted into the shaking of the constituent atoms in each surface. The transfer of kinetic energy per atom is referred to as "heat flow," and the temperature of the respective objects is raised, without a caloric fluid being present.

From Chapter 12 we know that energy comes in only two flavors: potential or kinetic. The energy of the air molecules in your room is essentially all kinetic. As the mass of any given air molecule is very small, the variation of the gravitational potential energy (given by the weight of the air molecule multiplied by its distance above the floor) across the height of the room is so tiny that we make only a very small error by ignoring it. The following discussion does not depend on the exact chemical composition of the atmosphere, so we'll refer to generic "air" molecules. The hot air underneath the Atom in fig. 21 has a greater kinetic energy than the colder air above him. Of course, this last statement is unscientific, for there is no such thing as "hot air" or "cold air," just air that is hot compared with some other object or cold compared with another reference, which for the sake of argument we'll take to be the Atom's body temperature.

The hot air beneath the Atom in fig. 21 warms and lifts him. When a speeding car collides with a stationary or slowly moving vehicle, the faster car will usually slow down and the slower car will speed up. Similarly, air molecules that have a higher kinetic energy than those in the Atom's body will, upon colliding with him, transfer some of their kinetic energy, on average, to the atoms in his costume, causing them to shake more violently than before. Since energy is conserved in the process, the kinetic energy (that is, the temperature) of the air will decrease following the collisions, while the temperature of the Mighty Mite will increase. The atomic nature of matter makes clear why, when two objects are in thermal contact, the net flow of heat is always from the higher-temperature object to the lower-temperature object, and never in the other direction. Similarly, when exposed to cold air, the atoms in the Tiny Titan's costume will be moving back and forth much faster than the air molecules that they collide with, so after the collision, the air will be moving faster at the expense of the atoms in his costume, cooling the Atom down.

When the Atom is lightweight enough to float on air currents (as illustrated in fig. 21), and if all of the air surrounding him is the same temperature, he is being bombarded in all directions with roughly equal force. In this case, no matter how small his mass, gravity will eventually pull him down to the ground. The thermal draft that keeps the Atom aloft arises from the hotter air molecules below the Atom that are moving faster than the cooler molecules above him. There are therefore more collisions per second beneath the Atom that push him upward than collisions above him that direct him toward the ground. Furthermore, the force needed to reverse the direction of a faster-moving molecule is larger than compared with slower-moving molecules. Forces come in pairs, so the force the Atom exerts on the air molecules, changing their direction, also pushes back on the Atom. There will therefore be a net unbalanced force on the Atom due to the fact that there are more collisions underneath him pushing him upward than above him driving him down. Of course, this force will not be smooth and uniform, but will in fact be discontinuous and noisy, and occasional statistical fluctuations will lead to him being pushed down rather than up. But over time, the average force from the thermal gradient will be to push him away from the hot source and toward the cooler region.

The Atom in fig. 21, therefore, gains both kinetic and potential energy. He warms up slightly (that is, the kinetic energy of the atoms in his body increase) due to the flow of heat from the hot-air current. He will also be raised to a greater altitude, increasing his gravitational potential energy, thanks to the work done on him by the net force exerted by the hotter air molecules. The First Law of Thermodynamics states that the net change in the Atom's total energy is the sum of the heat flow and the work done on him.

Another example: When a hot gas pushes against a piston in an automobile cylinder, lifting the piston and thereby, through a series of cams and shafts, causing the wheels to rotate, the energy of the hot gas that goes into moving the piston is termed "Work." The Industrial Revolution came about when scientists and engineers realized that during the transfer of energy through the flow of heat from hot to cold, productive work, meaning a force applied across a given distance, could be extracted. Prior to this, the development of simple machines such as levers and pulleys to amplify forces required the energy stored within humans, draft animals, wind, or waterfalls. The work supplied by humans or animals is converted from the stored chemical energy in the foodstuffs ingested. The potential energy in the food has to accomplish many other tasks, from maintaining body temperature to continuing metabolic functions, and so on. Consequently, the amount of energy available for pushing down on a lever is only a small fraction of that stored within the food. In contrast, releasing the stored potential chemical energy by burning coal or oil enables a more direct conversion into work. While not 100-percent efficient, it is far more advantageous than using living beings.

The First Law of Thermodynamics tells us that in a best-case scenario, with all losses and external noise removed, the total amount of work we can get out of any device is exactly equal to the heat flow (change in kinetic energy) that drives the machine. By the Principle of Conservation of Energy, it is impossible to extract more work than is available from the heat flow from a hot-to-cold source. The heading of this section alludes to the fact that the universe ensures that you can never win (that is, get more out than you put in).

Well, if winning is ruled out, why is breaking even so difficult? We can certainly convert all of our work into heat—when we rub our hands together. We do work, applying a force against friction over a distance, and the result is that our hands warm up. Why couldn't we make a perfect machine that extracts all possible heat and converts it to work, without waste? We'll see in a moment that the random motion of the gas atoms during the transfer of heat places strict limits on the amount of usable work we can obtain from any machine, no matter how cleverly designed.

THE SECOND LAW—YOU CAN'T EVEN BREAK EVEN

There is nothing in the Principle of Conservation of Energy, which underlies the First Law of Thermodynamics, that prevents or forbids the construction of a 100-percent-efficient machine, where the work created by the device exactly equals the heat energy put into it. In fact, if all we had to go on was the First Law of Thermodynamics, we would reasonably expect that machines must be 100-percent efficient, for we know that energy can never be gained or lost, but only transformed from one form into another. In order to understand what limits the conversion of heat into work, we must introduce a new concept that is complementary to, but as important as, energy.

This concept, called "entropy," is intimately connected to heat flow, and will give the Atom a very bumpy ride, even when he isn't floating along thermal drafts.

Whenever there is an explosion in the Justice League of America's satellite headquarters orbiting in outer space, or on the quinjet spaceship used by the Avengers, there is a violent outpouring of air into the low-pressure surroundings. Why? What compels the air to rush out of the opening in the JLA satellite? A common metaphor invoked to explain why air races into any region that is at a lower pressure is that "nature abhors a vacuum." And yet there would be no violation of Newton's laws of motion if all the air were to remain inside the JLA's satellite, even with the door left open, though admittedly such a situation is extremely unlikely. It is the random motion of the air associated with the air molecules' kinetic energy that underlies the explosive decompression on the satellite.

Imagine that the room next door to the one you are sitting in has had all of its air removed. As long as the door connecting the two rooms is kept hermetically sealed, you would never know that a perfect vacuum was waiting in the next room. The air molecules in your room are at a certain temperature and pressure, and are buzzing merrily along. This peaceful, stable scenario changes when the door from your room swings open into the vacuum room.

Instead of asking why the air would rush out of your room into the vacuum room once the door was opened, the better question is, why wouldn't it? Those air molecules that were moving toward the door, and would have bounced off of it had it remained closed, will now continue moving in a straight line into the vacuum room (per Newton's first law). However, only a small fraction of the air molecules in your room would have been heading toward the door right before it was opened. Some of the air molecules would be moving away from the door, where they would collide with other air molecules moving in different directions. It is conceivable, though implausible, that aside from those air molecules initially heading toward the now-open doorway, all of the remaining air molecules would continue to collide with one another, with no more passing into the second room. This is as likely as, with the door remaining closed, all of the air molecules, through random collisions, managing to always avoid the region right next to the door. You would not have to worry about suffocating if you sat next to the door, because a fairly constant fraction of the air is always heading toward you at any given moment. The air molecules occupy all regions available to them—there is no reason for them not to. Any particular air molecule may spend most of its time in one particular corner of the room, but on average, every volume is just as likely to have air in it as any other, just as in a well-shuffled deck of cards, any one of the fifty-two cards is as likely to be turned up at the top of the deck as any other.

Air molecules have no free will, and if they are going to collide and head toward the door when it is closed, they will do so when the door is open. The only difference, and it is a big difference, is that once the air molecules move into the vacuum room, there are initially no other air molecules for them to collide with. There are many, many more ways for the atoms in the first room to move into the vacuum room than there are for them to bounce off one another and never pass into the second room. "Entropy" is the term used to describe the number of different ways a given system can arrange itself. A brand-new deck of cards with all the cards arranged in numerical order by suit has low entropy (I always know what the top card will be), while the entropy is at a maximum once the deck has been thoroughly shuffled (I have a one in fifty-two chance of correctly predicting the top card). It is hard to be dealt four aces in a row from a well-shuffled deck, just as it is harder to know where a particular air molecule is if it can possibly be in two rooms rather than just one.

When physicists say that systems tend toward maximum entropy, all they are saying is that the most probable situations will be the ones observed. When Peter Parker pulls his socks out of the dryer after they have been tumbling for some time, it is possible that he will extract them two at a time in perfectly matched pairs, but don't count on it. There is only one way that could happen, but many more ways in which the socks can be mismatched. The dryer randomizes the socks, so every sock has an opportunity of pairing with any other sock. Issues concerning entropy only apply to such fully randomized situations, and there are obviously many examples of externally imposing order on a system (such as Aunt May manually sorting the socks).

When the door between rooms is opened and the atoms move into the second, initially vacant room, we say that the entropy of the air molecules increases, but what this basically means is that if something can happen, it will happen. There are many more ways for the air molecules in the room to be spread out uniformly, each sharing a nearly equal portion of the total kinetic energy of the room, than for any other scenario, such as one molecule having all the kinetic energy of the atmosphere and the rest of the molecules having none. Those things that are observed most often are those most likely to occur. Wanda Maximoff, the mutant Scarlet Witch, who was originally a villain in Magneto's Brotherhood of Evil Mutants in Marvel Comics' X-Men and later rehabilitated as a hero in the Avengers, had a "hex power," whereby she could gesture at an object and something untoward would occur. It was suggested in *West Coast Avengers # 42* that her power consisted of the ability to alter probabilities, so that a highly unlikely event would actually have a near-certain probability of happening. The above would suggest that her mutant talent involves the power to alter the entropy of a system, bringing about rare configurations (such as all the air moving over to one side of a room) much sooner than would reasonably be expected.*

In order to get useful work out of the compressed gas in a car cylinder or the steam in a boiler, it typically must expand from a confined region to a more spacious configuration. Think about the air molecules in the room, separated from the vacuum room. It would be nice if they could be arranged to push the door open on their own, without your having to do so manually. If we unlock the door between the two rooms, then there are many air molecules striking the door on one side, and none counteracting on the other, so that there will indeed be a net force on the door that could push it open. This is the same unbalanced force that pushes the Justice League members out of their satellite headquarters when there is a hull breach. This unbalanced force arises quickly—the time between collisions for air molecules at room temperature is less than a nanosecond (one billionth of a second).

Once the air has moved into the vacuum room, it has a higher

---

* Years later Wanda's mutant power was associated with "chaos magic," but this is just silly. No less an authority than Dr. Strange (the Sorcerer Supreme of the Marvel universe) has stated (*Avengers # 503*) that there is no such thing as "chaos magic."

degree of disorder, that is, its entropy has increased. Closing the door, the air will not again push the door open, unless I scoop all the air out of the second room, repeating the original configuration (one room at atmospheric pressure, the other room under vacuum). This takes effort, and when I count up all the energy expended in returning the system to its original state, I wind up using more energy than I gained when the door was pushed open. No matter how cleverly I arrange things, I can never convert all the energy of a system into useful work—there will always be some part that goes into increasing the entropy, which is not useful.

When the mixture of gasoline and oxygen is ignited in your automobile engine, it undergoes a chemical reaction, releasing heat (that is, the reactant products are moving faster than they had been before the explosive reaction). Only those faster-moving molecules that are heading in the right direction will displace the piston in the engine, leading to rotation of the tires. It would be much more useful, from the driver's point of view, if all of the molecules following the explosive combustion of gasoline and air were moving in a direct path toward the piston head, so that they all contribute to the lifting force that leads to tire rotation. But there is essentially only one way that the gas molecules can arrange themselves to be all moving in the same direction, while there are many, many more ways they can be moving in all possible directions, with only a fraction of them heading toward the piston head. Those molecules that are heading away from the piston are wasted, from the point of view of getting something useful out of the chemical reaction. Not only can you not get more energy out of a process than you put in, but the entropy constraint means that you will always get out less useful work than you used to set the system up.

This is the heart of the Second Law of Thermodynamics. No process can be 100-percent efficient (defined as all of the heat being converted into useful work), and in fact, most motors and engines rarely convert more than a third of their available energy into useful work. The Second Law of Thermodynamics is a harsh mistress, but there doesn't seem to be any way around it. Or is there?

Could I use the talents of the Atom to try to beat the Second Law of Thermodynamics? The air molecules in your room are characterized by a certain temperature, which measures the average energy of the air. The key word is "average"—not every single air molecule in the room has exactly the same kinetic energy.

Some are moving a little faster than average, while some poke along a little slower. The steam coming off a fresh cup of coffee is a reflection of the fact that not every molecule in a cup has exactly the same energy. Those water molecules in the coffee energetic enough to escape from the liquid state (more on such phase transitions in Chapter 15) form the clouds hovering over the coffee. The hotter the coffee is initially, the greater the steam forming over the liquid surface, as there are more water molecules at the high end of the kinetic-energy distribution. When you blow on your coffee to cool it off, you do not reduce the temperature of the coffee, because your breath is a frosty 98.6 degrees Fahrenheit. True—your breath is cooler than the hot coffee, but not enough to effect significant refrigeration. Rather, what you do is disturb the steam, pushing the most energetic water molecules away so that they are unable to be deflected back into the coffee. Once they are permanently removed from the coffee/steam system, the average energy (that is, temperature) of the remaining coffee decreases. This physical process is called "evaporation cooling"—it is the physics underlying the operation of refrigerators and is the reason why sweating is more effective in cooling you down if there is a strong breeze to carry the perspiration away.

The idea of the Atom getting around the Second Law of Thermodynamics is a variation on the concept of evaporation cooling. We'll start by having the Atom shrink down so that he is only several times bigger than an air molecule. He'll have a box with him, with a small, hinged door. In this example the Atom assumes the identity of "Maxwell's Demon," proposed by James Clerk Maxwell to test the Second Law of Thermodynamics. All the air molecules in the room are at the same temperature, which means that they cannot be used to generate a heat flow to power a machine. But now the Atom makes use of the fact that the temperature is an average measurement, and starts sorting out the air molecules, based upon their kinetic energy. Those air molecules coming toward him that are moving faster than the average he collects by opening the door of his box and trapping them inside (it's a thermally insulated box, so those molecules retain their kinetic energy once secured inside). Those that are moving slower than average, he ignores. Before long he has acquired a large number of air molecules that have a kinetic energy larger than the corresponding initial average value. Furthermore, by removing these faster-moving molecules, the average energy of the remaining air molecules decreases, just as when you blow on your steaming coffee. The Atom can now take these hotter-than-average molecules and, bringing them into thermal contact with some colder molecules, allow the net heat flow between them to power an engine, thereby getting useful work out of air that was initially at one average temperature.

Or he would, if we didn't have to worry about the Atom himself. He expends energy in opening and closing his box to sort and trap the energetic molecules. This energy must be included in any balancing of the total energy added to and extracted from a process. To neglect his expenditure of energy would be equivalent to saying that you are able to drive your car for just pennies a day, if you ignore the cost of the gasoline. When the heat and work contributions of the Atom's sorting of the air molecules are carefully accounted for, we find that by collecting the faster molecules, the Atom himself contributes energy to the remaining atmosphere, increasing its average kinetic energy, so that in the end there has been no net temperature differential. If you blow on your coffee and remove the steam but replace it with other molecules that are just as hot, you have not cooled your drink.

No matter how hard you try (and believe me, many have tried), there is only one way, discussed below, around the no-win scenario presented by the Second Law of Thermodynamics. Unfortunately, even that option is not available to us.

THE THIRD LAW–YOU CAN NEVER GET OUT OF THE GAME

If entropy considerations limit the amount of useful work we can extract from any process, whether it's a V-8 engine, a gas turbine, or the chemical reactions in your cells' mitochondria, then couldn't we just get around this problem by dealing with systems with no entropy? After all, it is conceivable, no matter how difficult it may be in practice, to have a system where all the atoms are in a precise, uniform configuration, so that there is no uncertainty regarding the location of any single element within it. Why can't I arrange my two systems that generate the heat flow that powers my engine to have no entropy, so that I don't have to worry about the Second Law?

The reason why this won't work is that the entropy of a substance and its internal energy (which could be available for heat
transfer) are related, such that we can't change one without affecting the other. The entropy of the air molecules in the room is a
measure of their random motion. If I lower the air's kinetic energy,
eventually the gas condenses into a liquid. The entropy of the liquid is lower than that of the same molecules in their vapor state,
because there is less uncertainty as to where any given molecule
might be (they're in the puddle on the floor, as opposed to spread
out throughout the room). But there are still chaotic fluctuations
in position and velocity of the molecules in the liquid state. Lower
the temperature of the liquid further, and eventually the average
kinetic energy of the molecules is insufficient to overcome the
attractive bonding between molecules, and the material freezes
into a solid. The chemical bonds between the molecules have preferred orientations, so the natural configuration of the solid will
be a particular crystalline arrangement, with all of the atoms or
molecules lined up in a certain way. At very low temperatures, all
of the atoms will be in their ideal crystalline spots, and we will
know the location of any given atom.

The entropy of any crystalline solid would therefore be zero,
except for the atoms' vibrations about their crystalline positions.
The solid will still have some temperature, no matter how low, so
the atoms in the crystal will be shaking back and forth. We will
truly have no uncertainty, and the entropy will be exactly zero,
only when all vibrations of every atom in the solid cease. The
fact that the entropy is zero only when the temperature is also
zero is termed the Third Law of Thermodynamics. In the zero-
temperature state, none of the atoms have any kinetic energy at
all. At this point we say that the solid has a temperature of
absolute zero degrees. We use the prefix "absolute" because no
matter what type of thermometer you use, it will read zero average kinetic energy at this point. Note that not even outer space is
this cold. There is background light and stray cosmic rays even in
the vacuum of space, and they carry energy. In fact, the radio-wave
background radiation that is a remnant of the Big Bang origin of
the universe has an energy characterized by an average temperature of 3 degrees above absolute zero. So, even outer space has a
temperature, and hence an entropy. The only way to beat the Second Law of Thermodynamics is to use systems with zero entropy,

but this can only be realized at absolute zero. But if everything in your machine is at zero degrees, how would there be any heat flow to power our engine? The three laws of thermodynamics almost conspire to prevent us from constructing perfectly efficient machines, and just like supervillains in a comic book, we must reconcile ourselves to inevitable loss.

Our discussion of entropy has relied so heavily on the fluctuations of the constituent atoms that it is striking that the Second Law was formulated long before most scientists were convinced that matter was indeed composed of atoms. From the mid-1800s onward, some scientists had been taking the atomic theory of matter more and more seriously, while others remained unconvinced of the reality of atoms. These critics felt that the idea of matter as composed of atoms was useful in simplifying many of the calculations about the properties of fluids and gases, but that it was nonetheless meaningless to ascribe physical reality to entities that were too small to ever be seen. Many of the elder statesmen of physics in the late nineteenth century, notably Ernst Mach (for whom the speed of sound in air, the Mach number, is named) held this view.

Nevertheless, the atomic hypothesis eventually won out, via the same time-tested strategy by which all revolutionary ideas succeed. As Max Planck, himself a young Turk of the quantum revolution (about whom we will have much more to say in Section 3), once remarked, "A new scientific truth does not triumph by convincing its opponents and making them see the light, but rather because its opponents eventually die and a new generation grows up that is familiar with it."

A key development that convinced these younger scientists that atoms were real, regardless of what the older establishment claimed, was the explanation for the jitter that small objects underwent due to their random bombardment from still-smaller atoms and molecules striking them from all sides. This phenomenon is termed "Brownian motion" after Robert Brown, the botanist who observed the excursions of a pollen grain in a drop of water using a new scientific instrument, the microscope. While Brownian motion had been known since 1828, it was not until 1905 that a satisfactory theoretical description was provided by another young upstart, Albert Einstein. Einstein was able to quantitatively calculate the excursions of a pollen particle due to the collisions with the water in which it was suspended, and also relate the magnitude of the fluctuations to the temperature of the surrounding medium. The close agreement between Einstein's calculations and experimental observations convinced many physicists that the atomic hypothesis was indeed correct. This work (a related aspect of which formed Einstein's Ph.D. thesis) may not be as revolutionary as his subsequent discovery of relativity published the same year. Nevertheless, his elucidation of the statistical nature underlying Brownian motion would ensure that Einstein would be well known to physicists even if he never got mixed up with relativity and quantum mechanics.

As the Atom shrinks down so that he is roughly the size of a pollen grain—say, a hundredth to a tenth of a millimeter, less than the diameter of a human hair—he will begin to experience the back-and-forth motion that Brown first observed. This size is critical: When he is larger, the average bombardment is negligible; when much smaller, he fits between the atoms in the air,* so that as long as he can avoid being struck, he will be fine. To go back to the example of the Atom floating on air currents, at one particular instant there may be more molecules striking him from below so that he will suddenly be pushed upward, while the following moment may see a downward thrust, perhaps not as severe as the last upward swing, or possibly even more extreme. This is a very slow way to get anywhere, and the Atom will need some Dramamine before too long.

One doesn't have to be as small as the Atom to directly experience Brownian motion. The random collisions of the air on our eardrums produce deflections that are just at the limit of our hearing. Sit in a soundproof room for about thirty minutes and your hearing will improve (just as your eyes' sensitivity to stray light increases when you've acclimated to a darkened room) enough for you to be able to detect the deflection of your eardrums by the motion of the air atoms. It is possible in a very quiet room to hear the background noise emanating from the entropy of the air—in essence, to hear the temperature of the room. Super-hearing—it's not just for Kryptonians anymore!

---

* So what is he breathing? Stay tuned for the answer, Fearless Reader!

MUTANT METEOROLOGY—

CONDUCTION AND CONVECTION

STAN LEE, head writer and editor of nearly all Marvel comics in the Silver Age, was fond of using radiation as a source of his heroes' and villains' superpowers. The bite of a radioactive spider gave Peter Parker the proportionate strength and abilities of a spider; the Fantastic Four were bombarded by cosmic rays (high-energy protons from as near as the sun and as far as other galaxies); Bruce Banner turned into the Hulk when he was belted by gamma rays (more radiation); and Matt Murdock was struck in the eyes with a radioactive isotope that literally fell off a truck, blinding him but endowing him with a "radar sense" and enhancing his other senses so that he could fight crime as Daredevil. After all of this radioactivity, Lee was tired of trying to come up with origins for bizarre superpowers. Hence, in 1963, when he co-created with Jack Kirby a new team of superpowered teens, the X-Men, he essentially threw in the towel, claiming that they were mutants, * and hence simply born with their strange abilities and attributes.

One of the original X-Men, first appearing in *X-Men # 1*,† was Bobby Drake, code-named Iceman. Drake's mutant power was the ability to lower the temperature of his body and his immediate surroundings to less than 32 degrees Fahrenheit. His body would

---

* Lee had originally wanted to call the series *The Mutants*, but his boss, Martin Goodman, thought the term too obscure for a comic book title.
† Later to be known as *The Uncanny X-Men*.

thereby acquire a protective coating of frozen water. As explained in *X-Men* # 47, Bobby does not generate the ice that covers his body or projects from his hands when he employs his mutant power. Instead, by lowering the temperature around him, he condenses the water vapor that is always present in the air.

By now, I don't need to belabor the point that any heat—that is, any kinetic energy—that Bobby is able to subtract from his surroundings must be compensated for by an amount of heat delivered somewhere else and, given the Second Law of Thermodynamics, the heat added is most likely greater than the heat removed. Refrigerators remove heat from an enclosed space, but this heat must then be deposited elsewhere. In addition, the motors in the compressors of the refrigerator require energy, and some of this electrical energy is not converted into useful work but takes the form of "waste heat." The waste heat of a refrigerator is ejected from its rear, usually placed against a wall. (If you ever want to heat your kitchen, just leave the refrigerator door wide open. As the appliance struggles to lower the temperature of the room, it will deposit more heat into the kitchen than it is able to remove, thanks to the Second Law of Thermodynamics.) Where Iceman deposits the excess heat generated when he lowers the temperature of his surroundings remains a mystery.

Iceman's body covering initially took the form of fluffy snow, and when he had gained further control over his power in *X-Men* # 8, he acquired an icy, crystalline appearance. The difference between snow and ice is in the arrangement of the water molecules that results when the water freezes into the solid state. Snowflakes are constructed from aggregations of water within clouds. When water molecules condense from the vapor phase, they release energy, warming the surrounding air. The lower-density, warmed air keeps the cloud aloft, as in a hot-air balloon. When too many water molecules come together, frequently coalescing around a dust grain in the cloud, they form a droplet. When the temperature in the cloud is above 32 degrees Fahrenheit, the droplet can fall from the cloud in the form of rain, converting its potential energy into kinetic energy. A frozen rain droplet is called sleet. The construction of a snowflake is a more delicate affair. A snowflake is created when water vapor slowly freezes around a dust particle.

The chemical properties of atoms dictate how they are arranged in a solid. The chemical forces between metal atoms makes them stack in a solid, like oranges in a grocery store display. Water molecules have a more V-shaped geometry, with an oxygen atom at the vertex bonded with two hydrogen atoms, protruding like a rabbit-ear antenna. The shape of the water molecule determines the geometry of its packing, which turns out to be a hexagon.

Chemical stacking explains the sixfold symmetry of snowflakes, but how do they form their characteristic lacy structure? In the snowflake-generating clouds that have a relatively low humidity, the water molecules must diffuse to the growing flake before they can be incorporated into its structure. The water molecules in the cloud are not being pushed in any given direction, but are undergoing Brownian motion as they fluctuate one way and another. This type of "random walk" is a very slow way to get anywhere, since you are just as likely to take a step away from your goal as toward it. When you hold your hand over a hot object, the warmth you feel is carried by air molecules that collided with air diffusing away from the high-temperature region. This method of conveying energy from one location to another is termed "conduction," and is fairly inefficient. In general, unless the object in question is glowing white hot, you typically have to place your hand very close before a significant energy transfer is detected.

Einstein's equation for how far a fluctuating atom moves in a given period of time, derived in his 1905 paper on Brownian motion, indicates that it will take a hundred times longer for the water in a growing snowflake to diffuse a distance of one centimeter than it will take for it to travel one millimeter. Consequently, those regions of the growing snowflake that extend farther out from the body of the flake will accrete water faster, because they shorten the distance the molecules must random-walk to reach the flake. The six pointed regions at the corners of the hexagon will therefore grow first by adding water molecules, and as they extend farther, they continue to grow more rapidly than neighboring regions. The exact details of how the flake develops—how the dendritic branches grow secondary offshoots, the role that the energy deposited by the diffusing water molecules plays in locally melting and then refreezing the growing flake—will all depend sensitively on the humidity and temperature within the cloud. The final structure of the flake will also depend on the unique details of how the flake forms around a dust particle, such that no two snowflakes will ever be exactly alike, though it is possible to find flakes that are strikingly similar. But at its heart, the beautiful symmetry and near-perfect order of a snowflake arises from the disordered, statistical fluctuations underlying Brownian motion.

* * *

As his mastery of his mutant ability progressed, Iceman was able to project "freeze rays" from his hands, icing up another person or object, or even creating a large mountain of ice beneath his feet. To get around while fighting such evil mutants as Magneto, the Blob, or the sinister, robotic Sentinels, Bobby would frequently generate an "ice slide" underneath his feet on which he would glide, as illustrated in fig. 22. Bobby would, in principle, create a large ice mountain underneath himself and then generate a ramp on which he would slide to his intended destination. Well, this in itself would not violate any physics principles, assuming, of course, that one could indeed control local temperatures in this manner and that there was sufficient humidity in the air to create all this ice. What is troubling, however, is the apparent stability of Bobby's ice slides, no matter how long they became. At some point, Bobby would get out in front of the center of mass of his ice slide—a point at which one would expect bad things to happen.

The "center of mass," also referred to as the "center of gravity," is the point at which an object behaves as if all of its mass, no matter how much or how unevenly distributed, were concentrated at this one location. A yardstick would have its center of mass located exactly at its center. You can balance the yardstick across your index finger, held so that it lies parallel to the ground, but only if your finger is underneath this midpoint. Placing your supporting finger too close to either end will cause the stick to rotate and fall. The center of mass depends on the distribution of matter in the object. A baseball bat, thicker and heavier at one end, will have its center of mass closer to the wider end than to the narrower handle.

To see why Bobby's ice slides cannot extend too far without crumbling, place a book on a table. The weight of the book is

© 1971 Marvel Comics

Fig. 22. *Scene from* Amazing Spider-Man # 92, *where the mutant X-Man Iceman battles Spider-Man due to a mistaken impression (such confusion that would lead two heroes to fight until they realized that they were on the same side occurred no more often in Marvel comics than once or twice—a month). It's no surprise that Spidey has never seen anything like Iceman's ice ramps, as they display questionable stability when the mutant hero ventures far beyond the ramps' centers of gravity.*

directed toward the ground and is balanced by a force from the table. The center of mass is in the middle of the book, and as long as it is on top of the table, the book is stable. But now slide the book near the edge of the table. At first a small fraction of the book can hang over the edge of the table with no problem. The portion of the book that's over the edge creates a twisting force, a torque. This unsupported weight of the book tries to rotate the book, but there is more weight over the table, trying to rotate the book in the other direction, so the book remains stationary. But as you slide the book farther, so that its center of mass is no longer above the table but over the side, the book will rotate and flip onto the floor.

Similarly, when Iceman's slides become too extended, he moves too far from the center of mass of the ice mountain he'd originally generated. Rather than flipping his ice ramp and mountain, it is more likely that the torque that results from him gliding along its edge becomes greater than the strength of the ice slide, and his slide should break off. In order to satisfy basic mechanics, Bobby Drake should continually reinforce the underside of his ice slides with ice pillars, in order to avoid getting too far in front of their center of mass. Sometimes the implausible mechanical stability of Iceman's constructs would be acknowledged, as indicated in fig. 23 from a backup feature "I, the Iceman" in *X-Men # 47.* If

Fig. 23. *A scene from* X-Men *# 47, where Bobby Drake (Iceman) talks directly to the reader about the mechanisms underlying his mutant freezing powers and gives a knowing wink that he recognizes the physical implausibility of some (OK, all) of his uncanny feats.*

any "wiseguy physics majors" wonder what keeps Iceman's ladders and ramps aloft, the answer, which no one can refute, is "a hulkin' helping of imagination!"

* * *

The X-Men comic in its first incarnation in the 1960s was not a major sales success, and original stories ceased publication in 1970. Five years later, new management at Marvel Comics decided to try a revival of the X-Men, and the *All-New, All-Different X-Men* (written by Len Wein and drawn by Dave Cockrum), as their debut giant-size issue was titled, was a financial success from the very beginning. Despite the promise implied by the title, some of the X-Men characters featured in this issue were from the original team, but many of the new characters that debuted, such as Ororo Munroe (Storm); Logan (Wolverine); Peter Rasputin (Colossus); and Kurt Wagner (Nightcrawler) would go on to become fan favorites.

Not everyone was impressed, however. Stan Lee has been

quoted as complaining about the implausibility of Storm's mutant ability to control the weather. He did not seem bothered by Colossus's ability to transform his skin into "organic steel" (I have absolutely no idea what this phrase means, or how this power could possibly work in the physical world) or Nightcrawler's teleportation power (well, if you give me enough "miracle exceptions," I suppose I could try to make this work), but "weather powers" seemed too implausible for the man who brought the world the Incredible Hulk and the Silver Surfer (if he travels through outer space, what exactly is he surfing on?).* But Stan should not be too quick to throw stones at superhero houses. The same mutant power manifested by one of Stan's own creations, Iceman of the original X-Men—the ability to generate and control thermal gradients—also enables Storm to influence meteorological phenomena.

At its core, the weather is simply a matter of the atmosphere absorbing energy in the form of sunlight, and is in fact so simple that it is nearly impossible to accurately predict. When one thinks of weather, terms such as "wind," "rain," and "snow" come to mind (particularly if, like me, you live in Minnesota). All are governed by spatial differences in temperature, driven by variations in the amount of sunlight energy absorbed by the atmosphere.

Spatial variations of the atmospheric temperature are associated with changes in the atmospheric density (the number of air molecules in a given volume). When a volume of denser air is adjacent to a dilute region, there will be a net flow of air from the high- to low-density spaces until the density in each volume element is equal. This flow of air can be understood simply on the basis of the entropy argument discussed in the previous chapter. If there is a constant input of energy, keeping one region at a lower density than another, then this flow of air—or wind—will persist. The wind can, in turn, move cloud cover, changing the spatial pattern of sunlight absorption, which changes the airflow trajectories that influence the cloud cover, and so on. Of course, the Earth's rotation determines the global direction of air circulation.

The ability to accurately predict the weather is therefore limited by the precision with which one knows, at some specified

---

* Cosmic rays?

time, the air speeds and temperatures everywhere in space. Furthermore, changes in temperature lead to airflow that in turn changes the absorbed sunlight, leading to new airflow patterns. There is a nonlinear feedback in place, such that any small uncertainty in our knowledge of the initial conditions quickly becomes magnified. In a linear system a small change in the input results in a corresponding small variation in the output, while for nonlinear systems, such as the weather, a small change could lead to a very large variation in the output. This has become popularly known as "the butterfly effect," whereby the beating of a butterfly's wings in Cleveland may, several weeks later, produce tornado-like conditions in Chile. Meteorologists do an excellent job of predicting the weather in the short term, but anything beyond a few weeks is intrinsically unreliable, regardless of the quality of the measurement systems.

A physically plausible explanation for Storm's ability to control the weather is that she is able to alter atmospheric temperature variations in space and time at will. The wind that allows Storm to fly, as illustrated in fig. 24, is created by a temperature gradient beneath her. Storm presumably uses her mutant power to make the region of air under her hotter than that above her. The temperature of the air is a measure of its average kinetic energy, so air at a very low temperature is moving much slower than warmer air. This cooler air is denser, and it will fall toward the ground. The faster-moving, less dense "hot" air molecules will occupy the space left vacant by the falling "cold" air molecules, simply because if they are zipping around at great speeds, colliding with one another, there are many more ways they can scatter into unoccupied regions than if they collide and always manage to stay near the ground. The average kinetic energy of the hot air molecules is large; consequently, the gravitational potential energy is only a small addition to its total energy. Once the hot air molecules are near the cold upper region, and the cold air molecules are near the ground, the lower molecules will gain energy from collisions with the heated ground, and the hot air will lose energy following collisions with the cool air above it. There will once again be a situation where hot air is on the ground and cold air is above it, and the cycle will continue.

This process is termed "convection" and such thermal convection rolls are an extremely efficient way to transfer energy from a

Fig. 24. Scene from X-Men # 145, *where the mutant Storm employs her ability to generate controlled thermal gradients to warp wind patterns, carrying her aloft on convection cells.*

hot source to a cold one through the thermal link of the air in the room.

The fraction of air that is composed of water vapor depends on the average kinetic energy (ambient temperature) and pressure of the atmospheric molecules. Colder air is denser and has less room to accommodate the water molecules. If Storm is indeed able to control the local temperature, then she can also vary the barometric pressure and humidity at will. It is not unreasonable that she would be able to cause localized rain- or snowstorms, or even generate lightning strikes, though her ability to control the exact position of the strike would be hampered by factors (such as the local charge buildup on the ground) outside of her control. All told, if Stan Lee was not bothered by his creation of Iceman, a mutant who could lower his own body temperature to below 32 degrees Fahrenheit and also project localized regions of lower temperature in his immediate vicinity, then a mutant such as Storm, who can control not her own temperature but that of the surrounding atmosphere, should not have been too great a stretch.

* * *

A final thought about a connection between genetic mutation and thermodynamics. According to Stan Lee, mutants, particularly those with dramatic superpowers, comprise an entirely new species, Homo superior, and are distinct from most comic-book readers, Homo sapiens. The process of speciation, by which new species develop, was put forth by Charles Darwin and independently by Alfred Wallace in the 1850s. In Darwin's original formulation of the theory of evolution, he proposed that speciation was a slow, gradual process and that several hundred million years were necessary to account for the current biological diversity. The only problem was that the physics of the time estimated that the Earth was only roughly twenty-million years old.

One of the foremost scientists of the nineteenth century, William Thomson (later honored as Lord Kelvin for his efforts in developing transatlantic telegraph cables) performed a thermal conductivity calculation that challenged Darwin's hypothesis. Thermal conductivity is a basic property of all matter, and it reflects the rate that heat is transferred in response to a given temperature difference. Metals have a very high thermal conductivity, such that they are able to carry away heat very efficiently for a particular temperature difference (for example a wet tongue at 98.6 degrees Fahrenheit and a metal lamppost below 32 degrees Fahrenheit in winter), while wood is a fairly poor thermal conductor. Making reasonable assumptions that the Earth was a sphere of molten rock at 7,000 degrees Fahrenheit when it initially formed, and knowing the thermal conductivity of rock, Lord Kelvin was able to determine how long it would take the Earth to cool to its present temperature. His conclusion, that the Earth was at least ten times too young to provide sufficient time for evolution's effects, was considered to be a near-fatal flaw in Darwin's arguments. Kelvin's understanding of thermodynamics was so highly regarded that the absolute temperature scale mentioned in the previous chapter, in which zero kinetic energy is recorded as zero degrees absolute, is named in his honor (and is now referred to as "degrees Kelvin"). There was nothing wrong with his calculation.

While Darwin could not refute Kelvin's result, he remained convinced of the validity of the theory of evolution, as it was able to account for too many biological phenomena to be completely wrong, despite Kelvin's objection. Darwin passed away in 1882. A few years later, radioactivity was discovered, and it was realized that there was an additional internal heat source within the Earth that Kelvin had not taken into account, simply because he (like the rest of the world) was unaware of its existence. This extra heat within the Earth would lengthen the time needed for the planet to cool to its present temperature. When Kelvin redid his calculation in 1905, now incorporating the energy provided by radioactive decay, he arrived at a minimum estimate of the Earth's age of several hundred million years, closer to Darwin's original proposal. The current determination of the age of the Earth is 4.5 billion years, easily old enough to provide a landscape for evolution to operate. Darwin went to his grave not knowing that Kelvin was mistaken, yet nevertheless maintained his belief in the correctness of his theory of evolution.

There are critics of evolutionary theory today who point out particular biological phenomena that the theory cannot currently explain, but this does not necessarily invalidate a scientific theory. For example, the motion of three masses interacting through their mutual gravitational attraction turns out to be so complicated as to defy analytical calculation, but this does not indicate that the theory of gravity is wrong. There are always gaps in our knowledge and many things we do not presently understand, but the only way we will change this situation is through critical thinking and the experimental testing of evidence. If you find the scientific method lacking in one aspect of science, then honesty would indicate that you should refrain from using its results in other parts of your life, which will certainly save you some money on doctor and electricity bills.

15

HOW THE MONSTROUS MENACE OF THE MYSTERIOUS MELTER MAKES DINNER PREPARATION A BREEZE—

PHASE TRANSITIONS

NOT EVERY SUPERHERO possesses powers and abilities far beyond those of mortal men. Some, such as Batman and Wildcat, bravely face down supervillains armed with nothing more than a good right hook and the courage to appear in public wearing their underwear on the outside of their clothes. Of course, Batman would try to even the odds somewhat by using his analytical brain, as highly trained as his body, to produce a fabulous array of crime-fighting weapons that he stored within his utility belt. Over at Marvel Comics, the engineer as superhero reached its apogee in *Tales of Suspense # 39*, featuring the debut of the invincible Iron Man. When electrical engineering genius and weapons manufacturer Tony Stark dons his flexible suit of red-and-golden armor, he has the strength of a hundred men, is able to fly using jets built into the soles of his boots, and can fire concussive "repulsor rays" from the palms of his gloves.

We will have much, much more to say about Tony and his Golden Avenger alter-ego when we get to Chapter 24, which is devoted to a discussion of solid-state physics. Right now I want to consider one of the charter members of the gallery of superpowered villains who would bedevil Iron Man time and again. This villain was one of the first to actually strike fear into Tony's shrapnel-damaged heart.* For if you are wearing a suit of iron, and

---

* The "shrapnel" reference will make sense in Chapter 24, when we discuss in depth Iron Man's origins. Stay tuned, Fearless Reader!

your only superpower stems from this armor, then one of your worst nightmares would involve a villain possessing a "melting ray," capable of dissolving iron like butter on a stovetop. Unfortunately for Iron Man, Bruno Horgan possessed just such a melting gun, and as the costumed criminal the Melter, he was only too willing to use it. When the Melter first appeared back in 1963, the notion of a melting ray seemed suitable only for comic books. As we'll now discuss, science and engineering have advanced to the point where such devices are commonplace. You probably have one in your home right now (you no doubt refer to it as a "microwave oven").

Before we can answer why solids melt when they become very hot, we need to address a more basic question: Why do atoms combine to become solids in the first place? It all comes down to energy and entropy. Under certain circumstances, two atoms may have a lower total energy when they are close enough to each other that their electrons' "orbits" overlap. When this happens, a chemical bond forms between the two atoms. This lowering of energy is not always very significant. If the two atoms are moving very fast when they are brought together, then their individual kinetic energies can be much greater than any lowering of energy resulting from the formation of a chemical union, and no chemical bond will form. It's easier to hook up a trailer to a towing hitch on the back of a truck if you slowly back onto the hitch, rather than smash into it at 100 mph. The slower the atoms are moving, the greater the chance that the resulting lowering of energy will prevail when they overlap, and they will remain linked, bonding to form a molecule.

What is true for two atoms will also hold for two hundred, or two trillion trillion, atoms. As the temperature of a gas is lowered, the average kinetic energy of each atom decreases and the greater the chance that the atoms, when they collide, will condense into a new phase of matter—a liquid. Adding thermal energy—heat—to the liquid reverses the process, and the fluid will boil and return to the vapor phase. Similarly, upon lowering the temperature of a liquid, a point is reached at which the atoms cannot glide past one another, and they become locked into a rigid, solid network. Pressure also plays a role in phase transitions. If I squeeze on the collection of atoms, I force them to remain closer to one another than they would ordinarily, changing the temperature at which a phase transition will occur.

If the atoms or molecules lower their energy when they link up and form a collective state as either a liquid or a solid, then it follows that one has to add energy to remove atoms. It takes quite a bit of energy to enable a water molecule to leave the liquid state. This energy comes from its surroundings, either the rest of the liquid, or the surface on which the water is in contact. In either case, when the water molecule leaves the liquid state and returns to the vapor state, that is, during evaporation, it leaves its surroundings with less energy. The average energy per atom, characterized by the temperature, is lowered when a water molecule evaporates from the liquid state. In this way, water molecules undergoing a phase transition from the liquid state to vapor cool the remaining cup of coffee, or the skin on which perspiration is drying.

What determines the exact temperature and pressure at which a phase transition takes place depends on the details of how the individual atoms link up when their electronic clouds overlap. To determine the temperature at which a phase transition such as melting or boiling occurs, we must do more than simply count up the energy needed to break each chemical bond that holds a solid or liquid together. We also have to take into account the large change in the randomness of the atoms—that is, their entropy. For a given internal energy, systems tend to increase their entropy, because all other things being equal, there are generally more ways to be in disordered configurations than in neat, ordered piles. The competition between lowering energy and increasing entropy leads to a fascinating collective phenomenon in which all of the atoms in a solid decide to melt at the same temperature. By the way, the bubbling we associate with boiling water in a pot arises from small irregularities in heating at the bottom of a typical saucepan. Individual points on the pan's bottom will be hotter than neighboring regions, and the liquid-to-vapor transition occurs first at these locations. The underwater vapor forms a buoyant bubble that rises to the surface.

To initiate the melting process, we must add energy to the solid. We can do this the slow, conventional way, by placing the solid in an oven, or the quick way, à la Bruno Horgan and his melting ray. In a conventional oven, the heating elements—either gas-flame jets or electrical coils—cause the average temperature inside the oven to rise. A solid placed in the oven, such as a nice roast, will come to the temperature of the oven as air molecules collide with the walls of the oven, pick up some extra kinetic energy, and then make their way to the roast. Striking the surface of the roast, these fast-moving air molecules transfer their energy to the meat. With a conduction oven, one must wait for the hot air molecules to randomly make their way from the hot walls to the cooler roast, while with a convection oven a fan generates circulation cells from hot to cold and back again (as in our discussion about Storm of the X-Men in the previous chapter). In either case, the surface of the roast warms up first, and one must wait, sometimes several hours, for the center of the meat to reach a higher temperature. As the internal temperature of the meat increases, the atoms shake more and more violently about their equilibrium positions. At a given temperature, the shaking of the connecting fibers and deposits of fat in the roast is sufficiently pronounced that these fibers undergo a phase transition and melt.* Since these were the tough, stringy tissues binding the muscle cells in the roast, melting them makes the meat more tender and easier to eat. This is the same principle utilized by the Flash when escaping from the solid blocks of ice in which Captain Cold would routinely entomb him.†

If you're in a rush but unable to vibrate at superspeed, there is another technique you can use. This involves grabbing hold of every single atom in the solid at the same time and shaking them back and forth very rapidly, using internal friction to cook all parts of the roast simultaneously. This is what a microwave oven—and the Melter's deadly ray gun—does.

---

* While the thermally driven chemical and structural changes during cooking can be quite complex, for our purposes the key step will be approximated as a melting transition.

† By vibrating back and forth at high speed, the Flash imparts kinetic energy to the ice crystals surrounding him. A vibration rate back and forth of 100,000 times per second, even if he could only flex half an inch, would correspond to a total kinetic energy $(\frac{1}{2})\ mv^2$ of 35 million kg-meter^2/sec^2. The melting transition of ice to water requires the addition of 336 kg-meter^2/sec^2 per gram of ice at 32 degrees Fahrenheit. With his excess kinetic energy, the Flash is able to melt a hundred kilograms of ice around him, freeing him to return Captain Cold to the Central City Jail.

Every atom in a solid is electrically neutral, with exactly as many positively charged protons in the nucleus as there are negatively charged electrons swarming about it. But the electrons are not always distributed around the nucleus in a perfectly symmetrical manner. Due to the vagaries of the probability clouds and the nature of the chemical bonds holding the atoms together, one side of the atom may have a little more electric charge than the other. In this case, the atom will be a bit more negative on one side and a bit more positive on the other, just as a bar magnet will have one end with a North magnetic pole and the other end will be the South magnetic pole. This charge imbalance is not very great, but it gives an applied electric field something to grab onto. Even molecules with perfectly symmetrical charge distributions can be polarized by an external electric field.

If a large enough electric field is applied across the solid, the imbalanced atoms line up with the field, just as the needle of a compass will rotate and point in the direction of an external magnetic field. If I now suddenly reverse the direction of the electric field, all the atoms will flip 180 degrees and point in the opposite direction. Changing the electric field back to its original orientation, the atoms will have to rotate once again. If I flip the direction of the electric field back and forth several billion times a second, the atoms are going to do some serious rotating. This energy of vibration will very quickly raise the average internal energy of each atom in the material and, in so doing, raise its temperature. As the external electric field penetrates deeper within the material (with a few exceptions), more atoms will move back and forth due to the oscillating electric field at the same time, not just those on the surface. This process is many times more efficient than waiting for the transfer of kinetic energy by the impact of hot air molecules. The frequency of oscillation of the alternating electric field is in the microwave portion of the electromagnetic spectrum, hence, this type of cooking device is called a microwave oven.

Microwave emitters (called magnetrons) were first created for radar applications during World War II. The cooking benefits of such a device were noted in 1945 when engineer Percy L. Spencer, studying the range of microwave energy emitted from a magnetron, noted that the candy bar he had in his pants pocket had melted. A follow-up experiment with popcorn confirmed the non-military usefulness for this device.

The easier it is for the atoms in an object to move back and forth and rotate with the oscillating electric field, the quicker the temperature of the object will rise. This is why liquids heat up faster than solids in a microwave. You can dig a deep hole in a large chunk of ice and fill it with water. Placing this "ice cup" filled with water in a microwave oven enables you to boil the water while the outside of the ice cup remains cold and solid. Don't leave the ice cup in the microwave for too long, however, as it will also melt due to the ministrations of the alternating electric field, and in much less time than it would take in a conventional thermal oven.

From the descriptions given in the pages of *Tales of Suspense* and Iron Man comics, can we infer that the Melter's weapon used the same principle underlying a microwave oven? Yes and no. Bruno Horgan first appeared in *Tales of Suspense # 47* as an industrialist competitor of Tony Stark's and was embittered when he lost a government contract to build tanks for the U.S. Army once the military discovered that Horgan was using "inferior materials." Stark's company then won the Army contract, despite the fact that there was an apparent conflict of interest in the information presented, as the report describing Horgan's use of inferior components was written by Tony Stark himself. Later, one of Horgan's laboratory testing devices (built with inferior parts) goes haywire while he is examining it and emits an energy beam that melts any iron it strikes.* When Horgan realizes that the "inspection beam" he has created is actually a melting ray, he redesigns the device into a compact, portable unit and, donning a hideous blue-and-gray costume (sadly reinforcing various stereotypes concerning the fashion sense of engineers), he decides to destroy his enemies and make himself supreme (sadly reinforcing various stereotypes concerning the ethical sense of modern industrialists). His initial success against both Stark Industries and Iron Man is short-lived. Horgan is dumbfounded when he discovers, at the story's conclusion, that his ray is no longer effective against the

---

* I cannot stress enough that, nearly without exception, one cannot randomly combine a collection of circuitry and power supplies into a box and "accidentally discover" that it is a fully functioning death ray (believe me–I've tried!).

Golden Avenger. The Melter is unaware that Tony Stark has sur-
mised his weapon's weakness: It only works on iron! Creating a
suit of "burnished aluminum" that appears indistinguishable
from his regular suit of armor, Tony is able to fight the Melter to a
standstill, and it is only the accidental melting of an iron drain
above a sewer system that enables Horgan to escape to fight
another day.

From this story, we must conclude that Horgan's melting ray is
not a portable microwave device. A microwave oven's oscillating
electric field grabs hold of any atom, while Horgan's weapon
works on iron (which contains twenty-six electrons) but not on
aluminum (with thirteen electrons). Later on (in *Tales of Suspense*
 # 90), Horgan's melting-ray gun would become even more specific,
with dial settings for stone, metal, wood, and flesh (yeesh!). This
specificity came to Tony's rescue when, while in civilian clothes,
he was shot in the chest by Horgan with this weapon, yet was
unhurt. Bruno Horgan did not know that Tony Stark was also Iron
Man, and was therefore unaware that Stark always wore his metal
chest plate underneath his shirt (in order to keep the shrapnel near
his heart at bay—see Chapter 24) and thus had the gun set to
"Flesh," when "Metal" would have been the correct setting.

Now, it is certainly true that when two atoms form a chemical
bond, the lowering of energy is unique to the particular atoms participating. Thus, every chemical bond has its own energy signature, and it is, in principle, possible to design a microwave-type
weapon that would be tuned to the chemical bonds in stone and
not those in metal. Similarly, tuning the resonant frequency to
water would make the beam effective against people (flesh) and
not inanimate objects (metal). Just such a microwave-based "heat
ray"—a weapon that induces extreme pain, similar to a second-
degree burn, while the beam is incident on a person—has in fact
recently been developed. The motivation for the development of
such a weapon is for use in crowd-control situations, as the heat
beam provokes a group of people to disperse from a given location
in order to avoid the burning pain of the ray. However, regardless
of how the frequency of the oscillating electric field is tuned, the
bonds between iron atoms and aluminum atoms in their respective metals are much too similar for a weapon designed for melting iron to not also melt aluminum.

Of course, all this discussion about the Melter and Tony Stark's metallic union suit raises a question that has long plagued modern man: If we can put a man on the moon, why can't we put metal in a microwave? The answer is that the free electrons in the metal may cause some serious problems. Metals have high thermal conductivities, and can cause fires when in contact with paper in a microwave oven. Applying an external electric field to the electrons that are able to roam over the entire volume of the metal does more than just push them back and forth as it does the fixed atoms.

Any metal in a microwave is an isolated object, and there is no place for these pushed electrons to go—hence, they can build up at one end of the metal. If there are sharp points or edges, this pileup of electrons can cause a large electric field to be created inside the metallic object. If this electric field becomes larger than 12,000 Volts per centimeter, it can cause a spark, as the air is no longer able to insulate the high-voltage metal from the wall of the oven, and tiny lightning arcs will emanate. Depending on the metal's curvature, the electric field induced may be less than the critical discharge level, while a sharp corner on a foil-covered stick of butter can be sufficient to create a spark that permanently scars the internal surface of the oven. (Personal note to my wife: Sorry, honey.)

ELECTRO'S CLINGING WAYS—

ELECTROSTATICS

UP TILL NOW, we have focused primarily on how forces change the motion of objects, and the force that has concerned us almost exclusively has been gravity. Whether slowing Superman down as he leaps or speeding Gwen Stacy up as she fell, it is gravity that has been invoked when a force F is needed in Newton's second law $\mathbf{F} = \mathbf{ma}$. But there are forces other than gravity in this—and the comic-book—universe.

Physicists have discovered that only four basic forces in nature are both necessary and sufficient to account for the wide range of complex physical phenomena observed. These forces are: (1) Gravity, (2) Electromagnetism, and (3) and (4), the unimaginatively named Strong and Weak forces.* The latter two only operate inside atomic nuclei. The Strong force binds protons and neutrons together in close proximity within the atomic nucleus, and without it, the positively charged protons would repel each other and no stable elements other than hydrogen could exist. The Weak force is responsible for certain forms of radioactivity (such as the nuclear decays that led physicists to suggest the existence of neutrinos, as mentioned in Chapter 12), and without radioactivity, few superheroes or supervillains would exist. Nearly every force we encounter in our everyday dealings, aside from gravity, is

---

* These four forces, in the form of the villains Graviton, Zzzax, Quantum, and Halflife, can be quite formidable when combined, as the West Coast Avengers found in "The Unified Field Theory."

electrostatic in nature. The forces generated by our muscles, the force the chair exerts on the seat of your pants to keep you from falling to the floor, the force exerted by the hot gases in your car engine's cylinders that lead to locomotion, all these and many others are, in the final analysis, electrical. It is thus time for us to consider the twin forces of Electricity and Magnetism, and see how these wonder twin powers activate into one single force properly termed "electromagnetism."

Very few superheroes have powers that are electromagnetic in origin. Two of the earliest Silver Age comic-book characters whose powers do utilize electricity and magnetism are Lightning Lad and Cosmic Boy. These heroes are from the future, and they first appeared in *Adventure # 247* (April 1958) when they, along with Saturn Girl, traveled back in time in order to recruit Super-boy into the Legion of Super-Heroes. Lightning Lad is able to create and discharge electrical bolts from his hands, while Cosmic Boy can control magnetic objects. The third founding member, Saturn Girl, possessed the superpower of mental telepathy, which we will argue later is intimately connected with electromagnetic-wave propagation. Consequently the three founders of the Legion are direct manifestations of electricity and magnetism theory in action.

The Legion hailed from the year 2958 (current stories take place in 3010) and was comprised of teenagers from different planets who each had a unique superpower. The concept of a club of teenaged superheroes in the future proved very popular with comic readers, and the Legion of Super-Heroes became a regular backup feature in Adventure Comics and eventually squeezed Superboy out of his own comic. The Legion membership grew over time, and currently boasts more than thirty heroes. All of the fundamental forces of nature, as well as several basic symmetries of physical laws, were pressed into service as the writers of Legion stories strained to develop a superpower for each hero. Legion-naires include Star Boy, who could make objects heavier while Light Lass could make them lighter, Element Lad, who was able to transmute one element into another (implying control over nuclear forces), and Colossal Boy, who could grow to great heights, while Shrinking Violet could miniaturize herself. Ferro Lad could transform himself into some sort of organic iron (an early, teen version of Colossus of the X-Men); as a kid I was profoundly shaken when he nobly sacrificed himself in order to destroy the Sun-Eater in *Adventure # 353.*

While only a few heroes draw upon electricity and/or magnetism as the source of their superpowers, supervillains frequently employ these fundamental forces of nature as they seek either financial gain or world domination (and occasionally both). In particular, in the next few chapters, we'll focus in turn on two such evildoers, Electro and Magneto (I leave it as a challenge to you, Fearless Reader, to determine which villain is associated with electricity and which one is involved with magnetism).

STATIC ELECTRICITY—NATURE'S MOST POWERFUL FORCE!

When comic-book readers first met Max Dillon in *Amazing Spider-Man # 9,* he was a highly skilled but self-centered electrical-utility worker. When a coworker was trapped atop a high-tension line, Max was cavalier about his fate until his foreman offered Dillon a $100 reward (in 1963 dollars, worth about $700 today) to rescue him. Freeing the unconscious colleague and lowering him to the ground with a cable, Dillon then received an unanticipated bonus when he was struck by lightning while grasping the high-tension lines. Just as in the case of Barry Allen (the Flash), Dillon not only didn't die or suffer any burns or neurological damage from this traumatic event, but he in fact gained the ability to store electrical energy that he could discharge at will in the form of lightning bolts.*

Dillon's accident may have changed his body, but it left his antisocial attitudes intact. Realizing that he now possessed fearsome electrical powers, he designed a garish green-and-yellow disguise, with a bright yellow lightning-bolt-themed mask, and embarked on a life of crime as Electro, as shown in fig. 25. Personally, if I gained mastery over such a powerful, fundamental force of

---

* Young readers during the Silver Age could be forgiven if they reached the conclusion that being struck by lightning, preferably in conjunction with some other hazardous activity, was one of the best things that could happen to them, second only to being exposed to massive doses of radiation.

© 1963 Marvel Comics

Fig. 25. *A scene from* Amazing Spider-Man # 9, *where the supervillain Electro simultaneously demonstrates an advanced concept in electromagnetism and a significantly less sophisticated fashion sense.*

nature, I don't think this would necessarily be the costume that I'd choose to wear in public. Perhaps if Max Dillon had not been such a rat, his friends might have gently provided some better fashion advice. But it is exactly such a pattern of bad choices that frequently leads these superpowered miscreants to a life of crime.

Dillon found that his body could store an electrical charge that enabled him to hurl lethal bolts of electricity. Stories featuring Electro would frequently show him charging up his body at some "abandoned" power station, standing between two transformer towers and letting the electric current flow through his body. (The fact that New York City had fully functioning power stations lying dormant throughout town, available as hideouts for various supervillains, surely accounts at least in part for the high utility bills city residents must now pay.) Fully charged, Dillon could project lightning bolts from his hands, though sometimes he would discharge through other parts of his body. Once his stored charge was depleted, he was basically powerless until receiving another charge. Essentially, the freak accident on the power line turned Max Dillon into a walking rechargeable taser gun.

What does it mean to have "electrical powers," such that one could hurl electrical bolts at the police and costumed super-heroes? Anyone who has shuffled his feet through shag carpeting on a dry winter day and then touched a metal doorknob has verified that matter is composed of electrically charged elements. Unlike the mass of an object, which is always positive, electrical charge comes in two varieties and is arbitrarily labeled "positive" and "negative." The expression "opposites attract" may or may not be a reliable guide in affairs of the heart, but it does accurately summarize the nature of the force between positively and nega-tively charged objects. Two objects with opposite charges will be pulled toward each other with an attractive force. Similarly, two objects that are electrically charged with the same polarity, either both positive or both negative, will repel each other. When a ship-ping box picks up an excess electrical charge due to random fric-tional rubbing, this charge can be transferred to the foam packing peanuts inside the box. All of the lightweight foam bits then have the same charge, repelling each other and flying into the air when the box is opened.

The negatively charged electrons in an atom are attracted toward the positively charged protons in the nucleus by the attrac-tive electrostatic force. The more protons there are, the larger the positive charge and the greater the force pulling the electron into the nucleus. However, the more electrons in an atom, the greater their mutual repulsion. These two forces—the attraction by the nucleus and the repulsion by the other electrons—tend to roughly cancel out, which is why a uranium atom with ninety-two elec-trons and an equal number of positively charged protons in its nucleus is approximately the same size as a carbon atom, with six electrons and six nuclear protons.

The attractive force between two oppositely charged objects, or repulsive force for two objects with the same charge, has, remark-ably enough, the same mathematical form as Newton's law of gravitational attraction described in Chapter 2. That is, the force between two objects that have electrical charges (charge 1 and charge 2) is given by the equation:

$$FORCE = K \, [(CHARGE \; 1) \times (CHARGE \; 2)] / (DISTANCE)^2$$

This expression, attributed to the eighteenth-century French scientist Charles Coulomb, is nearly identical to Newton's gravitational expression, except that instead of the charge of two objects, we multiplied their masses, and the constant wasn't called "k" but "G." Recall from Chapter 2 that Newton's law of gravity described the force between two masses (mass 1 and mass 2) by the expression:

$$FORCE = G \; [(MASS \; 1) \times (MASS \; 2)] \; / \; (DISTANCE)^2$$

Mathematically, these two expressions for Force are equivalent when "mass" is replaced by "charge" and the constant G is renamed as the new constant k. Because electric charge is not the same quantity as mass, the units of the constant k are different from the units of the constant G in order for both equations to have the units of a force.

More important than k having different units from G is the fact that the magnitude of k is very much larger than the magnitude of G. Consider a single proton in the nucleus of a hydrogen atom, orbited by a single electron a certain distance away. The attractive force of gravity pulls the electron in toward the proton, and there is an additional attractive force since the positively charged proton is pulled toward the negatively charged electron. The magnitude of the charge on the proton is exactly the same as that of the electron, where the charge of the proton is labeled positive by convention and that of the electron is considered negative. While they may have equal but opposite charges, the mass of the proton is nearly two thousand times larger than that of the electron. However, when k in Coulomb's expression is multiplied by the product of the electron's and proton's charge for a given separation in an atom, the resulting force is ten thousand trillion trillion trillion (a one followed by forty zeros) times stronger than the gravitational attraction. On the atomic scale, gravity is irrelevant, and matter is held together by electrostatics. There'd be no molecules, no chemistry, and no life without static cling.

If gravity is so much weaker than electrostatics, why does gravity matter so much for planets and people? Because it is always attractive. Two masses, no matter how big or small, will always be pulled toward each other due to gravity. While there is such a thing as antimatter, it has a positive mass, and therefore has a normal gravitational attraction to other matter. As far as anyone has been able to experimentally determine, gravitational attraction between masses is always attractive. Certain puzzling astronomical observations have recently been interpreted as indicating the presence of some sort of "antigravity" associated with a mysterious quantity termed "dark energy." However, this explanation is somewhat controversial, and at the time of this writing, scientists don't have the foggiest idea what dark energy is.

The situation with electricity is very different. The fact that electrical charges come in two different types—positive and negative—introduces the ability to screen out electric fields. An electron orbiting a proton feels an attractive pull. A second electron brought near this arrangement is pulled toward the proton but is pushed away from the first electron. Until the second electron comes very close to the proton, the sum of the pull and push cancels out, and there is no net force on the second electron. If we could as easily screen out gravitational attraction, then levitating devices such as the Fantastic Four foe the Wingless Wizard's antigravity discs would be commonplace. Regardless of whether it has a positive, negative, or no (neutral) electrical charge, all matter has a positive mass and feels an attractive gravitational pull from other matter. In this way, gravity always wins in the end and pulls everything together—even those objects that are electrically neutral.

But make no mistake, electrostatics is the stronger force. Consider the form of Coulomb's force equation stated earlier. If you had only 10 percent more negative charge than positive charge on your body, the force of electrostatic repulsion would be large enough for you to lift an office building that had a similar 10-percent excess negative charge. On the other hand, while the mass of the office building is much greater than your own, you are not gravitationally bound to the building, despite the occasional dictates of the workplace and your boss.

As he runs, the Flash should pick up an enormous static charge due to the friction between his boots and the ground. This friction is necessary in order for him to run, and an accumulation of excess charge as the Scarlet Speedster sprints is similar to what happens when we rub our feet along a carpet in the winter. The friction of our feet rubbing against the carpet, which is a violent process on the atomic scale, results in the transfer of electrons, which spread out over our bodies. These excess charges repel each other and don't want to stay on you. When you approach the doorknob, a path for the charges back to the Earth (which is able to take up a few more or less electrons without bother) becomes available. If the charge is large enough, the electrons will jump through the air, the same way a lightning bolt allows the excess charge in a thundercloud to discharge to the ground. When driving, your car will frequently pick up excess charge due to the friction between the tires and the road, which you can remove by touching the metal door once the car has stopped. The discharge is painful for two reasons: The surface area of your fingers is very small, so the current per area is large, and your fingertips have more nerve endings, so they are more sensitive to the current. Better to touch the car frame or the doorknob with your elbow, or you could adopt my strategy of draping your entire body over the hood of the car. The arch looks you'll receive will be a small price for the reduced pain (and you'll find that over time few people will want to park near you!).

This friction-induced electrostatic buildup (technically referred to as "contact electrification") was acknowledged in *Flash # 208*. The Flash had just finished saving the citizens of Keystone City,[*] yet again, from an attack by a subset of his Rogues Gallery, and was being thanked by a group of bystanders. In addition to requests for his autograph, one person patted the Flash's shoulder and, receiving a shock, noted, "Hey, check it out! His uniform is covered in static electricity!" This excess charge should in general discharge to the first metallic object near the Flash that was connected to electrical ground as soon as he stopped running. The fact that this contact electrification was only noticed in 2004 (and not in the previous fifty years of Flash comics) suggests that during the majority of his crime-fighting career, the Flash, in addition to possessing an ability to ignore air resistance and punishing accelerations, was similarly immune to electrostatic buildup.[†]

Returning to Electro, his electrical powers no doubt stem from

---

[*] This being the Wally West Flash, who based his operations out of Keystone, the Twin City to Central City, home of the Barry Allen Flash.
[†] Suggesting that Barry Allen's Flash costume was at the forefront of anti-cling dryer-sheet technology!

the fact that he is able to store very large quantities of a net electrical charge, either all positive or all negative, within his body. He can then discharge himself at will, in a similar fashion as the spark that leaps from your fingertip to the brass doorknob mentioned earlier. This is consistent with the fact that Electro needs to charge himself up before employing his powers, and if he lets loose with too many electrical bolts, he is, in essence, depleted and susceptible to a good right hook.

* * *

More than sixty years ago, a Swiss engineer's hiking frustration led to a technological innovation. George de Mestral's investigations into why burrs clung so tenaciously to his woolen hiking pants resulted in the invention of a fastener consisting of millions of tiny hooks and loops, and gave the world Velcro. More recently, Robert Full, Keller Autumn, and coworkers have discovered that the gecko lizard's ability to climb up smooth walls and ceilings can be traced to millions of microscopic hairs on the lizard's toes called "setae." But without miniature hooks in the walls or ceiling, what holds the fibers and the attached gecko in place? Static cling!

The fibers in a gecko's feet are electrically neutral, but the lizard does not need to shuffle across a shag carpet to cling to a wall, because he makes use of fluctuations of charge in his setae. The electrons in the fibers in the gecko's toes are constantly zipping around. Sometimes a few more electrons are on one side of the fiber, making that side slightly negatively charged, while other times a few less electrons are on that side, making it slightly positively charged. If the side of the fiber closer to the wall is, just for a moment, slightly negatively charged, then it will induce a slight positive charge in the wall (by repelling those electrons in the wall closest to the surface, exposing the positively charged ions) and an attractive force between the fiber and the wall will result. You would expect that this force (known as the van der Waals force) is very weak, and you would be right, which is why the gecko has millions of these fibers in each toe, so that the total attractive force can be large enough to support the lizard's weight.

Or possibly even Peter Parker's weight. Marvel's writers have suggested that Spidey's wall-crawling ability is electrostatic in nature—the 2002 *Spider-Man* film included a scene showing scores of microscopic barbed fibers sprouting from Peter Parker's fingers once he had gained his spider-powers. Turns out that both the comic and the movie are on to something. A report from the University of Manchester in England described the development of "gecko tape," which consists of millions of tiny fibers (the length of each fiber is fifty times shorter than the width of a human hair) that can provide a strong enough attraction to support a Spider-Man action figure from one's palm. A tape that makes use of the force arising from fluctuating charges can be, in principle, instantly used and reused, unlike a single-application adhesive that requires a curing time. The fibers on the tape must be very small in order to maximize the ratio of surface area to volume, since only the fluctuating excess charges on the surface of the fiber contribute to the attractive force. In order to provide enough force to support a grown person's weight, the density of microfibers must be very high, to compensate for the extremely weak force of each fiber. Whether these engineering challenges can be resolved remains to be seen. But if "gecko tape" ever becomes as common as Velcro, I, for one, will never wait for the elevator again!

SUPERMAN SCHOOLS SPIDER-MAN—

ELECTRICAL CURRENTS

LET'S NOW TAKE A CLOSER LOOK at those electric bolts emanating from Electro's hands. A large enough positive charge can pull electrons from very far away, even through miles of copper wire. The term for the pull exerted on electrons as they move through a wire is "voltage." Electrons are negatively charged, so a positive voltage pulls them in one direction while a negative voltage repels the electrons in the opposite direction. The "current" expresses the number of electrons moving past a given point in the wire per second.

Imagine a garden hose connected to an outdoor faucet. In this case, the voltage plays the role of the water pressure that pushes the water through the hose. The amount of water that comes out the end in a given time period is the current. The resistance of the hose arises both from kinks in the line and small holes along its length, from which some water can escape before making it to the end. The more defects in the hose, the greater the water pressure needed to maintain the same water flow (current) from the end of the hose. However, a faucet in a sink can have water flowing without a hose connecting the faucet to the drain—similarly, an electrical current can be pushed by a large enough voltage even in the absence of a wire. This is what happens when a spark jumps from your fingertip to the doorknob or from a cloud to the ground during a lightning strike. The greater the distance, the bigger the force needed to pull the charges, as the Coulomb electrostatic-force expression decreases with the square of the separation of the charges. A long garden hose, with various imperfections and holes, will have more resistance to water flowing through it than a similar short segment of hose. This is why you don't receive a static shock until your fingers are just about to touch the doorknob: Air is a pretty good electrical insulator, and it takes an electric field of more than 12,000 Volts per centimeter before the pull on the electrical charges is sufficient to make them jump the gap. When it does happen, it stings, and it's why you most definitely do not want to be zapped by Electro's massive discharges.

Alas, in *Amazing Spider-Man # 9*, where Spidey first tangles with Electro, the basic physics of electrical current flow is mangled. In one scene during their climatic battle, Spider-Man manages to deflect an electrical bolt that Electro has hurled at him by tossing a metal chair over Electro's head. "Anyone with any knowledge of science knows that anything metal can act like a lightning rod," Spider-Man lectures Electro, "as this steel chair is doing!" Actually, Spider-Man's mistaken understanding of how lightning rods function suggests that his allegedly advanced knowledge of science isn't all it's cracked up to be. The electrical bolt is shown arcing away from Spider-Man and chasing after the soaring chair—even though the chair is not electrically connected to anything! Why would Electro's lightning bolt be pulled toward the chair, metal or not, if once it reaches the chair there is nowhere for the electrical current to go?

When you turn on the water in the kitchen sink, the water flows from the tap to the drain, owing to the downward pull of gravity on the water. It's not that water is attracted to drains—if one were to install a drain on the ceiling above the faucet, the water would not arc upward, compelled to flow into the drain at all costs. When dealing with electrical charges, what gravity is for the water flow, the voltage difference is for the electrical current.

Electrical charge can't flow if there is no place for it to go. Actually, this is true for our water analogy as well. Want to know how you can overturn a glass of water filled to the brim and manage to not spill a drop? Do it when the glass is underwater! If there is no place for the water in the glass to go, it will stay inside the container.*

---

* Provided you ignore random collisions between the water molecules in the pool and those at the top of the glass that will cause those molecules to switch locations.

The same is true for electricity. Regardless of the magnitude of the net electrical charge an object possesses, it will not discharge if every other object surrounding it has exactly the same charge. Technically, the voltage that pulls or pushes electrical charges around is a measure of the "potential difference," defined as the difference in potential energy of a charge as it moves from one point to another. This is what makes Electro so dangerous (in addition to his daring fashion sense): He is able to control his potential difference relative to his surroundings at will, so he can decide when and where he will discharge the excess electrical charge he has stored up.

By applying a voltage across a conductor, I raise the potential energy of the electrons within the conductor, just as when lifting a brick over my head I raise its potential energy. The brick keeps this extra potential energy until I release it, at which point the potential energy is converted into kinetic energy and the brick speeds up as it falls. But this conversion cannot take place until I let go of the brick. Similarly, the electrons in a wire speed up and increase their kinetic energy in the form of an electrical current, in response to the voltage applied across the wire—but only if I close the switch in the circuit and the electrons have someplace to go. Just as the raised brick will keep its potential energy indefinitely until I drop it, the electrons cannot speed up in response to an applied voltage if the wire is not electrically connected to anything. Think again of a garden hose connected to a faucet. No matter how much I turn the faucet tap, absolutely no water will flow through the hose if it is completely sealed at the other end. Before any water can flow through the hose (a current) in response to the water pressure (voltage) at the faucet, I have to uncap the end of the hose so that the water can drain out. The technical way of expressing this is to say that in order for an electrical current to flow through a wire, it must be grounded.* The Earth, or "ground,"

---

* Strictly speaking, the wire does not have to be connected to "ground," but simply to another point at a lower potential than the starting point. But ultimately, the terminus of any current must be the ground. Following the water analogy, we can have a flow of water through a hose that is connected from one faucet at one end to another faucet at the other end. As long as there is a difference in water pressure between the two faucets, there will be a net flow of water, but for this to keep up indefinitely, the second, lower-pressure faucet must be able to eventually discharge its excess water down some drain (or "ground" back in the electrical situation).

is obviously a large object, with many electrical charges; consequently, it can take up extra electrons, or donate electrons to a wire without difficulty. This notion, that in order for a current to flow, it must have someplace of a lower voltage to go, is fairly reasonable, but Spider-Man seems to not have grasped it, while Superman, from his very first adventure, has demonstrated a solid understanding of electrical currents.

In Chapter 1 we mentioned the early exploits of the Man of Steel, as described in *Superman # 1*, before the world at large knew of his existence. In this story, Superman sought to learn the identity of the person bankrolling the Washington lobbyist who was bribing a senator with the goal of embroiling our country in the war in Europe. (Recall that this story occurred in 1939.) The secret employer of Alex Greer, "the slickest lobbyist in Washington," turns out to be Emil Norville, the munitions magnate (war being good for business, from Norville's point of view). For some reason, Greer initially refuses to divulge the name of his employer to this strange man wearing a blue-and-red long-underwear ensemble accessorized with a flowing red cape. In Chapter 1, I mentioned that Superman purposely falls from the top of a high building holding Greer, pretending that the fall will kill them both. Prior to this scene, in order to loosen Greer's tongue, Superman picks him up like a sack of potatoes and leaps atop some high-tension lines, as illustrated in fig. 26. Greer protests that they'll be electrocuted, but Superman finds the time to give the lobbyist a physics lesson. (Whether this lecture should be considered an additional part of Superman's efforts to psychologically torment and elicit information from the lobbyist, I leave to the reader to decide.) "No, we won't," the Man of Steel explains, since after all, "birds sit on telephone wires and they aren't electrocuted—not unless they touch a telephone pole and are grounded!"

Superman is exactly right. It is only when you touch a high-voltage wire and simultaneously grab the telephone pole (or touch another wire at a different voltage), and thereby provide a pathway for the current in the wire to flow to the lower voltage, that you have to worry. In this unfortunate situation, the flow of electrons (a current) passes through the conductor—namely, your body—connecting the two points.

The physics of grounding is well understood a thousand years from now, as indicated in "The Secret Origin of Bouncing Boy" in

Fig. 26. *A scene from* Superman # 1, *where the defender of truth, justice, and the American Way coerces information from a Washington lobbyist by giving him a "hands-on" demonstration of the principles of electrical grounding.*

Adventure # 301. Chuck Taine became a member of the teen-superhero association the Legion of Super-Heroes when he accidentally drank a super-plastic serum and consequently gained the power of "super-bouncing." Chuck was quite disappointed to be turned down for membership in the Legion, owing to the super-silliness of his power. Yet he made good use of his second chance to impress the Legion, when he managed to subdue a crook whose electrical gloves had stopped the Science Police and Saturn Girl of the Legion. Chuck knew that when he was bouncing and airborne, he could collide with the crook and not fear electrocution, as he was not grounded at the time. Perhaps it is not so surprising that this story correctly illustrates the principle of grounding, as it was written by Jerry Siegel, Superman's co-creator.

It seems that Superman could teach a thing or two about currents and grounding to Spider-Man. The "grounding goof-up" in *Amazing Spider-Man # 9* was not a solitary error, sadly, as *Amazing Spider-Man Annual # 1* (Feb. 1964) finds Spidey again facing off against Electro. This time, as an extra precaution, he *deliberately* attaches a wire to his ankle to ensure that he remains electrically grounded at all times! Promise me, Fearless Reader, that when next you fight a supervillain capable of hurling lethal lightning bolts at you, you will avoid a good, solid electrical connection to the ground!

The whole point of a lightning rod is not that it's made of metal, but that the lightning will strike the tallest feature on the building (the rod), and the electrical current is then carried from the rod by way of a wire safely to ground, thereby avoiding igniting a fire on the roof of the building. Lightning rods can also carry away excess charges in the atmosphere, decreasing the voltage difference and reducing the likelihood that the building will be struck by lightning in the first place. The static shock between your fingertip and the metal doorknob occurs only when your finger is very close to the door, since the shorter the distance, the less resistance the arc has to overcome. Similarly, the lightning bolt is trying to minimize the distance and resistance on its way to electrical ground. This is why you don't want to stand under a tree during a lightning storm, as you increase the chance that the lightning striking the tall tree will take a detour through your body— when alone in an empty field during a thunderstorm, one should lie flat on the ground in order to decrease the chance of being struck by lightning. If a building's lightning rod is not connected to ground, any electrical current entering the rod will find a higher resistance pathway to ground, through the roof and building, with concomitant damage to the structure.

Just such damage will surely be Spider-Man's fate when he intentionally connects himself to ground, thereby guaranteeing that all of Electro's electrical energy will pass through his body on its way to a lower potential state. Spider-Man's "spider-strength" will enable him to withstand some of the damage of the electrical strike, but grounding himself as he does makes the situation much worse than it needs to be.

It's not clear which of Spider-Man's co-creators, writer Stan Lee or artist Steve Ditko, should get the blame for these goofs. This ambiguity stems from the "Marvel method" of producing comic books in the 1960s. At Marvel's Devilish Competitors (as Lee would jokingly refer to DC Comics), a comic-book writer would generate a full script that detailed not only the captions and dialogue and thought balloons in each panel, but also what the artwork in each panel should look like. An editor would then go over the script, making changes as needed, and pass it along to the artist, who would draw the comic story as described in the script. The artwork would then be inked, colored, and lettered, using the dialogue and captions in the script, and the writer would commonly not see the story again until it was available for sale on newsstands. This system works fine as long as one has enough writers and editors to cover the number of comics produced per month, but at Marvel in the early 1960s, the number of writers and editors was low—namely one: Stan Lee. Lee was both the editor and writer of (in 1965, to pick a particular year) the *Fantastic Four*; *Spider-Man*; *The X-Men*; *The Avengers*; Captain America and Iron Man stories (both in *Tales of Suspense*); Dr. Strange, solo Human Torch stories, and Nick Fury, Agent of S.H.I.E.L.D.* (in *Strange Tales*); Giant-Man, the Sub-Mariner, and the Incredible Hulk (in *Tales to Astonish*); *Daredevil*; and *Sgt. Fury and His Howling Commandos* (a World War II comic). If the stories in the Marvel universe had a coherent structure and feel, this was no doubt due to the fact that there was a single creative voice guiding the varied comic books.

With so many stories being created every month, there was simply no way that Lee had the time to craft full scripts for all of these comic books. Meanwhile, the artists working for Marvel were freelancers, and would bring in the artwork for one issue, get paid, and then need to pick up instructions for the next issue's story line (if they weren't working, they weren't being paid). I should mention, by the way, that the artists working for Marvel at the time were some of the very best in the business and included such titans as Jack Kirby, Steve Ditko, Don Heck, John Romita, and Gene Colan. The talent of these artists is reflected in the fact that they were able to continue making a living through the comic-book Dark Ages of the mid-1950s, when the entire industry

---

* Supreme Headquarters, International Espionage, Law-Enforcement Division.

was on the verge of extinction thanks in part to the "Seduction of the Innocent" brouhaha. Consequently, they were experts in how to tell a story in graphic terms, and did not need a comic-book writer holding their hands with panel-by-panel instructions of what should be drawn on every page.

Stan Lee, therefore, hit upon a clever solution to the problem of not enough time and too much available talent: Let the artists tell the story. Lee would write up a brief synopsis, varying in length from a few pages to a few paragraphs,* describing what the latest issue's story line would be. In essence, he gave the artists a plot outline of the major points of the story, such as who the villain would be and what his powers were and how he obtained them, as well as how the hero would lose the initial skirmish with the villain, and finally the clever stratagem that would provide the hero's victory by the issue's conclusion. The artists would then go back to their studios and construct a graphic story that followed Lee's synopsis. When the finished artwork would return, Lee would then write the captions and dialogue in each panel, and the comic book would be ready to be sent off to the printer. Consequently, both Lee and the artists could legitimately be said to co-write or co-plot any given issue of a comic book created in the Marvel style. It is therefore on both Lee's and Ditko's shoulders that the blame for Spidey's ignorance of basic electrical-current theory must be placed.

Lee and Ditko may not have had a firm grasp of the concept of electrical grounding, but they understood that electricity plus water equaled a short circuit. The climax of Spider-Man's battle with Electro in *Amazing Spider-Man # 9* came when Spidey grabs a nearby fire hose, such as used to be commonplace in most professional buildings before the advent of ceiling-based sprinkler systems, and doused Electro with a heavy spray of water. As Spider-Man grabs for the hose and turns open the main pressure valve, he thinks "Say!! What kind of science major am I, anyway?

---

* Or sometimes even less. One legend has it that Lee's précis to Jack Kirby for the classic *Fantastic Four # 48*, where the FF encounter the cosmic being Galactus, read as follows: "The FF fight God." Kirby, reasoning that such an entity would have a herald to prepare the unlucky planet for the devourer's arrival, created the character the Silver Surfer. Lee first learned of the Surfer when Kirby's artwork for *FF # 48* arrived at the Marvel office.

Why didn't I think of this right away??" As he lets Electro have a full blast, he continues, "Water and electricity just don't mix!!!"

Well, as mentioned above, we are beginning to have some doubts about just what kind of science major Peter Parker is, but he is certainly correct that water and electricity don't mix. This is because city water, while technically electrically neutral, contains a high concentration of impurity ions. Ordinary tap water is consequently a pretty good conductor of an electrical current. Electro is at a high potential difference, which is why he is such a lethal threat to superheroes. By dousing him with water, Spider-Man essentially connects a wire between Electro and ground, allowing the large excess charge Dillon has stored to flow out of his body. This is one physics lesson that seems to be well learned in the Marvel universe. When Electro is bested by Daredevil in the second issue of that hero's comic, the police keep him drenched with a water hose (Electro, that is, not Daredevil) in order to safely transport him to the paddy wagon and the station house.

18

HOW ELECTRO BECOMES MAGNETO WHEN HE RUNS—

AMPERE'S LAW

I SUPPOSE THAT we should cut Lee and Ditko some slack for their goof described in the previous chapter concerning whether metals can carry an electrical current if they're not connected to ground. The rush to put out a monthly comic book, combined with the need to tell an exciting story, has certainly accounted for more than a few science blunders at both Marvel and DC over the years. As emphasized earlier, these comic-book tales were never meant to serve as physics textbooks. It is therefore all the more impressive that in the very same issue of *Amazing Spider-Man* that features Electro's debut, we see a perfect illustration of one of the fundamental properties of electricity.

At one point in the story, following a brazen daytime bank robbery, Electro is shown escaping from the authorities by climbing up the side of a building, as easily as Spider-Man. The panel is reproduced on p. 197 in fig. 25, where we see one observer exclaim, "Look!! That strangely-garbed man is racing up the side of the building!" A second man on the street picks up the narrative: "He's holding on to the iron beams in the building by means of electric rays—using them like a magnet!! Incredible!"

There are three feelings inspired by this scene. The first is wonder as to why people rarely use the phrase "strangely-garbed" anymore. The second is nostalgia for the bygone era when pedestrians would routinely narrate events occurring in front of them, providing exposition for any casual bystander. And the third is pleasure at the realization that Electro's climbing this building is actually a physically plausible use of his powers. Utility-pole man Max

Dillon (Electro) understands, as does the second passerby in the panel, that electric currents do indeed create magnetic fields. This phenomenon, termed the Ampere effect, was first noted by Hans Christian Øersted (for whom a unit of magnetic-field strength is named) and was fully explained by André-Marie Ampère (after whom the unit of electrical current, the "amp," is named). Why does Electro's control of electricity enable him to generate magnetic fields, and wouldn't fairness therefore dictate that Magneto, the mutant master of magnetism, be able to control electric currents at will? The answer to this question reveals a deep symmetry between electricity and magnetism, found in both comic books and the real world.

We first must begin by defining an "electric field." An electric charge at rest exerts a force on another electric charge. The farther away this second charge is, the weaker the force and, depending on its polarity, the second charge will be either pulled toward or pushed away from the first electric charge. We can therefore say that there is a "zone of force" surrounding this first electric charge. Another way to describe this "zone of force" is to say that around the first charge is an "electric field." A second electric charge brought near the first charge will experience a force as it is being pushed or pulled by the electric field of the first charge. The strength of the electric field depends on the magnitude of the electric charge at this point and varies with its distance from the first charge—very close to it, the force on a second charge is large, while as the separation increases, the force decreases as the inverse of the square of the distance (from the Coulomb expression). If the separation of the two charges is doubled, the force goes down by a factor of four, and if the separation triples, the force is only one ninth as large.

There is another field that is created by an electric charge, but only when it is in motion, called a "magnetic field." If an electrical current flowing through a wire is held near a compass needle, the needle will be deflected just as if an iron magnet had been brought near the compass (this was Øersted's discovery). In fact, two parallel wires carrying electrical currents will, depending on the direction of the currents, either be attracted toward each other or repelled away from each other, behaving just as two magnets would when oriented north to south poles (attracting) or south to south (repelling). The magnetic field generated by Electro's "electric rays" does indeed provide an attraction to the magnetic field of the iron beams in the building, enabling him to scale buildings or adhere to passing automobiles (as he did to attempt an escape in *Daredevil # 2*).

The force that arises between current-carrying wires is not electrostatic in nature. The wire is electrically neutral before the current flows, with the number of electrons in the atoms making up the wire evenly balanced by the same number of positively charged atomic nuclei. While a current is passing through the wire, an equal number of electrons enter at one end as leave at the other. The extra force between the wires, present only when carrying a current, is due to the magnetic field they create.

Why is this so? Why does an electric current create a magnetic field, which in all aspects is identical to that of an ordinary magnet? A key clue behind the phenomenon of magnetism is that it involves electric charges in relative motion. That is, the charges must be moving relative to each other.

If two electric charges are moving in the same direction at the same speed, then from the point of view of one of the charges, the other charge is stationary. In this case, the only force between the two charges, from their point of view, is electrostatic. To someone stationary in the lab, however, there is an extra force associated with the motion, called magnetism. That the magnetic force is connected with the relative motion of the electric charges suggests that there's a simple four-word explanation for the phenomenon of magnetism: Special Theory of Relativity. To explain how Einstein's theory from 1905 is able to account for magnetic fields will require more than four words, but we'll try to get by without mathematics.

I'll use a nice argument given by Milton A. Rothman in his excellent book, *Discovering the Natural Laws*, that illustrates how relative motion of charges creates a force, where the force is absent when the charges are stationary. Think about two very long train tracks lying next to each other, one with a large number of negative charges equally spaced exactly one inch apart, the other with an equal number of positive charges, also one inch apart. We're making this up as we go along, so we'll assume that these rows of negative and positive electrical charges extend for miles and miles, and this way we won't have to worry about running out of charge as they move along the track. We next bring in a test charge—a positive charge, for sake of argument—some distance from these lines of charges. This test charge will feel no net force, as it is pushed away from the line of positive charges as strongly as it is attracted to the negatively charged array. Now the two tracks start moving at the same speed in opposite directions, the negatives to the left and the positives to the right. If the test charge is stationary, then the same number of negative charges and positive charges, in a given length, pass by it, and there is still no net force. An extra force develops, however, if the positive test charge moves to the right at the same speed as the positive charges on the track, also moving to the right.

Back in Chapter 11, when we considered the relativistic effects of the Flash's great speeds, we discussed the property of the Special Theory of Relativity that, from the point of view of a stationary observer, the length of the moving object is contracted. From the positive-test charge's "point of view," moving along with the same speed and direction as the positive charges, it is stationary compared with this array of positive charges. The test charge therefore sees the positive charges on the track as still being spaced one inch apart. The array of negative charges moving in the opposite direction, on the other hand, will be contracted in length and will therefore be closer than one inch apart to the moving test charge. The electrostatic push and pull on the test charge is now unbalanced, and it will now feel a net attractive force. We give this extra force, present to the outside observer when charges are moving relative to each other, a special name: magnetism. From this argument it is clear that a moving object having no net charge (that is, electrically neutral) will not feel any extra force, which is consistent with the experimental observation that magnetic fields are created only by positive or negative currents.

Earlier we noted that the friction between the Flash's boots and the ground should transfer static charge to the Scarlet Speedster. Because moving electric charges create magnetic fields, it is puzzling that the Flash, whenever he sprints at superspeed, does not simultaneously generate an enormous magnetic field that would pull every iron object not nailed down (and quite a few that are) after him in his wake. We will have to chalk both this missing electric charge and corresponding magnetic fields to the efficacy of his "aura," that also enables him to avoid the deleterious effects of air resistance.*

It is indeed strange that magnetism is explained by invoking the special theory of relativity for moving electrical charges, as in general it's easy to ignore relativistic effects when the object in question is moving much slower than the speed of light. There's only a very slight error if we neglect relativity even when the object is moving at one tenth of the speed of light. Nevertheless, for electrical charges moving much slower than light-speed, there is still a relativistic effect through the creation of a magnetic field. The effect is smaller, to be sure. How much smaller? One can show mathematically that the upper limit of the magnitude of the magnetic field created by a moving charge is equal to its electric field divided by the speed of light. The speed of light is a large number; so, for a given electric field, the magnetic field associated with the moving charge will be quite weak, but it will be there nonetheless. Increasing the magnitude of the electrical current, either by moving more electrical charges, or having them move faster, generates a larger magnetic field.

An understanding of Ampere's Law, connecting electrical currents and magnetic fields, makes possible useful devices such as electromagnets. An electromagnet is a coil of wire wrapped around an iron magnet core. The current flowing through the coils creates a magnetic field, enhancing that from the iron alone. Just such a device is constructed in *Superboy # 1*, when the Teen of Steel faces off against a gang of thieves who are running around town in a fleet of personal tanks stolen from an Army-surplus depot. The "Smash and Grab Gang" of crooks (yes, that was really their name) then uses these tanks to break into banks and cause mayhem. While he could easily fly around and collect all the tanks by hand, Superboy decides on a more technological approach, as shown in fig. 27. "I'll only need that locomotive, a

---

* When that extradimensional imp Mopee bestowed superspeed and a protective aura to Barry Allen in *Flash # 167*, he knew what he was doing! Mopee disappeared at the end of the story before Barry Allen could ask him how Wally West (Kid Flash, at the time) acquired his aura. Nowadays, all of Flash's auras derive from the Speed Force, which makes as much physics sense as a tenth-dimensional imp.

Fig. 27. Superboy demonstrates a practical knowledge of electromagnetic theory, as he constructs a portable electromagnet (from Superboy # 1)

Fig. 28. A continuation of the scene from Superboy # 1, where the Teen of Steel uses his brains, in addition to his brawn, to capture the "Smash and Grab Gang."

dynamo from that power-house, and a few miles of wire," the Teen Titan explains to a recent victim of the crime gang. Superboy flies a large electrical dynamo to the empty coal car behind a large locomotive engine, and notes, "This dynamo will give the current I need when it's hooked up! Now for the windings—." In the next panel, we see him take "a few seconds to wind these miles of wire" as he loops them around and around the body of the locomotive. On the next page (fig. 28) we see the payoff as Superboy starts up the engine (presumably there is enough coal to get it going) and announces, "I've got the biggest electromagnet ever made, and one that can go places!" The train tracks conveniently pass not only through the center of town, but right by the vandals' tanks. "What happened? We're flying!" one crook cries out as his tank is drawn toward the magnetic locomotive. "It's that locomotive," says a more informed villain. "It's a magnet drawing our steel tanks!"

This is all perfectly correct, from a physics perspective. The dynamo is the source of the electrical current that passes through the few miles of wire around the engine. The current in the wire creates a magnetic field that projects from the center of the wire loop. If a magnetic material such as an iron locomotive engine is placed within the loop, it enhances the generated magnetic field. However, why the strong magnetic fields generated by Superboy's homemade electromagnet do not also attract the nearby parked cars, any loose iron in any of the buildings, or cause the steel wheels of the locomotive to seize up, thereby preventing them from rotating, remains a mystery.

19

HOW MAGNETO BECOMES ELECTRO WHEN HE RUNS—

MAGNETISM AND FARADAY'S LAW

THE VERY FIRST VILLAIN the X-Men encounter in *X-Men # 1* is Magneto, the mutant master of magnetism, whose superpower consists of the ability to generate and control magnetic fields. Magneto could hurl missiles at our heroes and deflect magnetic objects, yet would be powerless against a wooden baseball bat. In fact, even some metallic objects are immune to Magneto's power: He is able to pick up an automobile easily enough, but not a silver spoon or gold bracelet. What determines whether some materials are magnetic, even without an electrical current passing through them, and others not? Where does magnetism come from?

The Special Theory of Relativity may be ultimately responsible for the magnetic field created by an electrical current that involves the motion of electrical charges, but what about magnets made of iron? The magnets we use to hold grocery lists to the front of our refrigerators don't seem to have any moving parts, yet they still create magnetic fields. It turns out that relativity is ultimately responsible for the magnetism of a stationary hunk of iron, too.

Every proton, electron, and neutron in the universe has a tiny magnetic field associated with it. This field is barely noticeable compared with the Earth's magnetic field, or fields created by an electrical current. The electrons orbiting the nucleus can be (very roughly) considered tiny current loops that generate magnetic fields. But even without this "orbital" effect, there is still an internal magnetic field within atoms. Where does this minuscule intrinsic magnetic field of subatomic particles come from? The answer involves quantum mechanics, which we will discuss in the next section. A principle of the Special Theory of Relativity is that space and time should properly be considered on an equal footing as a single entity, called space-time. When this relativistic adjustment is made to the basic equation of quantum physics, the theory predicts that electrons should have a very small internal magnetic field, the theoretically predicted magnitude of which agrees precisely with the measured value. The internal magnetic field of electrons, protons, and neutrons is understood mathematically only in the relativistic version of quantum mechanics. Even for stationary matter, relativity turns out to be crucial for understanding magnetism. So, no Einstein, no relativity, and, hence, no magnetism. No magnetism, no magnetic iron, and, most importantly, no refrigerator magnets! Therefore, without relativity there is no way to keep our shopping lists from falling to the floor and lying there, discarded and unread. Without Einstein's towering achievement in theoretical physics, a slow and lingering death of starvation would await us all.

Normally, the small magnetic fields of electrons inside an atom like to pair up, just as when you bring two magnets together, they orient themselves to align at separate poles. When the magnetic fields inside an atom pair, there is no net magnetic field associated with the atom, just as an ordinary atom will have no net electric field, because the number of positive protons in the nucleus is balanced by an equal number of negatively charged electrons. Most materials, such as paper and plastic, are not magnetic, and even most metals, such as silver and gold, have all of their magnetic moments paired up.*

---

* This point was illustrated in *Atom # 3*, in which the Tiny Titan travels back to the past using a Time Pool discovered by physics professor Alpheus V. Hyatt. By shining a light of "all colors—even the finest shadings" on a small region in space, Prof. Hyatt created a small portal into other times. The opening of this Time Pool is only about six inches in diameter, so the professor dangles a magnet on a fishing line and tries to retrieve magnetic objects from the past (he is apparently unconcerned about what "butterfly effect" any small change in history will have on the present). The Atom, able to shrink small enough to fit through the portal, would occasionally hitch along on the dangling magnet and have adventures in the past, as in

If most materials do not have net magnetic fields because their atomic magnetic moments are paired up, then how is Magneto able to levitate himself and other people, as shown in fig. 29? The physical basis behind this trick is that Magneto is able to generate such a large magnetic field that he essentially polarizes the internal magnetic fields of our atoms, turning us, or any other object, into a magnet.

Before we begin this discussion of magnetic levitation, I first must stress that Magneto does not lift people through his influence on the iron in their blood. Let's leave aside the question of the effect of an inhomogeneous pressure on the veins and arteries in a person's body (this would get messy), and focus instead on the blood's magnetism. A few metals, such as iron and cobalt, have just the right configuration of non-paired electrons' internal magnetic fields, so that the atom has a net magnetic field. However, the iron in your blood occurs primarily in the form of hemoglobin, a protein used to collect and transport oxygen and carbon dioxide as you breathe. Hemoglobin is a very large molecule that consists of four large proteins (called globins, which look like folded worms) bonded together. Each of these proteins contains a large molecule termed a "heme" group, composed of carbon, nitrogen, oxygen, hydrogen, and iron. The iron atoms in the center of each heme molecule are chemically bonded to their neighboring atoms. There's another technical term for an iron atom chemically bonded to oxygen atoms: rust. As anyone who has dealt with scrap metal can confirm, rust makes for a very weak magnet. The common form of rust has three oxygen atoms bonded to two iron atoms (called "hematite") and is nonmagnetic, though four oxygen atoms bonded to three iron atoms (termed "magnetite") is magnetic. The magnetic field of the iron in hematite disappears when it combines with oxygen atoms, because the iron and oxygen chemically sharing their electrons pair up the remaining

the story involving the change from the Julian to Gregorian calendar mentioned in the Introduction. At the conclusion of the tale in *Atom # 3*, the Atom brings back a gold coin from Arabia circa 500 A.D. (the coin has an odd mix of Roman numerals and Arabic writing on it). The Atom holds on to the coin as the magnet is pulled back into the present day, noting, "The professor will be mystified at how this magnet held on to a gold coin!" And indeed, Prof. Hyatt was, and properly so.

Fig. 29. *Scene from* X-Men # 6 *(above) and* X-Men #1 *(left), as Magneto attacks (above) or escapes (left) from the X-Man the Angel (the one with the wings), illustrating Magneto's ability to levitate nonmagnetic objects, such as a boulder or himself, through the principle of diamagnetic levitation.*

uncanceled electronic magnets in the iron. Depending on whether the hemoglobin has picked up an extra oxygen molecule to bring to the cells, or is carrying a carbon-dioxide molecule to be exhaled, the iron can either have an uncanceled magnetic field or not. But at any given moment only a fraction of your blood is even capable of being affected by an external magnetic field.*

---

* In the motion picture *X-Men 2: X-Men United,* Magneto is able to overcome his guard and escape from his plastic cell only after his accomplice Mystique has injected a small quantity of magnetic metal into the guard's bloodstream.

Even when iron is not chemically bonded to oxygen atoms, it is possible that it will be nonmagnetic, if all of the individual atoms are not properly aligned. Ordinarily the atoms inside a piece of iron or cobalt will line up, forming small regions termed "domains," where all the iron atom's magnetic fields point in the same direction. However, entropy considerations lead to the domains pointing in different directions, so their combined magnetic fields cancel out. Heat up a bar of iron so that the atoms have a lot of thermal energy and are free to rotate, and then place it in a strong external magnetic field. The external field induces the majority of the domains to all point in the same direction, so that the piece of iron, when cooled back to room temperature, has a large net magnetic field. If you hit the magnetized iron bar with a hammer or heat it in an oven, you will cause the magnetic domains to reorient themselves randomly, with the effect being that the magnet will lose nearly all of its field strength. Some flexible, credit card–size refrigerator magnets have their magnetic domains aligned in little strips along their length. Rather than have all of their domains point in the same direction, it is easier to line them up so that one strip has its north pole pointing toward the refrigerator, while the adjacent strip has its north pole heading away from the fridge, and so on.*

Materials that form magnetic domains with magnetic fields pointing in the same direction are called "ferromagnetic" (so named after iron, the most famous example). Many atoms in solids have a very weak magnetic interaction with their neighbors, so if placed in a strong external magnetic field, they will align in the direction of the field but will randomize again at room temperature once the field is removed. These materials

---

* Try this experiment: Take two flexible credit card–size magnets and hold them so that their magnetic sides face each other. When they are oriented so that both long sides point in the same direction, you can slide them smoothly past each other. Now flip one magnet, so that its long side is at a right angle to the long side of the other credit-card magnet. You should notice a bumpy, stick-slip motion as you slide the magnets past each other. What you are feeling is the magnetic field of one card bumping over the walls between the stripes of magnetic domains on the other card.

(such as molecular oxygen, gaseous nitric oxide, and aluminum) are termed "paramagnetic." And there is a third class of materials in which, due to the nature of the interactions between adjacent atoms and the chemical ordering of the atoms, their atomic magnets (generated by electron orbits within the atoms) line up opposite an external magnetic field. If an external magnetic field is applied to these materials and the field's north pole points up, the atomic magnet's north pole rotates to point down. These materials are called "diamagnetic," and they try to cancel out any external magnetic field. Gold and silver are diamagnetic—if you are able to pick up your jewelry using a refrigerator magnet—someone has some explaining to do! Water molecules are also diamagnetic, and since we are primarily composed of water, so are we.

It is through our diamagnetism that Magneto is able to levitate himself and other people, as shown in fig. 29. In moderate-strength magnetic fields, the atoms in your body are not susceptible to being polarized. The diamagnetic interaction is weak, such that at room temperature, the normal vibrations of the atoms overwhelm the attempt to magnetically align them. In a very strong field, roughly two hundred thousand times greater than the Earth's magnetic field (and more than one hundred times larger than the field of a refrigerator magnet), the diamagnetic atoms in your body can be induced to all point in the same direction—opposite to the direction of the applied field. Just as two magnets repel if they are brought together so that their north poles are facing each other, the now magnetically polarized person will be repelled by the external magnetic field Magneto is creating—the very field that magnetically aligned the atoms in the first place. As Magneto increases the magnetic field he generates, the magnetic repulsion can become strong enough to counteract the downward pull of gravity. (That is, the upward force of the magnetic repulsion can be equal to or larger than the downward force of the person's weight, and there is then a net upward force on the person, lifting him off the ground.) It takes a very big magnetic field requiring a great deal of power to accomplish this, and the heavier the person, the larger the effort. But it can be done, and the High Field Magnetic Laboratory at the University of Nijmegen in the Netherlands has amusing images and movies on their website of

floating frogs, grasshoppers, tomatoes, and strawberries, demonstrating the reality of diamagnetic levitation.

* * *

If an electrical current generates a magnetic field, then could a
moving magnetic field induce an electrical current in a nearby
wire? The answer, as anyone who has read X-Men comics would
know, is yes. In previous battles with the X-Men, and occasionally
with puny humans, Magneto employed his mutant talent to transform any metal object into either an offensive weapon or a defensive shield. Magneto's power is most effective on metals that are
already magnetic. There are only three elements (iron, cobalt, and
nickel) that are magnetic at room temperature. Magneto can
manipulate a steel girder into any form he wishes through the iron
it contains, but his power would be limited on a gold wedding
band, unless he is willing to expend an extreme effort in polarizing
the normally diamagnetic material. But Magneto's real power lies
not in his ability to exert forces on other magnetic materials, but
in his control over electrical currents.

For example, the Mutant Master of Magnetism once constructed a computerized control panel that automated the power-
dampening fields that neutralized the X-Men's mutant abilities,
keeping them from interfering with his plan of conquest. To prevent the X-Men from deactivating the device, Magneto configured
it so that it has no buttons or knobs that can be reprogrammed.
Magneto controlled the panel by altering the electrical currents
that flowed through its circuits, affecting them through the magnetic fields that they create. Furthermore, by varying the magnetic field over the control panel, Magneto could cause these
currents to flow.

How would a varying magnetic field create an electrical current? The answer brings us back to the point that started our discussion of electrical currents and magnetic fields: relative motion.

Just as a magnet can be pushed or pulled when a second magnet
is brought near it, an external magnetic field can exert a force on
an electrical current. As described in the previous chapter, moving
electric charges generate a magnetic field that can attract or repel
other magnetic fields, whether created by another electrical current or by a refrigerator magnet. When the charges are not moving,
but are sitting in a wire placed in an external magnetic field, there

will be no force exerted on them.* What about if the external bar magnet is moved, while the charges remain sitting still in the wire? Assume that the magnet is moved toward the wire. From the magnet's point of view, it is not moving at all, but it is the wire that is moving toward it.

Magnetism is, at its heart, all about relative motion. If you were a blindfolded passenger in a car moving at constant speed in a straight line, how could you prove that when you arrived at your destination, it was the car that moved, and not the scenery? If you change your speed or direction, then you will feel a force associated with the acceleration, and this will tip you off that it is you doing the moving. But for uniform motion, you cannot really prove whether you or everything else is moving. All you can say for sure is that you moved relative to your surroundings.

Similarly, when moving a magnet toward a wire, from the magnet's point of view, it is stationary, and it's the charges (both the mobile electrons and the fixed positively charged ions) in the wire that are moving toward it. But moving electrical charges create a magnetic field that will interact with the field of the magnet. So, by moving a magnet near a wire, the magnet sees two electrical currents from the positively charged ions and negatively charged electrons. A force is exerted on the charges in the wire and the electrons are free to move in response to this force. In this way, Magneto is able to affect the direction of electrical currents in any device at will, though the precision by which he can guide them depends on how sensitively he can manipulate these magnetic fields.

If relative motion is the only factor that matters when considering whether a magnetic field affects electrical charges, then how about a situation in which the magnet is stationary, but the wire is moved past the magnet? Would that generate a force on the charges? Sure!

If I pull a wire through space, the electrons in the wire are moving, just as surely as if I kept the wire still and applied a voltage

---

* Careful readers will note that even if a wire is stationary (with respect to some observers), the electrons are in constant motion, zipping randomly to and fro, since they have a kinetic energy characterized by their temperature. What is important for the discussion above is that in the absence of an external voltage or a changing magnetic field, there is no net motion of the electrons in the wire, so they can be considered, on average, to be stationary.

across it. In either case, the electrons are moving past a fixed point at a certain speed. Relative to the magnet, it is as if there were an electrical current flowing by it, and we know that electrical currents and magnets interact. In this situation, the charges in the moving wire will feel a force that will induce them to flow. By dragging the wire through the external magnetic field, we convert the physical energy spent moving the wire into a form of electrical energy manifested by the electrical current. For a coil of wire, it does not matter whether the magnet is pulled through the loop or the loop is moved past the magnet. As long as there is a relative motion between the charges in the wire and the magnitude of the magnetic field threading the loop, then a current will be induced, even without an outside voltage. This mechanism may sound a bit far-fetched, but it is in fact how the electricity coming into your house is generated.

A power station, such as the one Electro employs to charge up for a night of crime, operates on the principle that when a magnetic field passing through the plane of a coil of wire is changed, a current is induced in the wire. This is called Faraday's Law, after Michael Faraday, who was one of the first scientists to introduce the concept of electric and magnetic fields. The direction of this induced current is such that it creates a magnetic field that opposes the changing external magnetic field. This is a consequence of energy conservation, as we'll explain in a moment. In certain circumstances, this current is termed an "eddy current," but it occurs whenever the magnetic field passing through a coil is increased or decreased.

Imagine a large magnet bent in the form of a broken ring, so that the north pole faces the south pole, with a coil held in the open gap between its north and south poles. Initially the plane of the coil is at right angles to the magnet's poles, so the magnetic field passes through the loop. If the coil is now rotated ninety degrees, the plane of the coil faces away from the poles, so the amount of magnetic field passing through the coil is very small. Another right angle turn, and now the coil again faces the poles, and the magnetic field passing through it is large again. A further quarter rotation, and the field through the coil is minimized again, and so on. For every change in the magnetic field passing through the coil, whether an increase or a decrease, a current is induced. The direction of the induced current flips back and forth as the coil rotates. What we have just described is an electric dynamo, and by continuously changing the magnetic field passing through the coil faces, an electric voltage will be generated in the rotating coils. There are tricks for converting an alternating current (referred to as AC) to a direct current (known as DC). There are many practical reasons that we won't go into for using AC to supply our electrical needs. In the United States the coils are made to rotate sixty times in one second, which is why in the States the AC power has a frequency of 60 Hz (Hz is an abbreviation for the unit of frequency "Hertz" and measures the number of cycles or rotations per second), while in Europe the AC current's frequency is 50 Hz.

A current flows when the magnetic field passing through the rotating coil changes. From a conservation-of-energy standpoint, we realize that it must take energy to rotate the coil, in order to create an electrical current that did not exist before. In *The Dark Knight Strikes Again # 1*—Frank Miller's dystopian view of the future of the DC universe in which superheroes are forced into servitude and Lex Luthor runs the country—a third of America's electrical power was supplied by coercing the Flash to continually run on a treadmill. Recall from Chapter 12 that the Flash has managed to find a loophole around the Principle of Conservation of Energy (presumably through his ability to tap into a "speed force"), so in Luthor's view, one might as well get some economic benefit from this suspension of the rules of physics. In our world, where we have yet to find a single exception to the conservation of energy principle, the energy that turns turbines and generates electricity comes from the same mechanism employed for making tea.

Nearly all commercial power plants generate electricity by boiling water. The resulting steam turns a turbine (a fancy term for a pinwheel) to which the coils of wire in the powerful magnets are connected. As the turbines rotate, so do the coils, and electrical current is produced. To boil the water, one either burns coal, oil, natural gas, or bio-mass. Alternatively, the heat generated from a nuclear reaction can boil water and turn the turbine. But it all just goes toward creating steam in order to turn a pinwheel attached to a coil between the poles of a magnet. The stored chemical energy in the coal, oil, or garbage has the same source as the chemical energy in the food we eat—that is, from plant photosynthesis. The light from the sun is a by-product of the nuclear-fusion reaction running in the solar core. (So, all electrical power plants could be viewed as nuclear plants or solar plants, depending on your political bent.)

The rotation of the blades in wind-generated electrical power results from temperature differences in the atmosphere arising from spatial variations in the sunlight absorbed by the atmosphere or reflected by cloud cover. Obviously, solar cells (to be discussed in Section 3) require sunlight in order to function. Similarly, hydropower, in which the potential energy of water in a dam or waterfall is converted into kinetic energy to turn a turbine, requires solar-driven evaporation, followed by condensation, to replenish the high ground with water. Aside from harnessing the tides, and geothermal power, in which the internal heat of the Earth is used to boil water, all other mechanisms for generating electricity involve the conversion of energy from the sun in one form or another. Clearly, without sunlight, none of us would be here. Perhaps the writers of *Superman* were onto something when they changed the source of Kal-El's powers from Krypton's excessive gravity to the light from our sun.

ELECTRO AND MAGNETO DO THE WAVE—

ELECTROMAGNETISM AND LIGHT

IN THE MID-1800S, the expanding American frontier may not have seen many costumed crime-fighters, but there was no shortage of heroes willing to fight for truth, justice, and the Western Way. A good thing, too, as a century later, Western comics would tap into an upsurge in the popularity of cowboys in the 1950s and thereby help keep comic-book publishers solvent during the superhero crunch precipitated by the campaign launched by Dr. Wertham's *Seduction of the Innocent*. *All-American Comics*, featuring the adventures of Green Lantern and the Justice Society of America, became *All-American Western*— starring the "fighting plainsman" Johnny Thunder (schoolteacher by day, gunfighter by night)—and *All-Star Comics* became *All-Star Western* and detailed the adventures of the Trigger Twins. In the DC universe, the scarred (physically and psychologically) rebel loner Jonah Hex traveled across the western United States, righting wrongs and saving widows. Similarly, Bat Lash and the Vigilante dispatched justice...well, vigilante-style. Over in the Marvel universe, Western comics were strictly kid stuff, with the Two-Gun Kid, Kid Colt, Ringo Kid, and the Rawhide Kid working essentially the same beat, moving from town to town every issue (though rarely bumping into one another), facing down rustlers and stagecoach robbers. Back in the real mid-nineteenth-century world, when cowboy lawmen were cleaning up Dodge, physicists were seeking to tame the wild frontiers of electricity and magnetism, in the process laying the foundation for our modern wireless lifestyle.

It was the Scottish physicist James Clerk Maxwell who, in 1862, made the monumental theoretical leap connecting electricity and magnetism and ushered in a new era of scientific advancement. The set of equations elucidated by Coulomb, Gauss, Ampère, and Faraday is nowadays known by the general title of "Maxwell's equations," in recognition of his utilizing them to provide a fundamental understanding of electromagnetic radiation. None of these scientists have ever starred in their own comic books, but without these heroes, we'd still be reading by candlelight.

In order to understand how an electric lightbulb works, think back to the water analogy we invoked earlier to explain electric currents: The water pressure of the faucet was analogous to an electrical voltage, while the amount of water per unit time flowing through a hose represented the electrical current. To indicate that the hose was not perfect, and that a finite pressure had to be continuously applied to maintain a steady flow through the hose, we suggested that the hose had both partially blocked regions as well as small pinholes along its length, through which water could escape and avoid participating in the main current flow. For a given hose's resistance, the greater the water pressure, the larger the water current. Alternatively, for a fixed water pressure, the larger the resistance, the smaller the current. These commonsense principles can be combined into a simple equation

$$\text{VOLTAGE} = (\text{CURRENT}) \times (\text{RESISTANCE})$$

known as Ohm's law, after Georg Ohm, another pioneering scientist in the early days of electromagnetism, for whom the basic unit of resistance (an Ohm) is named. The longer and skinnier the hose, and the more clogs and holes along its length, the larger its resistance to current flow. A large water pressure at one end of a long narrow, leaky hose will correspond to a bare trickle at the other end, several miles away from the faucet. This is why your jumper cables are relatively short and thick, so that the current supplied from one car battery is not degraded by the time it reaches the second battery.

The pinholes in the hose represent energy loss, and account, in our water analogy, for why a uniform water pressure (force) leads to a constant water current, and not an accelerating flow as Newton's second law ($F = m \times a$) would indicate. Copper wires obviously do not have holes through which the electrons leak out, but they do have resistance. At one end of the wire, the electrons feel a large force due to the accelerating voltage. Hence, they have a large potential energy. As they flow down the wire, their potential energy is converted to kinetic energy. The greater their kinetic energy, the faster the electrons will move down the wire, and the larger the current. Imperfections or impurities in the wire act as speed bumps, and the fast-moving electrons collide with these defects, transferring some energy to them and causing the atoms to vibrate (which is why the wires become hot). The balance between the kinetic energy gained by the electrons from an applied voltage and the energy transferred to the imperfections determines the current.

These imperfections are impurities or atoms that are out of crystalline alignment with the rest of the lattice. As such, they have their own cloud of electrons surrounding them, just as the other atoms in the wire do. When these imperfections shake back and forth after collisions with the electrical current, their electric charges oscillate.

Consider a variation on the swinging pendulum in Chapter 10, in which the mass attached to a thin string now also carries an electric charge. As the charged mass swings back and forth, it is an electrical current, but one for which the speed of the charge's motion continuously changes (high velocity at the bottom of the arc, zero velocity at the highest points of the swing). As such, it generates a magnetic field that is also changing in magnitude. However, a changing magnetic field induces an electric field. The oscillating charge will therefore continuously generate a varying electric field in phase with a changing magnetic field, radiating out into space. The faster the charged mass swings back and forth, the higher the frequency of the electric and magnetic oscillations created. Since the electric and magnetic waves have energy, the oscillating, charged pendulum will slow down, even if there is no air resistance. There's a special name for the oscillating electric and magnetic fields created by the simple harmonic motion of the charged pendulum: It's called light.

WHY THE X-RAY SPECS ADVERTISED IN COMIC
BOOKS ARE A TOTAL RIP-OFF

In a wire carrying an electrical current, colliding with impurity atoms, the oscillating electric charges in these atoms give off alternating electric and magnetic fields. The faster the electrons connected to the imperfections in the wire oscillate, the higher the frequency of the electromagnetic waves that are generated. Just by being at room temperature, all of the atoms (and their electrons) in the wire are vibrating at a rate of approximately a trillion cycles per second. Therefore any object at room temperature emits electromagnetic waves with a frequency of a trillion cycles per second. Electromagnetic waves with this vibration frequency are called infrared radiation. The greater the temperature, the faster the atoms in the object shake, and the higher the frequency of the emitted radiation.

Depending on how fast the charges in the atom are shaking—that is, how many times a second they move back and forth—the emitted electromagnetic waves can have a wavelength (a measure of the distance peak to peak of the wave) of several feet all the way down to the diameter of an atomic nucleus. In the first case, we call these long-wavelength electromagnetic waves "radio waves" (with a frequency of roughly a million cycles per second) and in the second case, these ultrashort waves are termed "gamma rays" (with a frequency of more than a million trillion cycles per second). Gamma rays have more energy and can therefore do more damage to a person than radio waves—as reflected in the fact that no one has ever gained superpowers by standing too close to an FM broadcast antenna. But they are both the same phenomenon. In order for the atoms in a wire, such as the thin filament inside a lightbulb, to emit electromagnetic waves that our eyes are able to detect—which we call visible light—the atoms must be shaking back and forth roughly a thousand trillion times per second.

We are finally in a position to understand why the sun shines. As mentioned in Chapter 2, the intense gravitational pressure at the center of the sun means that protons (hydrogen nuclei) collide frequently, so that some fuse together to form helium nuclei. The mass of a helium nucleus is slightly less than that of the two protons and two neutrons separately, and the mass deficit corresponds to a great energy output, through Einstein's expression $E = mc^2$. This outgoing energy balances the inward gravitational attraction, and the sun remains relatively stable as it burns through its fuel (and a lot of fuel it burns—600 million tons of hydrogen nuclei every second!). Part of the energy resulting from this fusion reaction is in the form of kinetic energy, and the rapidly moving, accelerating, charged helium nuclei emit electromagnetic radiation. Acceleration is the rate of change in velocity, and so every time the helium nucleus speeds up, slows down, or changes direction as it collides with other nuclei in the dense stellar core, it emits light. The light we see from the sun is very old, as it slowly makes its way from the center of the sun to the surface. It is difficult to see anything on a foggy night because the dense water-saturated atmosphere scatters the light in all directions. It's much denser within our sun, and the light generated from a nuclear-fusion reaction takes an average of 40,000 years to diffuse from the core to the surface of the sun.

Our eyes can see visible light because most of the electromagnetic waves from the sun that make it through the atmosphere are in this portion of the electromagnetic spectrum. When creatures without eyes evolved into creatures with eyes on this planet, the eyes they developed were sensitive to the type of electromagnetic waves that were most prevalent. Much fewer X-rays emitted by the sun reach us compared with light in the "visible" portion of the spectrum. Consequently, if our eyes were such that we could see only X-rays, we would live in a world of near-total darkness. Those creatures that do live in total darkness, such as sea life at the ocean's deep bottom where sunlight does not penetrate, do not waste genetic resources on superfluous eyesight or skin pigmentation, but rather rely on other senses to navigate.

Back in the Silver Age of comic books, unscrupulous salesmen, preying on the prurient interests of comic-book readers, sold "X-ray glasses" that promised the wearer would be able to see through solid objects such as clothing. Even if these X-ray specs employed a principle similar to "night vision" goggles, which translates infrared radiation into the visible (more on this in Section 3), there are not enough X-rays outside of a dentist's office to make this a useful product. And just try getting your money back from these companies!

Animals that are primarily nocturnal devote more of their

optical receptors to high-sensitivity rods, sacrificing color vision
by having fewer cones, in order to detect the few electromagnetic
waves present. But any animal or person that evolved "X-ray
vision" would spend most of its time bumping into things, and
would be at a distinct evolutionary disadvantage. The more electrons an atom has, the more strongly it scatters X-rays. This is
why X-rays can penetrate through the soft tissue of a living body
(which is mostly water) but are reflected from the much denser
bone. Presumably Superman is able to emit X-rays from his eyes,
which then penetrate through low-X-ray-absorbing matter before
being reflected and then detected by the Man of Steel. Those not
from Krypton can see only when light from an external source is
reflected into their eyes from an object. It costs the human body
energy and raw materials to develop optical-nerve cells sensitive
to low-wavelength light; consequently, there is little point in
developing an ability to detect the odd, occasional X-ray.

I CAN TELL THAT YOU'RE THINKING YOU WISH YOU HAD A TINFOIL HELMET

The leader of the mutant team of superheroes known as the X-
Men is the wheelchair-bound telepath Charles Xavier, also known
as Professor X. While his shattered spine may have left him unable
to walk, he was a formidable general for this team of "good"
mutants thanks to his ability to read and project his thoughts into
other people's minds. The physical basis underlying Prof.
X's telepathy (and that of his protégée Jean Grey, and Saturn
Girl of the Legion of Super-Heroes, as well as Aquaman's fish
telepathy, for that matter) is that time-varying electrical currents
can create electromagnetic waves detectable to someone who is
supersensitive.

All cells in our body have a function. Muscle cells exist to generate a force, whether it is for the flexing of the biceps or the pumping of the heart. Liver cells filter impurities from the bloodstream,
while stomach and intestinal cells put them there in the first place.
The role of nerve cells or neurons is to process information. One
way they accomplish this is by transmitting and altering electrical
currents. The charged objects that move from neuron to neuron are

not electrons, but calcium, sodium, or potassium atoms that are either missing one or more of their electrons or have acquired extra electrons (such charged atoms are termed "ions"). An accumulation of ions in one region in the brain creates an electric field that in turn coerces other ions in other neurons to move. The moving ions constitute a current that generates a magnetic field. Experiments by neuroscientists using sensitive electrodes placed within the brain can detect the electric fields generated by the motion of these ions, which typically vary randomly in time. Depending on where in the brain the electrode is located and what task the brain is performing, the electric fields recorded will assume a coherent wave form, oscillating through several periodic cycles before abruptly returning to the random background. Neuroscientists are beginning the difficult task of identifying the voltage variations and determining their significance (if any) with behavioral tasks. From such simple building blocks, the human mind, with all its vast complexity, is constructed.

While scientists are a long, long way from understanding how or whether the ionic currents in the brain lead to consciousness, there is one aspect of the neuronal currents that we may rely upon: that moving electric charges generate magnetic fields. In turn, because the ionic currents in the brain are continually changing direction and magnitude, the corresponding magnetic fields vary in time and create electric fields as well. The net effect is that very-low-frequency electromagnetic waves radiate out from the brain whenever electrical activity occurs. The wavelengths, amplitudes, and phases of these electromagnetic waves are determined by the time-dependent ionic currents from which they originate. The amplitude of these waves is extremely weak, such that their power is more than a billion times lower than the background radio waves that surround you at this very moment (ordinarily, the fact that we live within a sea of radio-broadcast signals is ignored, until one turns on a radio and can't tune in a particular station clearly). Yet the electromagnetic waves created by brain currents do exist, though their intensity is too weak to be observed unless the sensor is directly on the person's head. For certain powerful mutants such as Prof. X, or residents of the moon Titan in the thirtieth century (Saturn Girl), their miracle exception involves brains sensitive enough to detect the electromagnetic waves generated by others' thoughts. Of course, if you are wearing a metal helmet (a precaution taken by Xavier's evil step-brother, the Juggernaut; Magneto; and other farsighted X-Men foes) then your head is shielded so that the outgoing (and any incoming) electromagnetic waves are grounded out.

Aquaman's ability to talk to fish presumably functions in a similar manner. Given that sea water is roughly a million times a better electrical conductor than pure, deionized water, and a hundred trillion times better a conductor than air, Aquaman starts with a considerable advantage in being able to detect and send weak electromagnetic signals. It also helps that fish have a specific organ that generates weak electric fields and separate receptors to detect these fields. Fish use these organs as a form of radar, helping them navigate in brackish waters or at night. Some fish, such as sharks, skates, sturgeons and sawfish (the first type of fish he communicated with in *More Fun Comics # 77*) have extremely sensitive electroreceptors that can detect a voltage as small as a millionth of a Volt a centimeter away—such sensitivity is difficult to achieve in the laboratory! Consequently, it would be surprising if the Aquatic Ace were not able to talk to fish—apparently he possesses the necessary language skills to make his wishes known to such simple life forms thanks to his Atlantean heritage.

A stone tossed into a pond creates a series of ripples that grow weaker the farther one is from the source. The water molecules have a great deal of kinetic energy imparted from the falling stone. But as the ripple becomes larger and larger, the amount of energy in the water molecules is spread out along the growing circumference, so that for a stone dropped in the middle of the Pacific Ocean there is no noticeable disturbance at the California coast. In the same manner there is a decrease in the intensity of electromagnetic waves the farther one moves from their source. The fact that the intensity of electromagnetic waves decreases with distance from the source of the waves explains why, when Professor X needs to locate a particularly distant mutant, he uses an electronic amplifier of his mental powers, called Cerebro. First introduced in *X-Men # 7* as an automated mutant brain-wave detector, it was adapted in later issues to increase the sensitivity of Prof. X's telepathic powers. The recognition that in order to detect a distant electromagnetic signal, one would have to use an external amplifier, is consistent with the physical mechanism underlying Prof.

X's mutant power. This is also why radio and television broadcast stations use megaWatts of power to transmit their signals. Recall that a Watt is a unit of power, defined as energy (in Joules or kg-m2/sec2) per second, and a megaWatt is a million Watts. The more power a radio station can broadcast, the greater the intensity of the electromagnetic waves reaching a given remote antenna, and the stronger the signal received by the radio. Commercial radio stations generate their signals by oscillating charges up and down a big antenna. Your television set or radio does not employ Cerebro technology to amplify the distant signal, but it does make use of transistors for this function, which we will discuss in detail in Chapter 24.

* * *

Detecting electromagnetic waves created by someone's thoughts is one thing, but can one work backward to determine the neuronal currents that generated them—that is, can one actually read and interpret someone else's thoughts? Yes. Prof. X and Saturn Girl presumably do it the same way that "reverse-television" functions. Let me explain how this would work.

Television signals consist of electromagnetic waves sent out by a powerful transmitter that, upon striking a rooftop antenna, cause the charges to oscillate back and forth with a frequency and amplitude characteristic of the incident signal. The information encoded within the electromagnetic wave is then sent to the television set. (The following description applies to old style cathode ray tube television sets.) The heart of the set is the picture tube, which consists of a large glass surface onto which is evaporated a phosphorescent material that gives a brief flash of light when struck by an energetic electron. This glass face is one end of an irregularly shaped glass box. At the narrow end of the box is a wire, heated by an electrical current so that electrons are stripped from it. These now-free electrons are then directed, by way of metal "steering" plates at suitable voltages, toward the other end of the picture tube—that is, the end having the large face coated with phosphorescent material. By choosing the right voltage on the steering metal plates, the electrons can be directed to strike a particular region of the screen. The interior of the television tube is evacuated, so as to reduce the number of stray air molecules that could cause unwanted scattering of the electron beam.

Whenever the beam strikes the screen, depositing its kinetic energy into the phosphor material, it causes a flash of light to be emitted. The voltages applied to the steering plates are then adjusted, and the electron beam is now directed to another location on the screen, lighting another phosphor or leaving it dark if the beam is stopped. This continues until the electron beam has moved across the entire screen. A given array of lit and dark regions across the screen yields an image on the face of the television.

By slightly changing the image being projected onto the screen, the illusion of motion can be induced. If three different phosphors or colored filters emitting red, green, and blue light are used at the spot struck by three electron beams, then slight adjustments as to how much of each filter is illuminated in each location results in a color picture. The basic physics underlying television is that the information encoded in the electromagnetic wave contains a set of instructions for the magnitude and timing of the voltages to be applied to the steering plates.

The varying electron beam in the picture tube gives off its own set of electromagnetic waves, different from the waves that were received by the antenna, but related to the television image. A sensitive antenna placed near this monitor could detect these electromagnetic waves and, with the appropriate software, reconstruct the image that the electron current is intended to create. This "reverse television" is a very inefficient way to have two sets showing the same image, but it would be one method by which a person could read the information projected onto a computer monitor without directly hacking into the computer.* Or send information from one brain to another.

What about Prof. X using the power of his mutant mind to control the actions of someone else? Recent experiments suggest that this situation may not be so far-fetched. It has been demonstrated that not only can we detect the weak magnetic fields created by ionic currents in the brain, but that the reverse process is also

---

* In 1985, the British Broadcasting Corporation actually demonstrated that they could monitor, from a significant distance, the images being watched on homeowners' television screens. It has recently been shown that LCD television screens are also susceptible to this form of remote viewing.

possible. Neuroscientists have developed a research tool called transcranial magnetic stimulation (TMS). In this procedure, a randomly varying magnetic field is applied to a human test subject's head, providing electrical stimulation to selected regions of the cerebral cortex. The subject's reaction time and ability to initiate a voluntary hand movement are disrupted by the application of the external magnetic field.

The ability to control the actions of others, using just the power of one's mind, is not limited to mutants and heroes from Saturn's moon Titan. In fact, I have been known to demonstrate just such amazing mental powers. Often my lectures can induce my students to either exit the lecture hall at a great speed or to fall into a deep and profound slumber!

SECTION 3

MODERN PHYSICS

21

JOURNEY INTO THE MICROVERSE—

ATOMIC PHYSICS

COMIC-BOOK READERS HAVE a reasonable expectation that the hero will triumph by the end of the story. The fun, therefore, lies in the challenges that must be overcome as each month's tale races toward its conclusion. Not to put too fine a point on it, but the better the supervillain, the better the story—which is probably one reason the *Fantastic Four* was such a popular comic book in the early sixties. Certainly, the artwork by Jack Kirby was a major factor, as was Stan Lee's plotting and characterization of the intrepid quartet. But if a superhero is only as good as his or her nemesis, then the Fantastic Four achieved greatness in issue # 5 (July 1962) when they became "Prisoners of Doctor Doom."

Victor von Doom was a scientific genius, second only to Reed Richards, the leader of the Fantastic Four. Richards and von Doom attended the same college, both on "science scholarships" (in the fantasy world of comic books, institutions of higher learning compete for academic scholars the way our real-world universities vie for athletes). Von Doom was expelled when one of his "forbidden" scientific experiments went disastrously awry, blowing up the lab and scarring his face. Hiding his disfigurement behind a metal faceplate, he designed a high-tech suit of armor that rivaled Iron Man's and began a long campaign of world conquest as Doctor Doom. Of course, not having finished his degree, von Doom is not really a doctor, and it is most likely his bitterness about his A.B.D.* status, along with his desire to humiliate Reed Richards,

---

* All But Dissertation.

that drives his evil ambitions. Unlike the villains in DC stories in the 1960s, who would inevitably be captured and turned over to the police at story's end, the Fantastic Four never seemed to be able to battle Doctor Doom to better than a draw. Of course, because Doom was the dictatorial ruler of the small European nation of Latveria, it was always a bit vague as to who the proper authorities were that one could hand an evil head of state over to.

More to the point, Doom's pride was so great that he would rather face near-certain death than incarceration. Consequently, a typical battle with Doctor Doom would end with Doom being lost in space, or marooned in another dimension, or trapped in time— all fates he had intended for the Fantastic Four. At the climax of the aptly titled "The Return of Doctor Doom" in *Fantastic Four # 10*, Doom was struck by a reducing ray he had planned to use on the Fantastic Four. The story ended with Doom shrinking to nothingness, but this would not be the last we would see of him. Six issues later in *Fantastic Four # 16*, the Fantastic Four journeyed to "The Micro-World of Doctor Doom," where they learned that Doom had survived his shrinking ordeal. At some point in his reduction, Doom entered a "micro-world—a world that might fit on the head of a pin." Later, in *Fantastic Four # 76*, Reed, Ben, and Johnny ventured into an entire microverse, that is, a universe (or at least a galaxy) of micro-worlds. The microverse was depicted as residing within a stain on a microscope slide in Reed Richards' laboratory. This, at least, removed the need to explain the enormous coincidence in *Fantastic Four # 16* of Doctor Doom and the FF standing exactly above one such micro-world just as they began to shrink.

If the micro-world that Doom encountered and subsequently conquered could indeed fit on the head of a pin, then its diameter at the equator is approximately 1 millimeter. For comparison, the Earth's diameter is thirteen thousand kilometers. One kilometer is one million millimeters long, so the micro-world is thirteen billion times smaller than the Earth. Recall that in Chapter 5 we discussed the difficulties involved in reducing an object in size. The micro-world cannot be six billion times denser than our planet, unless it is composed of white-dwarf-star matter. The fact that Doctor Doom, the Fantastic Four, and the inhabitants of this micro-world are able to walk around normally suggests that this is not the case. Both Doctor Doom and Reed Richards seem as smart on the micro-world as they do at their normal size, and the Thing is not any less strong, so it is unlikely that atoms are removed from them upon shrinking. We must therefore regretfully conclude that the micro-world of Doctor Doom is much like his other "master plans"—impressive in principle but lacking in execution.

* * *

If it's that hard to construct a world to fit on the head of a pin, what are we to make of the Atom's adventure with "The Deadly Diamonds of Doom" in *Atom # 5*? In this story, a diamond artifact found on Mount Pico in the Azure Islands is brought back to Ivy Town by an archeologist friend of Ray Palmer (alter ego of the Atom), whereupon a strange ray beam emanates from it, turning people and house cats into diamond statues. "Even though it appears solid" thinks Professor Palmer as he reaches for his size and weight controls, "there are vast gulfs of space between the atoms that comprise the diamond!"

True enough. Most of an atom is indeed empty space between the positively charged nucleus and the average location of the negatively charged electrons. As the Tiny Titan shrinks down to subatomic length scales, he discovers an entire other planet, inside an atom! I'm at a loss to suggest what this other planet could be composed of. It certainly can't be made of atoms, since it is smaller than the electrons that reside in the diamond artifact. Certainly the discovery that there could be entire civilizations residing within the atoms of ordinary matter would garner Palmer a Nobel Prize, at the very least, along with worldwide fame and fortune. Such is the hallmark of this hero that he never even considers reporting this scientific discovery, nor any of the other micro-worlds he encounters in *Atom # 4, # 19*, the *Justice League of America # 18*, or *Brave and the Bold # 53*.

While comic-book claims that there are micro-worlds within atoms is pure fantasy, the region within an atom as elucidated by quantum mechanics is no less strange. Within the "empty spaces" inside an atom are the "matter-waves" associated with an electron's motion. These matter-waves are the key to understanding atomic physics.

WHAT DO YOU DO WHEN EVERYTHING YOU KNOW IS WRONG?

It is now time for us to delve into the world of atoms. Things will get physics-y here for a few pages, but bear with me. We'll get back to comic books soon enough, but some background is needed in order to understand why at least some physicists take the notion of parallel universes and an infinite number of Earths seriously.

At the end of the nineteenth century, there was a growing body of experimental evidence that indicated that the physical principles described in the preceding chapters failed at explaining the behavior of atoms and light. For example, physicists were stymied trying to explain why hot things glowed. Place an iron poker in a roaring fireplace and, as it warms, it initially glows red and eventually gives off white light. Thanks to Maxwell's theory of electromagnetism discussed in the previous chapter, physicists understood that the oscillating electric charges in each atom, shaking back and forth as the poker became hotter, emitted light, and that the faster the atoms shook back and forth, the higher the frequency would be of the resulting electromagnetic radiation. Back in the 1800s, scientists had devised very clever techniques for measuring both ultraviolet and infrared light, which are at the upper and lower ends of the visible electromagnetic spectrum, bracketing the narrow slice of light that our eyes can see. Consequently, they could accurately measure exactly how much light a hot object emitted at any given wavelength as its temperature was increased. They discovered two surprising things. First, the fraction of light emitted at a specific wavelength depends only on the temperature of the object, and not on any other characteristic. Regardless of an object's material composition, shape, or size, the only thing that determined the spectrum of light emitted was its temperature. Second, the total amount of light emitted was not infinite and also depended only on the temperature. This second point was the first falling domino that ultimately led to the development of quantum mechanics.

The fact that the light from a hot object depends only on its temperature prevents us from getting something for nothing. If two objects made of different materials at the same temperature emitted different radiation spectra, there would be a way to have a net transfer of energy between them and hence useful work, without any heat flow. While this would be a convenient violation of the second law of thermodynamics, it turns out that this does not occur for just that reason. A practical benefit of the fact that the emitted light spectrum depends only on the temperature is that we can use the intensity of emitted light as a function of wavelength to determine the temperature of objects for which normal thermometers are useless. This is how the surface temperature of the sun (roughly 11,000 degrees Fahrenheit) and the background microwave radiation remnants of the Big Bang (3 degrees above absolute zero) are measured—through observations of the spectrum of light they produce.

The second discovery, that the energy emitted by a glowing object is not infinite, was not really a shock to physicists. What they found disturbing was that Maxwell's electromagnetic theory predicted that the amount of light energy emitted should increase without limit! Calculations using Maxwell's theory correctly predicted how much light would be emitted at low frequencies, in exact agreement with observations. As the frequency of light emitted by a hot object increased into the ultraviolet portion of the spectrum, the measured light intensity reached a peak, and at higher frequencies decreased again back to a low value, which is what one would expect from both conservation of energy and common sense. However, the curve calculated from Maxwell's equations and thermodynamics indicated that the intensity would become infinitely high for light above the visible portion of the spectrum. This was labeled the "ultraviolet catastrophe," though it was only a "catastrophe" for the theorists doing the calculations. Many scientists checked and rechecked the calculations, but they could find nothing wrong with the math. Apparently, there was something wrong—or rather, incomplete—with the physics.

Maxwell's equations had worked so well in all other cases (they led to the invention of radio in 1895 and would eventually enable the development of television as well as all forms of wireless communication), that it was doubtful that there was something fatally wrong with them. Rather, scientists concluded that the problem must lie with applying Maxwell's theory to the shaking atoms in a glowing object. Again, many tried to find an alternative approach, some different theory that could account for the observed spectrum of light emitted by a glowing object. Here is where the fact that the spectrum depends only on the temperature of the object becomes important. If the theory of electromagnetism could not account for the behavior of one or two exotic pieces of matter, well, that would be somewhat awkward, but not earthshaking. The inability to explain a property shared by all matter was downright embarrassing, and something had to be done.

In 1900 the theoretical physicist Max Planck, recognizing that desperate times called for desperate measures, did the only thing he could to explain the spectrum of light emitted by a glowing body: He cheated. He first determined the mathematical expression that corresponded to the experimentally obtained glow curve. Once he knew what formula he needed, he then set out to find a physical justification for it. After trying various schemes, the only solution he could come up with that gave him the needed glow-curve formula involved placing restrictions on the energy of the atoms that made up the glowing object. Planck essentially proposed that the electrons in any atom could only have specific energies. From the Latin word for "how much," this theory was called "quantum physics." The separation between adjacent energy levels was in practice very small. And I mean really small: If the energy of a well-hit tennis ball is 50 kg-meter^2/sec^2, then the separation between adjacent energy levels in an atom is less than a millionth trillionth of a kg-meter^2/sec^2. This should provide some perspective the next time you hear a commercial boasting that the latest innovation in an automobile design or a laundry detergent represents a "quantum leap."

Planck had to introduce a new constant into his calculations, an adjustable parameter that he labeled "h." He assumed that any change in the energy of an atom could only take on values $E = hf$, or $E = 2hf$ or $E = 3hf$, and so on, but nothing in between (so the atom could never have an energy change, say, of $E = 1.6hf$ or $17.9hf$), where f is the frequency characteristic to the specific atomic element. This is like saying that a pendulum can swing with a period of one second or ten seconds to complete a cycle, but that it was impossible to make the pendulum swing back and forth in five seconds. Planck himself thought this odd, but found it necessary in order to make his calculations come out right. He intended to let the value of h become zero once he obtained the correct expression for the spectrum for a glowing object. To his dismay, he discovered that when he did this, his mathematical expression went back to the infinite energy result from classical electromagnetism. The only way to avoid this nonsensical infinite result is to say something equally nonsensical (at least to scientists at the time), that the atoms cannot take on any energy value they want, but must always make changes in discrete steps of magnitude $E = hf$. Since h is very, very small ($h = 660$ trillion trillion trillionths of a kg-meter^2/sec), we never notice this "graininess" of energy when we deal with large objects such as baseballs or moving automobiles. For the energy scale of an electron in an atom, it is quite significant and absolutely cannot be ignored.

The fact that the energy of electrons in an atom can have only discrete values, with nothing in between, is indeed bizarre. Imagine the consequences of this discreteness of energy for a car driving down the highway at 50 mph if Planck's constant h were much larger. The quantum theory tells us that the car could drive at a slower speed of 40 mph, or at a faster speed of 60 mph, but not at any other speeds in between! Even though we can conceive of the car driving at 53 mph, and calculate what its kinetic energy would then be, it would be physically impossible for the car to drive at this speed, according to the principles of quantum physics. If the car absorbed some energy (say, from a gust of wind), it could increase its speed to 60 mph, but only if the energy of the wind could exactly bridge the difference in kinetic energies. For a slightly less energetic gust, the car would ignore the push of the wind as if it were not there and continue along at its original speed. Only if the energy of the wind *exactly* corresponded to the difference in kinetic energy from 50 to 60 mph, or 50 to 70 mph, would the car "accept" this push and move to a higher speed. The transition to the higher velocity would be almost instantaneous, and the acceleration during this transition would likely do bad things to the car's occupants. This scenario seems ridiculous when translated to highway traffic, but it accurately describes the situation for electrons in an atom.

Is there any way to understand why the energy of an electron in an atom has only certain discrete values? Yes, actually, but first you must accept one very strange concept. In fact, all of the "weirdness" associated with quantum physics can be reduced to the following statement: There is a wave associated with the motion of any matter, and the greater the momentum of the object, the shorter the wavelength of this wave.

When something moves, it has momentum. The physicist Louis de Broglie suggested in 1924 that associated with this motion is some sort of "matter-wave" connected to the object, and the distance between adjacent peaks or troughs for this wave (its wavelength) depends on the momentum of the object. Physicists refer to an object's "wave function," but we'll stick with "matter-wave" as a reminder that we are referring to a wave associated with the motion of a physical object, whether an electron or a person.

This matter-wave is not a physical wave. Light is a wave of alternating electric and magnetic fields created by an accelerating electric charge. The wind-driven ripples on the surface of a pond or the concentric rings formed when a stone is tossed into the water result from mechanical oscillations of the water's surface. Sound waves are a series of alternating compressions and expansions of the density of air or some other medium. In contrast, the matter-wave associated with an object's momentum is not like any of these waves, but in some sense, it is just along for the ride, moving along with the object. It is not an electric or magnetic field, nor can it exist distinct from the object, nor does it need a medium to propagate. Yet this matter-wave has real physical consequences. Matter-waves can interfere when two objects pass near each other, just as when two stones are thrown into a pond a small distance apart and each creates a series of concentric ripple rings on the water's surface that form a complex pattern where the two rings intersect. If you ask any physicist what this matter-wave actually is, he or she will give a variety of mathematical expressions that always boil down to the same three-word answer: I don't know. For once, our one-time "miracle exception" applies to the real world, rather than the four-color pages of comic books!

Unless an object is moving at nearly the speed of light, its momentum can be described as the product of its mass and its velocity. A Mack truck has more momentum than a Mini Cooper if they are both traveling at the same speed, since the truck's mass is much larger. The Mini Cooper could have a larger momentum if it were traveling at a much, much higher speed than the truck. Physicists typically use the letter "p" to represent an object's momentum, since obviously the p stands for *mo*-mentum.* The wavelength of this matter-wave is represented by the Greek letter lambda (λ). The matter-wave's wavelength was proposed by de Broglie (and experimentally verified in 1926 by Clinton Davisson and Lester Germer) to be related to the object's momentum by the simple relationship (momentum) times (wavelength) equals a constant, or $p \times \lambda = h$ where h is the same constant that Planck had to introduce in order to account for the glow curve of hot objects.

The fact that the product of an object's momentum and the matter-wave's wavelength is a constant means that the bigger the momentum, the smaller the wavelength of the matter-wave. Given that momentum is the product of mass and velocity, large objects such as baseballs or automobiles have very large momenta. A fastball thrown at one hundred mph has a momentum of about 6 kg-meter/second. From the relationship $p\lambda = h$, since h is so small, the wavelength (the distance between successive peaks in the wave, for example) of the matter-wave of the baseball is less than a trillion trillionths of the width of an atom. This explains why we have never seen a matter-wave at the ballpark. Obviously there is no way we can ever detect such a tiny wave, and baseballs, for the most part, are well-behaved objects that follow Newton's laws of classical physics.

On the other hand, an electron's mass is very small, so it will have a very small momentum. The smaller the momentum, the bigger the matter-wave wavelength will be, since the product is a constant. Inside an atom, the matter-wave wavelength of an electron is about the same size as the atom, and there is no way one can ignore such matter-waves when considering the properties of atoms. When the DC Comics superhero the Atom shrinks down to the size of an atom, he should see some rather strange sights. At this size he is smaller than the wavelength of visible light so, just as we can't see radio waves, whose wavelength is in the range of several inches to feet, the Atom's normal vision should be

---

* Seriously, you wouldn't want to represent momentum by the letter "m," for that would lead to confusion as to whether the "m" referred to momentum or mass. I'm not really sure why the letter "p" has been associated with momentum, but it seems to have stuck.

inoperable, and he will be roughly the same size as the matter-waves of the electrons inside the atom. It is suggested in his comic that at this size the Atom's brain interprets what he sees as a conventional solar system conception of the atom, for he has no other valid frame of reference to decipher the signals sent by his senses.

Imagine an electron orbiting a nucleus, pulled inward by the electrostatic attraction between the positively charged protons in the nucleus and the electron's negative charge. As the electron travels around the nucleus, only certain wavelengths can fit into a complete cycle. When the electron has returned to its starting point, having completed one full orbit, the matter-wave must be at the same point in the cycle as when it left. As weird as the notion of a matter-wave is, it would be even harder to comprehend if when the wave left it was at a peak (for example), and after having completed one full orbit, was now at a valley. In order to avoid a discontinuous jump from a maximum to a minimum whenever the wave completed a cycle, only certain wavelengths that fit smoothly into a complete orbit are possible for the electron. This is not unlike the situation of a plucked violin string, with only certain possible frequencies of vibration. Because the wavelength of the matter-wave is related to the electron's momentum, this indicates that the possible momenta for the electron are restricted to only certain definite (discrete) values. The momentum is in turn related to the kinetic energy, so the requirement that the matter-wave not have any discontinuous gaps after finishing an orbit leads us to conclude that the electron can only have certain discrete energy values within the atom.

These finite energies are a direct result of the constraint on the possible wavelengths of the matter-waves, which in turn are due to the fact that the electron is bound within the atom. An electron moving through empty space has no constraints on its momentum, and consequently its matter-wave can have any wavelength it likes.* A piece of string can have any shape at all when I wiggle one end, provided that the other end is also free to move. But if the string is clamped at both ends, as in the case of a violin string,

---

* Therefore our earlier analogy with an automobile driving on a highway that can only have speeds in multiples of 10 mph would be valid only for a car doing laps on a closed track.

then the range of motions for the string are severely restricted. When I now pluck the clamped string, it can only vibrate at certain frequencies, determined by the length and width of the string and the tension with which it is clamped. There is a lowest fundamental frequency for the string and many higher overtones, but the string cannot vibrate at any arbitrary frequency once it is constrained in this manner.

Likewise, the electron is held in an orbit by its electrostatic attraction to the positively charged nucleus. If "plucked" in the right way, a matter-wave for the bound electron can take on a higher energy value. When the electron then relaxes back to its lower fundamental frequency, it must do so by making a discrete jump. Energy is conserved; consequently, the electron can only lower its energy when returning to the lower frequency level by giving off a packet of energy equal to the difference between its higher energy level and the lower level it is relaxing to. Because the energies available to the electron are discrete, well-defined values similar to the overtones possible for a clamped string, this jumping from one energy state to another is termed a "quantum transition" or a "quantum jump." The discrete packet of energy given off by the electron when making this transition is typically in the form of light, and a quantum of light energy is termed a "photon" (a concept introduced by Albert Einstein, again in 1905—a busy year for him and physics—though the term "photon" was not coined until 1926 by Gilbert Lewis).

If a glass tube is filled with a gas such as neon, and an electrical current is passed through the gas, the energetic electrons of the current will sometimes collide with the neon atoms. When the energy of the energetic electrons is just right, the neon atoms can be excited into a higher energy state. After the collision, the excited neon atoms will relax back to their original lower energy configuration, emitting a photon of light that has the frequency (hence, color) that corresponds to the energy difference between its starting and final states. This is why neon lights have their identifiable color. By changing the type of gas in the tube, different colors of light can be selected. You could do this with any gas, but only certain elements have a transition within the visible portion of the light spectrum. If the atoms suffer highly energetic collisions, such that many higher energy states are excited, then many discrete wavelengths of light will be given off when the different overtones all relax back to the fundamental level. Different elements have differing arrays of overtones and fundamental frequencies, just as different strings on a violin or guitar will have different vibrational modes depending on the length, width, and tension. Two identical violin strings clamped with the same tension will have the same range of possible frequencies when plucked. Similarly, two identical atoms will have the same spectrum of emitted light when they relax from an excited state.

In this way, the spectrum of wavelengths of light emitted by an energetic atom is unique and can be considered the fingerprint of the particular element. The lighter-than-air element helium was discovered by the detection of its characteristic spectrum of light coming from the sun (the word "helium" is derived from Helios, the Greek sun god). By careful comparison with the spectrum of light emitted by hydrogen and other gases, scientists concluded that this array of wavelengths must arise from a new element that at the time had never been found terrestrially. Fortunately for the Macy's Thanksgiving Day Parade, underground pockets of helium were eventually discovered.

The notion that there is a wave associated with the motion of any object, and that the wavelength of this wave is inversely proportional to its momentum, is weird, but by accepting this mysterious concept, we gain an understanding for the basis of all of chemistry. Bring two atoms close enough together, and they may form a chemical bond, and in so doing create a new basic unit, the molecule. Why would the atoms do this? The negatively charged electrons in the first atom certainly will repel the negatively charged electrons in the second atom. Before quantum mechanics, there was no satisfactory fundamental explanation for why the universe didn't just consist of isolated elemental atoms.

The driving force underlying the bonding between atoms is the interactions of the matter-waves of the electrons from the different atoms. When the two atoms are held very far apart, the matter-waves of the atomic electrons do not overlap. When the atoms are brought close enough together so that the electron clouds around each atom intersect, the electronic matter-waves from each atom begin to interfere, forming a new wave pattern, just as two stones tossed into a pond create an intricate pattern of ripples that is very different from the pattern that would be created by each stone separately. In most cases this new pattern is a high-energy, discordant mess, similar to the sound resulting from a clarinet and violin played simultaneously by novices with no musical training or talent. In these cases, the two atoms do not form a chemical bond and do not chemically interact. In a few special cases, the two matter-waves interact harmoniously, creating a new wave pattern that has a lower-energy configuration than the two separate matter-waves. In these special cases, the two atoms can lower their total energy by allowing the matter-waves to interact in this manner, and once in a lower-energy state, it requires the addition of energy (called the "binding energy" or "bonding energy") to physically separate them. In this way, despite the considerable repulsion between the negatively charged electrons, the two atoms are held together in a chemical bond, owing to the wave-like nature of the electrons.

These arguments about discrete energy levels in an atom arising from those particular orbits that correspond to an integral number of wavelengths of the electron matter-wave seem so reasonable that it's a shame that they're not really correct. The electron cannot literally be considered to move in a circular or elliptical orbit around the positively charged nucleus, despite the appealing analogy to our solar system. For one thing, the electron would be constantly accelerated as it bends onto a curved path. As argued in the previous chapter, an accelerating electric charge emits electromagnetic waves that carry energy, so if the electron continuously radiates light in its orbit, it would lose kinetic energy. Eventually, the electron should spiral into the nucleus in a little less than a trillionth of a second, so no elements should be stable, and hence there would be no chemistry and no anything if the electrons really moved in curved orbits.

Nevertheless, the notion that only certain wavelengths are allowed, with corresponding discrete energy levels, is still a valid one, even if the picture we employed to get there can only be considered a useful metaphor and not a literal description. Instead of thinking of the electron as a point particle that moves in a circular orbit with a particular matter-wave associated with it, the full quantum theory of Heisenberg and Schrödinger, to be discussed in the next chapter, tells us that there is a "wave function" for the electron. For a plucked violin string, it makes no sense to ask at what position on the string the wave is. Similarly for the electron in an atom, its matter-wave extends over the atom, and we cannot specify the electron's position or trajectory more accurately than this. Electrons only emit or absorb light when moving from one wave pattern to another in the atom. As we'll see in the next chapter, these matter-waves are also responsible for a crisis on infinite Earths!

22

NOT A DREAM! NOT A HOAX! NOT AN IMAGINARY TALE!—

QUANTUM MECHANICS

THE ORIGIN TALE in *Showcase # 4* describing how Barry Allen gained his superspeed powers and became the Silver Age Flash featured a symbolic passing of the torch from the earlier Golden Age of superheroes. Just prior to being struck by the lightning bolt that simultaneously doused him with exotic chemicals, and that naturally would enable him to run at the speed of light, police scientist Allen relaxed with a milk-and-pie break in his lab while reading *Flash Comics # 13*, featuring the Golden Age Flash on its cover. After the freak accident gave Barry his superpowers, his immediate thoughts turned to how he could use his superspeed to help humanity. Taking his inspiration from the Flash comic book he had been reading before he was struck by lightning, he donned a red-and-yellow costume and began his crime-fighting career as the Silver Age Flash (though he referred to himself simply as the Flash, not realizing that he was himself a comic-book character in the dawning Silver Age of superheroes). In a turn of events that nowadays would be described as "postmodern," and back then was simply considered "pretty neat," it was established in the Flash comic books of the 1960s that the Flash character of the 1940s (who wore a different costume and gained his superspeed in a different, though equally implausible, chemical accident) was a comic-book character in Barry Allen's "reality."

The Golden Age Flash (whose secret identity was Jay Garrick) was considered fictional, as far as the Silver Age Flash was concerned, until September 1961. That month saw the publication of the classic story "Flash of Two Worlds" in *The Flash # 123* (fig. 30), where it was revealed that the Silver Age Flash and the Golden Age Flash both existed, but on parallel Earths, separated by a "vibrational barrier." In this story, the Silver Age (Barry Allen) Flash accidentally vibrated at superspeed at the exact frequency necessary to cross over to the Earth on which his idol, the Golden Age (Jay Garrick) Flash, lived. Once he realized that he was in the world of the Golden Age heroes, Barry met Jay and introduced himself. "As you know," the police scientist explained, "two objects can occupy the same space and time—if they vibrate at different speeds!" Apparently Barry Allen was a better forensic scientist than he was a theoretical physicist. Regardless of their vibrational frequency (and as we saw in Section 2, atoms in a solid

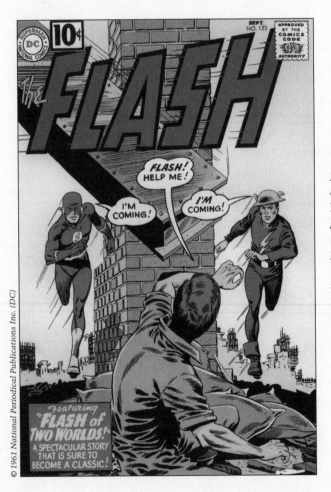

Fig. 30. The cover to The Flash # 123, *giving comic-book fans their first hint that there were at least two worlds beyond their own.*

do vibrate simply because they have some finite temperature), there is no way that two objects can be in the same place in space and time (unless we are discussing massless quantities such as light photons).

The writer of the "Flash of Two Worlds" story was Gardner Fox, who had also written many of the Golden Age Flash comics. He proposed a mechanism to explain how the Silver Age Flash could read comic books featuring the Golden Age Flash on this second Earth, while simultaneously providing insight into his work habits. As Barry hypothesized, "A writer named Gardner Fox wrote about your adventures—which he claimed came to him in dreams! Obviously when Fox was asleep, his mind was 'tuned in' on your vibratory Earth! That explains how he 'dreamed up' the Flash!"* This crossover meeting between the Silver Age and Golden Age Flash was a hit with comic-book fans, and the Silver Age Flash would more and more frequently cross the vibrational barrier to Earth-2. (The world on which the Golden Age Flash resided, though it appeared first chronologically, was labeled Earth-2, while the world of the Silver Age was designated Earth-1. Our world, Fearless Reader, in which all superheroes exist solely as fictional characters in comic books, is called Earth-Prime.) Eventually, the Silver Age Justice League of America of the 1960s, consisting of the Flash, Green Lantern, the Atom, Batman, Superman, Wonder Woman, and other superheroes, met and had an adventure with Earth-2's Justice Society of America from the 1940s, whose membership contained the Flash, Green Lantern, the Atom, Batman, Superman, Wonder Woman, and others. So popular was this meeting of the two superteams, that it quickly became an annual tradition. But the Justice League and Justice Society could visit each other's Earth only so many times before the novelty wore off. Soon the Justice League branched out and visited other Earths, such as Earth-3, where the evil analog of the Justice League of America had formed the Crime Syndicate of America (presumably to distinguish themselves from their European criminal counterpart). Captain Marvel—that is, Billy Batson,

---

* While Barry's explanation isn't actually all that exciting, it was the common practice in Silver Age comic books that every sentence that wasn't a question be punctuated with an exclamation mark!

who could become a superhero by shouting "Shazam!"—and the rest of his supporting cast inhabited Earth-S, and were in due time paid a crossover visit by the Justice League of America.* Earth-X, Earth-4, and others soon followed, and eventually the phrase "multiverse" became appropriate to describe the seemingly endless number of alternate universes that abounded.

The issues of the Justice League of America that describe the meeting of the Silver and Golden Age heroes always carried titles such as "Crisis on Earth-2" or "Crisis on Earth-X." The story lines eventually became so convoluted, with so much alternate history to keep straight, that in 1985, DC Comics attempted to normalize the multiverse. The yearlong miniseries describing this simplification process was called "Crisis on Infinite Earths." With a vast housecleaning of continuity under way, the writers and editors at DC Comics used this opportunity to weed out poor sellers from many of the less-popular worlds and bring all the heroes from the better-selling titles together on one Earth (coincidentally, the Earth-1 of the Silver Age heroes). Consequently, the "Crisis on Infinite Earths" miniseries is noteworthy to comic-book fans for the deaths of the Barry Allen Flash and Supergirl (both of whom died heroic deaths struggling against the evil tyrant who threatened to destroy the Silver Age Earth-1) and the removal of Superboy from Superman's history. Unlike most comic-book fatalities, both Barry Allen and Supergirl have remained dead for the most part (though they have both returned to continuity between the writing of the first and second editions of this book), while crimefighting adventures have begun to creep back into Clark Kent's teen years once again.

As ridiculous as all of the above may sound, the concept of an infinite number of parallel worlds may be one of the strangest examples of comic books getting their physics right!† Just four

---

* By this time DC had won its lengthy lawsuit against Fawcett, Captain Marvel's publisher, for copyright infringement on their Superman character. Contesting the lawsuit had contributed toward Fawcett's near bankruptcy, and DC was able to purchase the rights to the Captain Marvel character. Apparently, now that they owned the copyright to the character, DC was no longer concerned that publication of a Captain Marvel comic book would unfairly compete with and diminish the economic viability of Superman comics.

† This time, the exclamation point is justified!

years prior to the publication of *Showcase # 4,* the notion of an infinite number of parallel, divergent universes was seriously proposed as an interpretation of the equations of quantum mechanics. To reiterate: Some scientists believe that the concept of parallel universes is a serious, viable construct in theoretical physics. Current theories indicate that if such alternate Earths exist, they would be more like those described in the Marvel comics universe, where slight changes in a character's history (such as those presented by the Watcher in stories such as "What if Gwen Stacy Had Lived?") lead to diverging worlds that can never be visited by our reality, regardless of one's vibrational frequency.

GREAT MINDS REALLY DO THINK ALIKE

So far throughout this book, we have discussed what physicists term "classical mechanics." To understand the "mechanics" of something means you can predict its statistics (for example, the largest angle at which a ladder may be placed to remain balanced against a wall) and its motion (such as the speed with which the ladder falls when it begins to slip), once the external forces acting on it are identified. The fundamental equation that governs how a macroscopic object will move as the result of an applied force is our old friend Newton's second law of motion, $\mathbf{F} = \mathbf{ma}$. For the motion of large objects such as cars, baseballs, and people, the dominant forces are gravity and electromagnetism. Even when we considered electricity and magnetism, we still made use of $\mathbf{F} = \mathbf{ma}$, where the F in the left-hand side of this equation was either Coulomb's force of attraction or repulsion between electric charges or the force of a magnetic field on a moving electrical charge. The aspect of "quantum mechanics" that justifies its separation from "classical mechanics" as a distinct branch of physics is that when one considers electrons and atoms, and their corresponding "matter-waves," $\mathbf{F} = \mathbf{ma}$ suddenly doesn't cut it anymore.

After a great deal of effort trying to "fix" classical physics for atoms (essentially tweaking Newton's laws without overturning them completely), physicists reluctantly concluded that a different type of "mechanics" applied inside of atoms. That is, a new equation was needed to describe how atoms would respond to external forces. After roughly twenty-five years of trying one form or another for this equation, nearly simultaneously, Werner Heisenberg and Erwin Schrödinger obtained the correct form for an atom's equivalent of $\mathbf{F} = \mathbf{ma}$.

Have no fear: There is no chance that we will discuss either the Heisenberg or Schrödinger approach in any mathematical detail. In a few pages I'll give you the Schrödinger equation, but that will only be so that we can gawk at it as if it were some exotic zoo animal. Heisenberg's treatment of quantum physics involves linear algebra, while Schrödinger's makes use of a complex partial differential equation (for simplicity, we will focus on Schrödinger for the remainder of this chapter). To fully explain their theories would break the pact we have maintained up till now that nothing more elaborate than high-school algebra would be employed in these pages (the algebra we have used in this book has the same relationship to linear algebra as a housefly does to a house).

Nevertheless, there are two points about mathematics worth making here. The first is that, unlike the case of Isaac Newton discussed in Chapter 1, who had to invent calculus in order to apply his newly discovered laws of motion, both Heisenberg and Schrödinger were able to make use of mathematics that had already been developed at least a century earlier. The mathematical branches of linear algebra and partial differential equations that Heisenberg and Schrödinger employed to describe their physical ideas had been invented by mathematicians in the eighteenth and nineteenth centuries and were well established by the time they were needed in 1925.

Frequently, mathematicians will develop a new branch of mathematics or analysis simply for the pleasure of constructing a set of rules and discovering what conditions and principles then logically follow. Occasionally, physicists later discover that in order to describe the behavior of the natural world under investigation, the same tools that previously existed solely to satisfy mathematicians' intellectual curiosities turn out to be indispensable. For example, Einstein's task in developing a General Theory of Relativity in 1915 would have been much more difficult if he hadn't had Bernhard Riemann's theory of curved geometry— developed in 1854—to work with. This scenario of physicists making tomorrow's advances by using yesterday's mathematical tools has been repeated so often that physicists tend to not think about it too much.

The second point about the theories of Heisenberg and Schrödinger is that while they employ different branches of mathematics and look very different, careful inspection by Schrödinger in 1926 demonstrated that they are mathematically equivalent. Because they describe the same physical phenomena (atoms, electrons, and light), and are motivated by the same experimental data, it is perhaps not very surprising that they turn out to be the same theory, even though the mathematical languages used to express them are vastly different.

Schrödinger and Heisenberg, completely independent of each other, developed different descriptions of the quantum world in the same year. The notion that some ideas become "ripe" for discovery at certain moments in history is found time and again and is not confined to theoretical physics. Of course, simple mimicry accounts for much of the similarity in television programs or Hollywood movies, just as the breakthrough of Superman in Action Comics led to a rapid proliferation of superhero comic books by many other publishers, including National Comics, hoping to bottle lightning again. There are, however, well-documented cases of movie studios or television networks simultaneously and independently deciding that it is time for the reintroduction of a particular genre, such as the pirate movie or the urban doctor drama. This synchronicity also occurs in comic books, as in the example of the X-Men and the Doom Patrol. In March 1964, DC Comics published *My Greatest Adventures # 80*, featuring the debut of a team of misfit superheroes (Robotman, Negative Man, and the obligatory female teammate, Elasti-Girl) whose freakish powers caused them to be shunned by normal society. They were led by a wheelchair-bound genius named the Chief, who convinced them to band together to help the very society that rejected them, frequently fighting their opposites in the Brotherhood of Evil Mutants. Three months later, comic fans could buy *X-Men # 1*, published by Marvel Comics, where they would meet a team of mutants (Cyclops, Beast, Angel, Iceman, and the obligatory female teammate, Marvel Girl) whose freakish powers caused them to be shunned by normal society. These superpowered teens were led by the wheelchair-bound mutant telepath Professor X, who recruited and trained them to help the very society that rejected them, frequently fighting their opposites in the Brotherhood of Evil Mutants.

Despite some profound differences (Professor X is bald and clean-shaven, while the Chief has red hair and a beard), the striking similarities in concept have led many comic-book fans to wonder whether the X-Men were modeled after the Doom Patrol. However, interviews with the writers of both comics and the research of comic-book historians indicate that it is more likely that the near simultaneous appearance is a coincidence. The long lead time needed to conceive, write, draw, ink, and letter a comic book prior to its printing and newsstand distribution suggests that the X-Men were well into production by the time the Doom Patrol first appeared.

Another publishing synchronicity is the case of the moss-covered muck-monsters *Swamp Thing* at DC (written by Len Wein) and *Man-Thing* at Marvel (co-written by Gerry Conway), which appeared within one month of each other in 1971. Both Wein and Conway insist that their creations were not influenced by the other, and the fact that they were roommates at the time is purely coincidental.

* * *

If the behavior of objects on the atomic scale is governed by the matter-waves that accompany their motion, then what atomic physics needed was a matter-wave equation that described how these waves evolve in space and time. In 1925, Schrödinger (fig. 31) essentially guessed the right mathematical expression.

With the Schrödinger equation, scientists had a framework with which they could understand the interactions of atoms with light. This was Schrödinger's motivation for developing his matter-wave equation. A generation later, armed with the insights about the nature of matter made possible by the Schrödinger equation, a new class of scientists developed the transistor and, separately, the laser, as well as nuclear fission (as in atomic bombs) and nuclear fusion (in hydrogen bombs). The paths to both the transistor and the laser were difficult ones, and it was only with the guidance of quantum theory that these devices were successfully developed. A generation later still, the CD player, personal computer, cell phone, and DVD player—to name only a few—

Fig. 31. *Erwin Schrödinger, theoretical physicist, Nobel Laureate, and ladies'
man, thinking of the Schrödinger equation—the foundation of quantum
mechanics and our modern technological lifestyle.*

would be created. None of these are possible without the transis-
tor or the laser, neither of which are possible without the
Schrödinger equation. It is small wonder that until fairly recently,
Schrödinger's portrait was on the 1,000 Schilling note in his native
Austria, for he can truly be considered one of the architects of the
lifestyle that we in the twenty-first century take for granted.

A moment ago, I said that Schrödinger "guessed" the form of
the matter-wave equation. Perhaps "guessed" is too strong a word.
Erwin Schrödinger used considerable physics intuition to develop
a new equation to describe the behavior of atoms. Mere mortals
may never know how exactly someone such as Newton or
Schrödinger does what they do. The insight that leads to a new
theory of nature is perhaps more powerful than that of artistic creation, for a new physical theory must not only be original, but also
mathematically coherent and accord with experimental observations. The most elegant theory in the world is useless if it is con-
tradicted by experiments.

While we may not know how Schrödinger did what he did, we

do know where and when he did it. Historians of science inform us that Schrödinger developed his famous expression in 1925, while staying at a Swiss Alpine chalet that he borrowed from a friend during a long Christmas holiday. Furthermore, while we know that his wife was not staying at this chalet, we also know that he was not alone. Unfortunately, we do not know which of Schrödinger's many girlfriends was with him at the time.

At this point, the reader may wish to reexamine Schrödinger's portrait in fig. 31. We may have a new answer to the question, "Why is this man smiling?" Certainly Erwin does not strike one as a traditional ladies' man. If one ever wondered whether there might be some mathematical expression that would make one attractive to the opposite sex, the Schrödinger equation might be a good starting point. Even the brief overview of quantum physics presented in this chapter will no doubt enhance your romantic desirability. This is in addition, of course, to the irresistible sex appeal that accompanies an encyclopedic knowledge of superhero comic books!

SCHRÖDINGER'S CAT OF TWO WORLDS

Schrödinger's equation is the $F = ma$ for electrons and atoms. Newton's second law, when the external forces F are specified, describes the acceleration a, and from this the velocity and position of an object can be determined. Similarly, Schrödinger's equation, given the potential energy of the electron through the term V, enables the calculation of the probability per volume ψ^2 of finding the electron at a certain point in space and time. Once I know the probability of where the electron will be, I can calculate the average location or momentum of the electron. Given that the average values are the only quantities that have any reliability, this is actually all that should be asked of a theory.

The emphasis on average quantities in quantum physics is different from our consideration of averages in the earlier discussion of thermodynamics (Chapter 13). There we spoke of the average energy per atom in an object—characterized by its temperature— simply because it was convenient. In principle, if we had enough time and computer memory, or were superfast like the Flash or Superman, we could keep track of the position and momentum of every air molecule in a room, for example. We could thereby calculate the instantaneous force on the walls per unit area, which would convey the same information as a determination of the pressure. In quantum systems, on the other hand, the wavelike properties of matter set a limit on our ability to carry out measurements, and the average is as good as we'll ever get.

What is it about the wavelike nature of matter that makes it so difficult to accurately measure the precise location of an electron in an atom? Think of a clamped violin string with a fundamental vibration frequency and several higher harmonic tones. Assume that the string is vibrating at a given frequency, but one that we cannot hear. If the vibrations were so fast that we could not see the string vibrate back and forth, how would we verify that the string is indeed vibrating? One way would be to touch the string and feel the vibrations with our fingers. If our fingertips were sensitive enough (like Daredevil's), we could even determine the exact frequency at which the string had been vibrating.

I say "had been vibrating," because once we have touched the string, it will no longer be oscillating at the same frequency as before. It will either have stopped shaking altogether or will be vibrating at some different frequency. Perhaps we can determine the vibration frequency by bringing our fingers near to, but not in direct contact with, the string. In this way we can sense the vibrations in the air caused by the oscillating violin string. In order to improve the sensitivity of this measurement, we need to bring our fingertips very close to the string. But then the air vibrations will bounce from our fingers and ricochet to the string, providing a feedback that can alter its vibratory pattern. The farther away we hold our fingertips, the weaker the feedback, but then our determination of the vibratory frequency will be less accurate.

The matter-wave oscillations of an electron within an atom are just as sensitive to disturbances. Measurements of the location of an electron will perturb the matter-wave of the electron. Much has been written about the role of the "observer" in quantum physics, but it's no more profound than when you try to look at something smaller than the probe you are using to view it—you will disturb what you are trying to see.

Quantum theory can provide very precise determinations of the average time one must wait before half of a large quantity of nuclear isotopes has undergone radioactive decay (defined as the "half-life") but is not useful for predicting when a single atom will decay. The problem with single events is best illustrated by the following challenge: I take a quarter from my pocket and am allowed to flip it once, and only once. What is the probability that it will come up heads? Most likely your gut instinct is to answer 50 percent, but you'd suspect a trick. And you'd be right—it is a trick question. To those who say the chance of getting heads is fifty-fifty, I say: Prove it. And you can't, not based upon a single toss, as long as we live in world containing two-headed quarters. If you toss the coin a thousand times (or toss one thousand coins once) you would find that for a fair coin, it would land heads-side up very close to 50 percent of the time. But probability is a poor guide for single, isolated events. Yet probability is all that the Schrödinger equation offers. This did not sit well with many older physicists who were accustomed to the clockwork precision of Newtonian mechanics, and they proposed a conceptual experiment that would open a Pandora's box (a box that contained a cat).

They posed the following situation: Consider a box, in which is placed a cat and a sealed bottle of poison gas, along with another, smaller box that contains one single radioactive isotope. The radioactive element has a half-life of one hour, which means, according to quantum mechanics, that after one hour, there is a fifty-fifty chance that it will have decayed. A by-product of this nuclear decay is the emission of an alpha particle (otherwise known as a helium nucleus), and the bottle of poison is arranged such that it will break open if struck by this particle. So, after one hour, there is a 50-percent chance that the cat is dead, having succumbed to the poison vapors released when the bottle was struck by the alpha particle, and a 50-percent chance that the bottle remains undisturbed, leaving the cat alive and well. According to Schrödinger's equation, prior to the one-hour time limit, the cat can be meaningfully described only as "the superposition (or average) of a dead cat and a live cat." Once the hour has passed and the lid is opened and one looks inside, the "average cat's wave function" collapses into one describing either a 100-percent-live or 100-percent-dead cat, but there is no way to know which will be observed before the lid is opened. If the walls of the box are transparent, you can never be sure that the light from the outside has not disturbed the decay process (recall that observing quantum systems can sometimes alter them). This interpretation has been found wanting by many physicists (despite the fact that recent experiments on entangled quantum states of light, as described in *JLA # 19*, the latest version of the Justice League of America, suggest that this is exactly what does occur), and a great deal of thought and argument has gone into attempting to resolve the intellectual unpleasantness associated with Schrödinger's Cat. One provocative solution to this problem, described below, enables the Flash and Superman to travel to alternate Earths.

In 1957, the physicist Hugh Everett III argued that when the cat is sealed in the box, two nearly identical parallel universes exist: one in which at the end of the hour the cat is alive and another in which it is dead. What we do when we open up the box does not involve collapsing wave functions, nor is the cat 50-percent dead and 50-percent alive before we take a look. Rather, all we do at the end of the hour is determine which of the two universes we live in—one where the cat lives or one where the cat dies. In fact, for every quantum process for which there are at least two possible results, there are that many universes, corresponding to the different possible outcomes. The two Earths reflecting the two possible outcomes of a particular quantum event will each evolve in different ways, depending on the myriad additional quantum events that occur following this initial branching point. If the bifurcation of the Earths occurred recently, then a particular Earth may be similar to our own world. If the separation occurred a long time ago, then during the intervening time, there would be many opportunities for subsequent quantum events to have outcomes different from what was observed in our world. The history of this second Earth may be very much like ours, then, but there is also the possibility of dramatic differences.*

Hence, quantum theory provides a physical justification for both the "What If?" tales in the Marvel universe and the alternate Earths in DC comics. On one Earth, Jay Garrick inhaled "hard water vapor" in a chemistry lab accident, gaining the gift of superspeed with which he fought for justice as the Flash with his

---

* An excellent recapitulation of the above discussion of the Schrödinger Cat thought-experiment can be found in *Animal Man # 32*. Though, for best effect, start at the beginning of this story arc with issue # 27.

teammates in the Justice Society of America. On another Earth, police scientist Barry Allen was doused with an array of chemicals while simultaneously being struck by lightning, leaving him with the gift of superspeed, with which he fought for justice as the Flash with his teammates in the Justice League of America. On another Earth a super-speedster committed crimes as the evil Johnny Quick with his teammates in the Crime Syndicate of America. There are in principle an infinite number of Earths, corresponding to all possible outcomes of all possible quantum effects, though a basic tenet of this theory is that ordinarily there can be no communication between these multiple alternate Earths. Ordinarily. Apparently, for someone able to vibrate at superspeed like the Flash, travel between these many worlds could occur as often as readers kept buying such stories.

To physicists, Hugh Everett III's proposal led to a very different crisis of infinite Earths. The many-worlds solution to the Schrödinger's Cat problem represented to most physicists an example of the cure being worse than the disease. Nevertheless, there is nothing logically or physically inconsistent with this theory, and no one has been able to prove that it is incorrect.

Physicists who considered it intellectually unsatisfying to say that a complete theory of nature can predict only probabilities were not heartened by the notion that the theory actually described an infinite number of alternate universes. The "many-worlds" model has been considered the crazy aunt of quantum theory since its publication, and has been locked in the metaphoric attic until very recently. It was never taught to me, for example, when I studied quantum mechanics in college and again in more detail in graduate school. I discovered the "many-worlds" model completely by accident when, as a graduate student, I came across a copy of Bryce DeWitt and Neill Graham's 1973 book *The Many-Worlds Interpretation of Quantum Mechanics*, left abandoned in a graduate student's office. In a successful attempt to procrastinate doing my homework, I picked up this strange book, began reading it, and thereupon learned that somewhere there was another James Kakalios who was actually finishing his assignment on time (not that this knowledge did me any good).

Though few physicists give the "many-worlds" model the time of day, there is one class of theoretical physicists, some of whom are supporters of this idea: string theorists.

WHY SUPERMAN CAN'T CHANGE HISTORY

In the years following the development of the Schrödinger equation, scientists have developed techniques to describe how the electron's matter-wave interacts with quantum versions of electric and magnetic fields (a process called "Quantum Electro-Dynamics" or QED) and how the matter-waves of quarks inside a nucleus behave (a process termed "Quantum Chromo-Dynamics" or QCD). A remaining goal of theoretical physics is to understand how to unite the physics of large-scale massive objects, governed by gravitational physics, with the quantum world. There is a perfectly good theory for gravity, namely Einstein's General Theory of Relativity. There is an excellent theory to describe the quantum nature of electrons (QED). Combining these theories into one coherent whole has proven beyond the abilities of any scientist up until now. The closest that theorists have come to a quantum theory of gravity is something called "string theory."

A gross oversimplification of string theory is that it suggests that matter is itself a wave, or rather a vibration of an elemental string, and that these "strings" are the basic building blocks of everything in the universe. In its current state, many physicists are skeptical about string theory. Their first objection is that, in order for the equations to balance, string theory works only in eleven dimensions (ten spatial and one time). This is somewhat awkward because, as near as we can tell, we live only in three spatial dimensions, and no one has ever encountered additional dimensions.* To address this discrepancy string theorists have suggested that there really are eleven dimensions, but seven of these spatial dimensions curl up into little balls, with a diameter less than a billionth trillion trillionths of a centimeter, a length scale labeled the Planck length, though some versions posit the existence of very large dimensions. Another drawback of string theory is tied to this extradimensional notion: Probing length scales so small requires correspondingly higher energies than

---

* Except such imps as Mr. Mxyzptlk, Bat-Mite, Mopee, and possibly Dr. Strange, when he was warned to "Beware Triboro! The Tyrant of the Sixth Dimension" in *Strange Tales # 129.*

current and next-generation particle accelerators can achieve. Without the verification provided by experimentation, the only criterion to determine whether the equations are on the right track is mathematical elegance. This may be dangerous; while it is true that the equations of classical mechanics (electricity and magnetism) and quantum mechanics do indeed have a certain mathematical beauty, there is no a priori reason to believe that nature really cares whether we find the equations pretty or not. Nevertheless, string theory is presently the only likely candidate for a quantum theory of gravity, and only further study will determine its success.

Physicists investigating quantum gravity have invoked the many-worlds interpretation in order to resolve logical inconsistencies in their calculations involving time travel. Recently, some scientists have claimed that time travel is not physically impossible, though it is highly unlikely to ever actually be accomplished. The problem with time travel into the past is set forth in the famous "grandfather paradox." Essentially, if one could indeed travel back in time, it would be possible to murder your grandfather when he was a young man, before your own father was conceived. In this way you would prevent your own birth, but the only way you could have prevented it is if you had first been born. In order to find a way around this conundrum, modern theoretical physicists have dusted off Hugh Everett III's many-worlds interpretation. If there are indeed an infinite number of alternate parallel universes, then (the theorists argue) when you travel backward in time, the severe distortions in space-time necessary to make this journey would also simultaneously send you to a universe parallel to your own. You are therefore free to kill as many grandparents as ammunition allows, without fear of altering your own existence, because your own grandfather is safe in the past in your own universe, undisturbed by the havoc you are wreaking in alternate past worlds.

These modern theoretical notions were actually anticipated in the 1961 adventure "Superman's Greatest Feats" in *Superman # 146.* In this story, Superman agrees to travel into the past as a favor for Lori Lemaris, a mermaid from the sunken city of Atlantis with whom he had a "special relationship" (while she was a girl and was his friend, Lori was not Superman's girlfriend). Lori beseeches Superman to prevent the sinking of Atlantis, which occurred millions of years ago. Superman argues that all of his previous attempts (presented in earlier issues of Action Comics and *Superman*) to change history have failed, but Lori's pleading (and what appear to be bedroom eyes) convinces Superman to give it a try. Given that it takes great effort and a velocity larger than 1,100 feet per second to break the sound barrier (the effort, as discussed in Chapter 4, is due in part to the work one must do to push the air out of his way), it was proposed in DC comics that with an even-greater effort and a much faster velocity, one could pass through the "time barrier." (Both the Flash and Superman, each capable of these necessary speeds, could travel back and forth through time as their story lines required.) Superman thus zooms to at least 8,000,000 B.C.E. and reaches nearly the exact moment when the advanced civilization of Atlantis, which resides on a small island

Fig. 32. *Superman travels through time and saves Abraham Lincoln from being shot by John Wilkes Booth. Or does he?*

off the shore of what appears to be a coastal resort, is about to succumb to "giant waves caused by a colossal undersea earthquake."
Superman races to another island a safe distance from the undersea quake, which is the home to *another* advanced civilization.
Why we have never heard about this other ancient civilization is
not addressed. Superman borrows some "strange metal" from
buildings about to be torn down on this other island and fashions
an enormous crane with which he lifts the entire island of
Atlantis, depositing it onto a third, secure deserted island, where
it is spared by the earthquake. (Let's not even get into what this
"strange metal" could possibly be composed of that would give it
a tensile strength sufficient to lift an island.)

Amazed that, unlike all previous attempts, he was able to successfully change the course of history, Superman decides to make
various pit stops on his return trip to his own time period, using
this opportunity to "fix" various historical events. He saves the
Christians from being eaten by lions in the Roman Colosseum,*
takes Nathan Hale's place as he is about to be executed by the
British, prevents Custer's massacre at Little Big Horn, and drops
by Ford's Theater on April 14, 1865. As shown in fig. 32, as John
Wilkes Booth is about to assassinate President Lincoln, he utters
"Sic semper—Ulp!" as hands that can crush diamonds close upon
his pistol. Superman is now like a kid in a historical candy store,
and decides to try to save the population of his home planet, Krypton. Because he loses his superpowers under the red light of Krypton's sun, Rao (by this time the explanation for his amazing
abilities had been attributed to Earth's yellow sun), Superman
decides to build a fleet of spaceships from sunken Earth naval vessels and send them to Krypton in order to enable everyone to
escape to another world. Using his telescopic vision, he watches
his parents disembark on a new planet with an infant Kal-El. At
this point Superman realizes that he has stumbled upon a paradox,
for if his parents never sent him to Earth as a baby, how is he able
to save them now?

Returning to his present, in 1961, the Man of Tomorrow is sur-

---

* Contrary to popular opinion, most of the Christians devoured in ancient
Rome were killed in the Circus Maximus, not the Colosseum.

Fig. 33. *Superman, from the same story as fig. 32, now realizes that history has remained unchanged despite all of his time-traveling "Greatest Feats."*

prised to discover that history books are unchanged, shown in fig. 33. Lincoln was indeed shot at Ford's Theater and Nathan Hale and General Custer are similarly described suffering their Superman-less fates. Superman can't understand how this can be, since "surely, the [history] books are truthful." Ahem. Retracing his path through the time stream, Superman comes across an alternate Earth (fig. 34), in which the history books give proper credit to Superman's role in correcting the past's "mistakes."

Alas, Superman discovered in 1961 what theoretical physicists have rediscovered in 2001—that time travel is only possible via the many-worlds interpretation of quantum mechanics. Superman did indeed accomplish these amazing feats, altering the course of

Fig. 34. *Superman in 1961 discovers what quantum theorists have only recently hypothesized—that travel through time must of necessity also involve transport to alternate, parallel universes.*

history—but in an alternate universe, not in his own (see fig. 34). A similar phenomenon occurs in the Marvel Comics *Avengers #267*, where the evil time lord Kang the Conqueror is revealed to have single-handedly created a vast number of alternate Earths as a by-product of his frequent time-traveling in order to defeat his superhero foes. Still another example of comic books being ahead of the physics curve.

THROUGH A WALL LIGHTLY–

TUNNELING PHENOMENA

IN ADDITION TO BEING ABLE to run at amazing speeds, the Flash is said to possess total control over every one of his molecules' motions. This vibratory control was put to use in *Flash # 116, Flash # 123* (see fig. 30), and many times since. Matching his body's vibrations to the vibrational frequency of the atoms in a wall, it was argued by the writers of Flash comics that he could pass through a solid wall without harm to either himself or the wall. However, it is not true that the only reason we cannot walk through solid walls is that we vibrate at a different frequency than the atoms in the wall. As discussed earlier, the average rate at which our atoms vibrate is simply a reflection of our temperature. Our bodies are typically roughly within 40 percent of the temperature of a wall, so our atoms' vibrational frequency is already pretty well matched to those in the wall.

Nevertheless, there is a quantum mechanical phenomenon termed "tunneling" that predicts that an object, under the right circumstances, can pass through a solid barrier without disturbing either the barrier or itself. This very strange prediction is no less weird for being true. In this way quantum mechanics informs us that electrons are a lot like Kitty Pryde of the X-Men, who possesses the mutant ability to walk through solid walls (as shown in fig. 35), or the Flash.

Schrödinger's equation enables one to calculate the probability of an electron moving from one region of space to another even if common sense tells you that the electron should never be able to make this transition. Imagine that you are on an open-air handball

Fig. 35. *A scene from* X-Men # 130, *showing Kitty Pryde (not yet a member of the X-Men) as she employs her mutant ability to walk through walls to sneak up on the White Queen of the Hellfire Club.*

court with a chain-link fence on three sides of the court and a concrete wall along the fourth side. On the other side of the concrete wall is another identical open-air court, also surrounded by a fence on three sides and sharing the concrete wall with the first court. You are free to wander anywhere you'd like within the first court, but lacking superpowers, you cannot leap over the concrete wall to go to the second court. If one solves the Schrödinger equation for this situation, one finds something rather surprising: The calculation finds that you have a very high probability of being in the first open-air court (no surprise there) and a small but nonzero probability of winding up on the other side of the wall in the second open-air court (huh?). Ordinarily the probability of passing through a barrier is very small, but only situations for which the probability is exactly zero can be called impossible. Everything else is just unlikely.

This is an intrinsically quantum mechanical phenomenon, in that classically there is no possible way to ever find yourself in the second court. This quantum process is called "tunneling," which is a misnomer, as you do not create a tunnel as you go through the wall. There is no hole left behind, nor have you gone under the wall or over it. If you were to now run at the wall in the other direction, it would be as formidable a barrier as when you were in the first open-air court, and you would now have the same very

Fig. 36. *Scene from* Flash # 123, *where Jay Garrick, the Golden Age Flash, demonstrates the quantum mechanical process known as "tunneling." The matter-wave of an object has a small but nonzero chance of passing through a solid barrier. The faster the object approaches the barrier, the greater the transmission probability. As Jay correctly notes, the barrier is unaffected by the tunneling process.*

small probability of returning to the first court. But "tunneling" is the term that physicists use to describe this phenomenon. The solutions of Schrödinger's equation that predict tunneling phenomena find that the faster you run at the wall, the larger the probability you will wind up on the other side, though you are not moving so quickly that you leap over the wall. This is no doubt how the Flash, both the Golden and Silver Age versions, is able to use his great speed to pass through solid objects, as shown in fig. 36. He is able to increase his kinetic energy to the point where the probability, from the Schrödinger equation, of passing through the wall becomes nearly 100 percent.

Consider two metals separated by a vacuum. An electron in the metal on the left is like a person in the first open-air handball court. Instead of a concrete wall, a thin vacuum separates this electron from the second metal, which can be considered another open-air court. Solving the Schrödinger equation for this situation, one finds that the electron in one metal has a small but nonzero probability of finding itself in the second metal. The electron does not arc across the vacuum gap and does not have enough kinetic energy to escape from the metal on its own. (This is a good thing. Otherwise, all objects would be continually leaking electrons all over the place, and static cling would be one the most gripping problems of the day.) Rather, the electron's matter-wave extends into the separation space, decreasing in magnitude with distance in the gap. A similar phenomenon occurs with light waves moving from a denser to a less-dense medium. Under conditions for which the light wave should be totally reflected at the interface, there is still a small diffraction of light into the less-dense medium. The diffracted wave's magnitude decreases the further into the less-dense medium it progresses. Since the square of the electron's wavefunction represents the probability of finding the particle at a point in space and time, a finite magnitude for the "matter-wave" indicates that there is a nonzero probability that the electron will be found in the second metal. If the gap is not too large (compared with the electron's matter wavelength, which in practice means roughly less than one nanometer), then the matter-wave will still have an appreciable magnitude in the second metal. Let us be clear, the electron on one side of the barrier moves toward the obstruction, and most times simply reflects off the wall. If a million electrons strike the barrier then, depending on its height and width, 990,000 electrons might be reflected and 10,000 would wind up on the other side.

When the separation between the two metals is too large, then the chance of tunneling for even the most energetic electrons becomes very, very small. A person's momentum is large, so our matter-wavelengths are very small—much less than a trillion trillionths of the width of an atom and much smaller than the width of the concrete wall separating us from the second open-air handball court. Nevertheless, if you were to run toward the concrete wall, there is a very tiny probability that your matter-wave will arrive on the other side of the wall. The greater your kinetic energy, the larger will be your chance of tunneling. Those who doubt that this is indeed possible are invited to begin throwing themselves at concrete walls right now, and to persevere in their attempts no matter how discouraging the initial results.

Electrons in a solid rattle around at a rate of more than a thousand trillion times per second. Consequently, in one second they have a thousand trillion opportunities to tunnel through a barrier. Send enough electrons against a barrier, and if the height of the wall is not too high nor the thickness of the separation too large, an appreciable fraction will indeed tunnel to the other side. Not only has the phenomenon of quantum-mechanical tunneling been verified for electrons but it is the central principle behind a unique type of device called a "scanning tunneling microscope" that enables one to directly image atoms. As shown in fig. 37, when a metal tip is brought very close to, but not touching, a metal surface, it can intercept the electron clouds surrounding each atom on the surface. When the electrons tunnel from an atom to the metal tip, an electrical current is recorded in a meter connected to the tip. Whether or not tunneling occurs is very sensitive to the separation between the atoms on the surface and the scanning metal tip. A change in distance of just the width of an atom can change the probability of tunneling by a factor of more than a thousand. By moving the tip slowly over the surface and carefully measuring the current at each location, the position of each atom on the surface can be mapped out.

Just such an image can be seen in fig. 38, which shows the location of carbon atoms on the surface of a crystal of graphite (more commonly known as "pencil lead"). The gray scale is not real (carbon atoms aren't really black or white, or any color for that matter), but is used to represent the magnitude of the current recorded in the tip at any position, which in turn reflects the electron density at each spot. Fig. 38 shows us that the carbon atoms in graphite form hexagons much like the six-sided plates that make up a snowflake. The fact that the carbon atoms form a hexagonal lattice implies that a crystal of graphite consists of sheets of

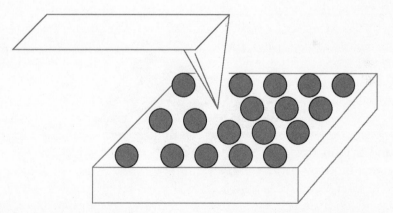

Fig. 37. *Cartoon showing the basic mechanism of a scanning tunneling microscope. A fine metal tip is brought very close to a conducting surface. By very close, I mean within a few atoms' widths. When the tip passes over an atom on the surface, the electron probability clouds of the atom may lead to quantum-mechanical tunneling into the tip. When the tip is right above an atom, the tunneling probability is high, and the current in the tip is large. In this way the atoms of the surface can be scanned and imaged.*

carbon atoms, as in fig. 38, lying atop one another. Building a three-dimensional crystal out of such two-dimensional sheets, the solid essentially stacks each sheet atop the other like the thin layers in a puff pastry. The planes in solid graphite are so loosely held together that you can easily peel them apart with just your hand, simply by scraping a pencil point along a sheet of paper. The fact that this form of solid carbon makes a better writing implement than if all carbon atoms had four equally strong bonds (otherwise known as "diamond") can be inferred directly from this atomic image. We'll have more to say about the nature of carbon bonds in Chapter 25.

In the next chapter we will discuss the physics of transistors and diodes, and I'll give away the punch line now and tell you that these semiconductor devices are essentially valves that regulate and amplify the flow of current. One way this current can be controlled is through the tunneling process. Normally, when two conductors are set close to each other, separated by a thin insulating barrier, no current can flow from one conductor to the other. By applying a voltage across this sandwich structure, the effective height of the wall separating the electrons of one region from the other can be varied. As noted, the tunneling probability is a very sensitive function of this barrier height. In this way, the tunneling

y-axis

x-axis

Fig. 38. Scanning tunneling microscope image of the surface atoms of graphite, the form of carbon used in pencil lead. Each white spot indicates a region of space where the tunneling current was high for that tip location (see fig. 37). The hexagonal lattice of carbon atoms is readily apparent. The grayscale is used to indicate the intensity of the tunneling current. The y-axis extends for 1nm while the x-axis is 0.5nm long. Courtesy of Dr. Laura Adams and Prof. Allen Goldman at the University of Minnesota.

effect is used to modulate the flow of electrons across the device. These "tunneling diodes" are integral components of cell phones, as well as many other solid-state devices. Quantum-mechanical tunneling is therefore not an esoteric theoretical novelty, or useful solely in atomic microscopes. Many of the products we associate with our current lifestyle would not be possible without tunneling being a reliable phenomenon.

When we apply the laws of quantum physics to large objects like Kitty Pryde of the X-Men (fig. 35), we find that tunneling is still possible, but very unlikely. How unlikely? Assuming Kitty's mass is 50 kilograms (such that her weight is 110 pounds—one of her code names was Sprite, after all), then even if she could run at a wall as fast as she could a million times a second, it would take longer than the age of the universe before she could expect to quantum-mechanically tunnel through to the other side. Clearly the one-time miracle exception comes into play in a big way here. With our improved understanding of physics, we can now more accurately describe Kitty Pryde's mutant power as being able to alter her macroscopic quantum wave function, increasing her tunneling probability to near 100 percent at will. Quite useful when one has locked the keys inside the car.

A long-standing puzzle in comic books is that if Kitty Pryde can walk through walls, why doesn't she fall through the floor at the same time? How, when she is "phasing" and immaterial, can she walk? In *X-Men # 141*, it was argued that while phasing, Kitty actually walks on a surface of air, and is not in actual contact with the floor. While immaterial in her phasing mode, she is therefore unaffected by any trapdoors opening beneath her. Assuming for the moment that she can indeed walk on air—that is, that somehow the air provides enough resistance to the backward force of her feet that it supplies a forward thrust, propelling her forward— the question still arises as to how her partially material foot can follow her body through a wall.

If the mechanism by which she is able to pass through solid barriers is indeed quantum-mechanical tunneling, then it is perfectly reasonable that she would not slide through the floor. When an electron tunnels from one side of a barrier to the other, it conserves energy in the process. If it has a certain amount of energy on one side of the barrier, it has the same total energy after the tunneling process is complete. In fact, tunneling occurs only when the energy of the object is exactly the same on both sides of the barrier. The ratio of kinetic to potential energy can change when tunneling across a barrier from one allowed region to another. But Kitty has shown that even when not walking through a wall, she can "phase" as bullets or energy blasts pass through her. In this case she is "tunneling" as she stands still and lets the "barrier" move through her.

Technically, she cannot walk while tunneling, as she may not increase her energy by pushing against any object, whether the object is the solid floor or a cushion of air. But at the same time, she cannot lose any energy either. All she needs to do is walk normally as she approaches a wall, turn on her mutant power of maximizing her tunneling probability, and she will glide through the partition with the same speed as she had when she neared it. For those times when she desires to phase through a floor, such as in *Astonishing X-Men # 4*, where she actually phases through nearly a hundred feet of solid metal to reach an underground laboratory, she would jump up slightly while in her corporeal mode, and then right before her feet touch the ground, activate her mutant tunneling ability. She would continue her motion with the last kinetic energy she had while solid, and descend with a steady velocity. It's probably safer for her if she keeps her mutant tunneling power activated until she is near the floor in the lower room, and avoids becoming material near the ceiling, where she would have to deal with her now-large gravitational potential energy.

24

SOCK IT TO SHELLHEAD–

SOLID-STATE PHYSICS

IF ANY SUPERHERO DEMONSTRATES the value of a physics education, and in particular the study of semiconductors, which is part of solid-state physics, it is Iron Man, who wore a high-tech suit of armor to fight for justice. The heart of Iron Man's amazing offensive and defensive capabilities was a modern (in 1963) technological marvel: the transistor.

The transistor truly is a revolutionary device, as its ability to amplify and modulate voltages has had a profound impact on all of our lives. Initially, the transistor was used solely to duplicate the functions of vacuum tubes, so that radios and television sets could become lighter and more efficient. As scientists and engineers learned how to make transistors smaller and smaller, their use for mathematical calculations led to the development of electronic computers. The solid-state transistor is the fountainhead of nearly every electronic device in use today. At this point, we have enough physics under our belt to understand how this remarkable device works. Before delving into semiconductor physics, let's review the events that led to Shellhead's* comic-book debut.

The Cold War loomed large in the monthly adventures of Silver Age comic books of the late 1950s and early 1960s. In DC comics, fighter pilots with the "right stuff" figured prominently in several

---

* The cylindrical helmet of his original suit gave Iron Man the nickname "Shellhead," which has clung to him for the past forty years, regardless of how streamlined and stylish later versions of his helmet became.

offerings. When an alien Green Lantern crash-landed on Earth and was at death's door in *Showcase # 22*, he instructed his power ring to seek out someone brave, honest, and fearless to whom he could bequeath the ring and power lantern. The ring selected an American test pilot, Hal Jordan. (In DC's *Showcase # 6*, another fighter pilot, Ace Morgan, led the Challengers of the Unknown, while the test pilot Larry Trainor was transformed into Negative Man during a flying mishap, as relayed in *My Greatest Adventures # 80*.) The Marvel Age of comics began in 1961 (the same year that both Russian and American astronauts first traveled to space), when four adventurers—a scientist, his girlfriend, her teenage brother, and a former fighter pilot—took an unauthorized rocket flight through a cosmic radiation belt in order to beat the Communists "to the stars." The cosmic rays they absorbed would turn this quartet into the Fantastic Four.

The "red menace" of Communism continued to rear its head in Marvel superhero comic books fairly often. The Hulk, who appeared a year after the debut of the Fantastic Four, owed his existence to a Commie spy, when physicist Dr. Robert Bruce Banner was exposed to an overdose of gamma radiation from a detonating gamma bomb. His assistant (who was in reality a Communist agent—you would think that a research assistant named Igor Starsky would have been a tip-off when he received his security clearance, but never mind) deliberately did not stop the countdown, hoping to eliminate America's leading bomb expert, but instead created the Hulk.

The Soviet presence in early Marvel comics was faced by the Human Torch, Ant-Man, Spider-Man, the Mighty Thor, and the Avengers. But none of the mighty Marvel superheroes of the early 1960s were so closely allied with the Cold War as the invincible Iron Man. *Tales of Suspense # 39* introduced the brilliant inventor and industrialist Tony Stark to the Marvel universe. Using his mastery of transistorized technology, Stark developed many new weapons for the military as part of his effort to help the United States win the fight against Communism in Indochina. Not content to merely test his new weapons in the lab, Stark accompanies an inspection team into the jungles of Vietnam, so that he could more accurately assess the effectiveness of his inventions. Alas, we soon see why more CEOs do not take such a hands-on approach to quality control. A booby trap kills the military advisers traveling with Stark, and leaves Tony himself with a piece of metal shrapnel lodged in his chest, dangerously close to his heart. To make matters worse, he is captured and brought to the hidden camp of Wong-Chu, "the red guerrilla tyrant." A doctor at Wong-Chu's camp determines that the shrapnel is actually migrating and that in a few days it will reach Stark's heart and kill him.

Wong-Chu offers Stark a deal: Work in my weapons development lab (this is apparently a pretty well-equipped guerrilla camp) in return for surgery that will save your life. Stark agrees, intending to use his brief remaining days to create some sort of weapon to both save his life and combat his captor. Along with the brilliant physicist Professor Yinsen, also a captive of Wong-Chu, Stark constructs a metal chest plate that, once fully electrically charged, will prevent the shrapnel from reaching his heart. Realizing that he will need offensive and defensive weapons if he and Yinsen are to escape the guerrilla camp, the chest plate becomes part of an iron suit, which contains an array of transistorized weaponry. The suit is completed moments before the shrapnel can claim Stark's life, though Prof. Yinsen is killed when he decoys their captors away from Stark as the chest plate finishes charging up. Yinsen's death is avenged, and the other prisoners in the camp are freed as Iron Man defeats the Communist warlord. (The 2008 film *Iron-man* would faithfully follow this origin storyline, substituting Middle Eastern terrorists for Communists.) Making his way back to the States through the jungles of North Vietnam (a story not fully told until years later in *Iron Man # 144*), Tony Stark would continue to use his technological suit of armor to defend America against Communist aggression.

And boy, did Shellhead attract the Communist foes! At times he seemed to have a "red magnetism" field built into his suit of armor. Borne out of the Vietnamese conflict, Iron Man would battle more Communist villains in his first four years than nearly all other Marvel superheroes combined. Iron Man would face the Red Barbarian (*Tales of Suspense # 42*); the Crimson Dynamo (*Tales of Suspense # 46* and *52*—a Russian power station built in the form of a suit of armor designed to defeat Iron Man); the Mandarin (*Tales of Suspense # 50, 54, 55, 61, 62*—a Chinese warlord modeled after the Sax Rohmer pulp-fiction villain Fu Manchu, who possessed ten deadly rings of power); and the Titanium Man (*Tales of Suspense # 69–71*—a Soviet-built stronger version of Iron

Man created to defeat our hero in televised conflict, thereby proving to the world the superiority of Communism over capitalism). Throughout all this, Tony Stark maintained the fiction that Iron Man was a separate person, hired by Stark to serve as his bodyguard. Given the number of times Communist agents tried to kidnap Stark or steal his research plans, this was not such a far-fetched cover story.

As if the constant battles with costumed villains were not distracting enough, Stark was continually called to testify before Senate committees, who insisted that it was his patriotic duty to turn the Iron Man technology over to the military. Little did Senator Byrd (not the long-serving senator from West Virginia), who led the investigations into the connections between Iron Man and Stark Industries realize that the secret of Iron Man's success—the transistor—was already in the public domain. The transistor was developed by three physicists in 1947 at Bell Laboratories in Murray Hill, New Jersey—the research lab of utility Bell Telephone. In order to facilitate the adoption of this new technology, Bell Labs ran seminars for other firms interested in using transistors, instructing them in the details of the new field of solid-state physics. It's not enough to build a better mousetrap; you must also make sure that the mice know about it!

CLOTHES MAKE THE MAN

Following his first appearance, Iron Man's suit would undergo nearly constant modifications, both cosmetic and significant. The suit was gray in *Tales of Suspense # 39*, but by the very next issue, Stark decided to change the color to gold, so as to make more of an impression on women. You would think that a multimillionaire industrialist who looks like Errol Flynn (the model upon which artists based their drawing of the mustachioed Tony Stark) would not have to worry about whether his secret identity as Iron Man was attractive to the ladies, but it is presumably just such attention to detail that brought Tony Stark his success. Within a year, the suit would be redesigned again, now as a more form-fitting yellow-and-red ensemble shown in fig. 39 that, with minor variations, would persist to recent times.

The weapons that were distributed throughout the suit would

© 1963 Marvel Comics

Fig. 39. Panels from a bonus section "All About Iron Man" in Tales of Suspense *# 55, providing a schematic of the 1960s Golden Avenger, and a tutorial stating that the charge to power supplies feeding his transistors is depleted the more he uses them.*

also undergo near-constant upgrades. Initially Stark had "reverse magnetism" projectors in the palms of his gloves, but these were soon modified to "repulsor rays" that were essentially "force beams." A large recessed disc on his chest housed a "variable power spot-light" that evolved into a "uni-beam" (I'm not really sure what this did). He originally had a radio antenna extending from his left shoulder, but improvements in wireless transmission and reception technology enabled this function to be incorporated within the body of his iron suit.

The armor itself is still quite heavy, even in its more flexible, streamlined form. The only way that Stark is able to walk in this iron suit, and to lift objects weighing up to several tons, is through the application of "tiny transistors within his armor which increase his power tremendously!" And he's going to need this extra power. In order to estimate how much the armor weighs, assume that the Iron Man suit is one eighth of an inch thick and has the average density of iron, which is roughly 8 grams per cubic centimeter. The surface area needed to make the suit can be approximated by taking Tony's trunk as a cylinder, his head as another, smaller cylinder, and his arms and legs as smaller, longer cylinders. If Tony is about six feet tall and his suit jacket is a 50 regular, then his total surface area is roughly 26,200 square centimeters. The volume of iron within his suit is found by multiplying the surface area by the armor's thickness of ⅛ inch, or 0.32 centimeters, yielding a total volume of metal of roughly 8,400 cubic centimeters. To determine how much it weighs, we multiply this volume by the density (8 grams/cm^3) and obtain 67,000 grams, or 148 pounds, excluding the weight of all of the transistorized circuitry. Tony Stark would frequently carry all of this in his briefcase (with one dress shirt covering the armor as camouflage (see *Tales of Suspense # 55*), except for the chest plate, of course, which he wore constantly in order to keep the shrapnel from worming its way to his heart. Consequently, simply by lugging nearly 90 pounds of armor around in this briefcase, Tony would have developed considerable upper-body and bicep strength as a side benefit of being Iron Man.

The suit's weight leads to the question of how his jet boots enable Iron Man to fly. If the suit weighs nearly 150 pounds, and Stark himself tips the scales at 180 pounds, then his boot thrusters have to supply a downward force of 330 pounds (equal to a mass of 150 kg), just for Iron Man to hover in the air. Presumably, these jets use a chemical reaction to violently expel the reactants from the soles of his boots. As every action is balanced by an equal and opposite reaction, this downward force leads to an upward push on Iron Man, keeping him aloft. If he wants to accelerate, then his boots have to provide even more force, as only the force in excess of his weight provides an acceleration ($\mathbf{F} = \mathbf{ma}$).

Tony would frequently need to travel from his Stark Industries plant on Long Island to Avengers Mansion in the heart of Manhattan, a distance of approximately fifty miles (as the Golden Avenger flies) in a time of ten minutes. This corresponds to an average speed of 300 mph, which is nearly half the speed of sound! (In the film *Ironman*, he dueled with fighter jets traveling at least twice that fast!) Ignoring the energy that Shellhead must expend to push the air out of the way as he travels at this speed, this implies that for a kinetic energy $(\frac{1}{2})mv^2$ at this speed, his armor requires at least 1.37 million kg-meters^2/sec^2 of energy. When Iron Man needed to travel great distances, he would eschew the boot jets and make use of motorized roller skates that were built into his boots. Not only would it be more fuel efficient, since energy does not have to be expended counteracting gravity to keep Shellhead in the air, but whenever he is decelerating, he can use the rotational energy and an alternator to recharge his internal batteries, as in an automobile. In this way, Tony Stark anticipated recent hybrid-engine automotive technology.

In the seventies, Iron Man had gone green, and his armor was now coated with a thin layer of solar cells, enabling him to recharge whenever he was in direct light. The energy from the sun on an *average* day in the United States is roughly 200 kg-meter^2/sec^2 per second over an area of 1 meter^2. We have just calculated the surface area of the Iron Man suit to be 26,200 centimeters^2, which means that the amount of energy per second striking Iron Man is 262 kg-meter^2/sec^2 (at any given moment only half of his available surface area can be facing the sun) while his jet boots require an expenditure of power of more than a million kg-meter^2/sec^2. If the solar cells are 50-percent efficient in converting the energy of the sun into stored energy in Tony's battery packs (and most commercially available solar cells have a conversion efficiency of only about 10 percent), Iron Man would need nearly three hours to soak up enough sunlight for this one trip. We have considered neither the energy needed to run the suit's internal air-conditioning unit (pushing air out of the way at 300 mph will make any person inside a metal suit a tad sweaty), nor whether he has to fire his repulsor rays during the flight. In the normal course of a typical day in the life of Iron Man, he will expend energy much faster than he can recharge his storage batteries using solar cells.

To give the writers of the Iron Man comic credit, their concern with mechanisms for Tony Stark to recharge his armor's storage batteries implies recognition of the Principle of Conservation of Energy. From his very first appearance in *Tales of Suspense # 39*, it was always acknowledged that running the mechanized Iron Man suit required large amounts of energy and that the greater his expenditure of power, the faster the drain on the energy reserves he might carry on his person. Not only was a ready supply of electrical energy necessary to animate his jet boots and activate the servo-motors that enabled him to move in the suit and that increased his strength, but his chest plate needed electrical energy as well in order to protect his heart from the shrapnel he carried with him ever since that fateful day in Vietnam. The 1960s Iron Man would occasionally have to drag himself dramatically along the ground after a particularly energy-exhausting battle, searching for an electrical outlet in order to recharge his battery reserves.

Even after he made the transition to solar-powered battle armor, Stark's suit could run dry in an emergency. In *Iron Man # 132*, Tony drained every last erg of energy (one erg is one ten millionth of a kilogram-meters^2/sec^2) from his suit in an exciting, no-holds-barred battle with the Incredible Hulk. Tony focused all of his suit's stored energy into one final punch, and accomplished what had previously been impossible: Iron Man knocked the Hulk unconscious. But the cost to Stark was high. With absolutely no power to move his suit, Stark was trapped, unable to move within his now rigid shell of armor. To make matters worse, the protective covering over his eye and mouth slits had been engaged, to shield Stark from a previous explosion. Tony therefore faced suffocation once the air contained in the suit was used up. It would take all of the following issue for Ant-Man, forcing his way in through the exhaust port in Iron Man's boot jet, to travel the length of the armor, avoid the suit's internal protective mechanisms, and disengage the faceplate's protective covering.

HE FIGHTS AND FIGHTS WITH REPULSOR RAYS

Of all of Iron Man's weapons, the most effective are his "repulsor rays," which are fired from discs on the palms of his armored gloves. Back in his first appearance in *Tales of Suspense # 39*, the first version of this glove-based repulsion weapon was a "reverse magnetism" ray used to fight his way out of Wong Chu's prison camp. Wong Chu's guards, finding that small-arms fire bounced harmlessly off the iron suit that the intruder wore, responded by preparing to shoot bazookas and throw grenades at the Yankee invader. Fig. 40 shows that as they fetch the heavy weaponry, Tony takes the time to "reverse the charge on this magnetic turbo-insulator and use a top-hat transistor to increase its repelling power a thousandfold!" As the rays are emitted from his hand, deflecting the weapons, he exclaims, "There! Reverse magnetism—it works like a charm!" In fact, it would have to work like a charm, because there's no way it could work using solid-state physics.

Fig. 40. *Iron Man, in his first appearance in* Tales of Suspense # 39, *fights his way out of a Vietnamese prison camp using a top-hat transistor and a "magnetic turbo-insulator."*

There's only one aspect of the scene summarized above that is physically correct, and that involves the "top-hat transistor." There is no such thing as a "magnetic turbo-insulator"; this is just technobabble. The "turbo" modifier is just to make these insulators sound cool. There are magnets that are nonmetallic—that is, they are electrical insulators, but still generate a large magnetic field—and devices called "top-hat transistors" do indeed exist. They are so named because they look like small cylinders, about the size of pencil erasers (this was back in the early 1960s, long before the microminiaturization of transistors enabled millions of such devices to be fabricated on a chip measuring only a few millimeters on each side), with a small disc at their base at which the electrodes extended, and this makes them look a little like the top-hat playing piece in a Monopoly game set. The panel showing Tony Stark employing such a device to amplify the current to his "magnetic turbo-insulator" is physically plausible. But the second-to-last panel, in which he then employs said device to deflect the grenades and bazooka shots using "reverse magnetism" is not.

While every electron, proton, and neutron inside every atom has an intrinsic magnetic field, the natural tendency of magnets to line up north pole to south pole has the effect of canceling out the magnetism of most atoms. Any magnetic field that Iron Man would create using a powerful electromagnet in the palm of his glove would only be effective if: (1) the grenades being tossed at him were for some reason already magnetized and (2) they were all perfectly thrown so that their north poles were all pointing in the same direction and (3) the magnetic field created by Iron Man's hand was also oriented so that the north pole was directed toward the incoming grenades and not the south pole, which would have the effect of accelerating the weapons toward him. It is unlikely that Tony Stark could always count on his opponents to cooperate with suitably magnetically oriented weapons.

Iron Man's reverse-magnetism ray actually has a better chance of working on nonmagnetic objects! Recall our discussion in Chapter 19 concerning Magneto and the phenomenon of diamagnetic levitation. Unlike metals such as iron or cobalt, for which the internal atomic magnetic fields align in the same direction, many materials, including water, are diamagnetic. In this case, when they're in an external magnetic field, the atomic magnets orient themselves to oppose the applied field. Thus, the very process of trying to magnetize the object leads to a repulsive force. Iron Man's reverse magnetism could repel objects, but only if they were diamagnetic, and it would not work on many metallic objects that are either ferromagnetic or paramagnetic (which align with an applied field). Magneto creates these large magnetic fields through his mutant power, but Iron Man must do it the old-fashioned way, using electromagnets (similar to the one constructed by Superboy in Chapter 18). As Iron Man does not carry around with him an electrical dynamo like the one used by Superboy, a few shots of this reverse-magnetism ray would drain his batteries faster than a fight with the Hulk. Furthermore, the recoil of such weapons is considerable. When supplying a large force against a target, they will induce an equal and opposite force against the gun and the shooter holding it. Tony Stark was clever to build his repulsor rays into his gloves. By locking the servo-motors that enable his armored arms to move, his iron suit provides a large and rigid inertial mass to take up the recoil whenever he fires this glove-based weapon.

While "reverse magnetism" may not be physically practical, hand-held pulsed-energy weapons have begun to make the transition from comic-book fantasies to military research facilities. Certainly these weapons cannot be the same "magnetic repulsors" as Iron Man uses, for the reasons argued above. The energy needed to generate a magnetic field large enough to deflect an object using only diamagnetic repulsion is so large that it would be more effective to employ conventional weaponry. Nevertheless, "pulsed" energy systems are under active development by the military. By generating a large voltage inside the weapon that can be rapidly discharged in a thousandth of a second, the power (change in energy per time) can be quite high. This electromagnetic pulse, if directed at a target, would deposit this energy in a localized region faster than the heat could be safely dissipated away. High-intensity laser beams delivered in extremely brief pulses are used in physics laboratories to nearly instantaneously melt a small region of a solid's surface, and in principle the same process could be employed in an offensive capability. The big drawback is once again the energy requirements of such a weapon. If one must carry a miniature power plant around in order to fire such a pulsed energy weapon, the element of surprise in any combat situation would be lost.

SOLID-STATE PHYSICS MADE EASY

What is a transistor? What is this electronic device that, at least according to Stan Lee, is endowed with miraculous abilities that enable Iron Man to successfully fend off the Mandarin, the Crimson Dynamo, and Titanium Man? A short answer is that transistors are valves that regulate the flow of electrical current through a circuit. Such answers are easy to remember, but they tell us nothing about how transistors actually function. The first question we should address is: What exactly is a semiconductor, which is neither a metal nor an insulator? We hear a great deal about how we are living in the "Silicon Age," but what is so special about silicon? In the next few pages I will try to condense more than fifty years of solid-state physics as I answer these questions.

Silicon is an atom, a basic element of nature, just like carbon, oxygen, or gold. A silicon atom's nucleus has fourteen positively charged protons and (usually) fourteen electrically neutral neutrons, and to maintain charge neutrality there are fourteen negatively charged electrons surrounding the nucleus. These electrons reside in the "quantum-mechanical orbits" that, as discussed in chapters 21 and 22, arise from the wavelike nature of all matter. The possible "electron orbits" are specific for each element, and determine the allowed electron energies.

Quantum mechanics enables us to calculate, via the Schrödinger equation, the allowed "orbits" of the electrons in an atom, and knowing how many different possible orbits an electron can have in an atom is like knowing the number and arrangement of chairs in a classroom (stay with me here; this classroom metaphor is going to be useful in explaining metals, insulators, and semiconductors). The chairs only represent possible or virtual classes; it is not until the students enter and take their seats that the class is real. If only one student comes in and takes a seat, this is like having only one electron in a possible quantum-mechanical orbit. We would call this class hydrogen, in analogy with the atom that has only one electron in its neutral, stable form. If there were two students sitting in the class, we would have helium, fourteen students would make up silicon, and so on. The first students to enter the class take the seats at the front of the room, close to the blackboard, in our hypothetical example. The last students to enter take seats near the back of the auditorium, far from the blackboard (where the positively charged nucleus will be). This arrangement, with the closest seats filled by students, describes the lowest-energy configuration. For a carbon atom with six electrons, the closest orbits are occupied. If the carbon atom gains some energy, say, from absorbing light, some of its electrons will then occupy higher-energy orbits (seats farther from the front).

Whether a material is a metal, a semiconductor, or an insulator depends on the energy separation between the highest level filled with an electron and the nearest available unoccupied level. In the classroom analogy, the solid can be thought of as a very large auditorium with many rows of seats, provided by the constituent atoms that make up the material. There will be an empty balcony that contains an equal number of seats. If the electrons sitting in the lower-energy orchestra seats* are to conduct electricity when a voltage is applied across the solid, then they gain extra energy. But they can move only to a higher energy if there is an empty state for the electron to be promoted into (recall the discussion of quantized energy levels in Chapter 21). The electrical properties of any solid are determined by the number of electrons residing in the lower orchestra seats and the energy separation between the lower occupied seats and the next empty ones in the balcony.

The difference between insulators and metals is clear in this analogy. An insulator is a solid in which every single seat in the orchestra is filled, while a metal is a material for which only half of the seats in the lower level are occupied. In a metal, there are a large number of empty seats in the orchestra available to an electron, and the application of a voltage, whether big or small, can accelerate the electrons to higher energy states (which correspond to carrying an electrical current). Metals are good electrical

---

* The lowest-energy levels are the first to be filled with electrons. Strictly speaking, every pair of electrons in an atom gets its own auditorium (they pair up due to their intrinsic magnetic fields, north pole to south pole). It is the last electrons to get placed in available levels that determine the chemical reactivity and electronic properties of the corresponding solid, and it is the auditorium containing these electrons that we consider here.

conductors because their lowest occupied auditorium seats for electrons are only half-filled. For the insulator, every seat is occupied, and absent promotion to the balcony, no current will result when a voltage is set up across the material. If I raise the temperature of an insulator, providing external excess energy in the form of heat, some of the electrons can rise into the previously empty balcony. In the balcony there are many empty seats for the electron to carry a current, but this will last only as long as the temperature is elevated. If the temperature is lowered, the electrons in the balcony will descend and return to their low-energy seats in the orchestra.

If the insulator absorbs energy in the form of light, it can immediately promote an electron to the balcony. When the electron returns to its seat in the orchestra, it has to conserve energy and thus gives off the same amount of energy that it previously absorbed. It will either do this by giving off light of the same energy as initially absorbed, or the electron can induce atomic vibrations (heat). This is why shining light on an object warms it up—the electrons absorb the energy of the light, but then can return the absorbed energy in the form of heat. If the energy of the light is insufficient to promote an electron from the highest filled orchestra seat to the lowest empty balcony seat, the light is not absorbed. In this case the lower-energy light is ignored by the electrons in the solid, and passes right through it. Insulators such as window glass are transparent because the separation between the filled orchestra and empty balcony for this material is in the ultraviolet portion of the spectrum, so visible light with a lower energy passes right through. On the other hand, metals always have available empty seats to absorb light even in the half-filled orchestra. No matter how small the light's energy, an electron in a metal can absorb this energy and then return it upon going back to its lower-energy seat. This is why metals are shiny—and good reflectors. They always give off light energy equal to what is absorbed, and there is no lower limit to the energy of light they can take in.

A semiconductor is just an insulator with a relatively small energy gap (compared with the energy of visible light) separating the filled lower band from the next empty band. For such an energy separation, a certain fraction of electrons will have enough thermal energy at room temperature to be promoted to the balcony. When electrons are excited to the upper deck, the material

now has two ways to conduct electricity. For every electron promoted to the higher energy band that is able to conduct electricity,
an empty state is left behind. The empty chairs in the previously
filled orchestra can be considered as "positive electrons" or
"holes," and can also carry electrical current. If an electron adjacent to an empty seat slips into this chair, then the empty spot has
migrated one position over. In this way we can consider the hole
to move in response to an external voltage, and also carry current.
Of course, the original electrons will eventually fall back down
into the orchestra, filling the empty seats they left behind (though
not necessarily their original seats). When certain semiconductors
absorb light, there are enough excited electrons in the upper band
and holes in the lower band to convert the material from an insulator to a good electrical conductor. As soon as the light is turned
off, the electrons and holes recombine, and the material becomes
an insulator again. These semiconductors are called "photoconductors" and are used as light sensors, as their ability to carry an
electrical current changes dramatically when exposed to light.
Certain smoke detectors, television remote controls, and auto-
matic door openers in supermarkets make use of photoconductors
for their operation.

Semiconductor devices are typically constructed out of silicon
because it has an energy gap conveniently just below the range of
visible light. Furthermore, it is a plentiful element (most sand is
composed of silicon dioxide) that is relatively easy to purify and
manipulate. There are times when the physical constraints of the
size of the energy gap in silicon limits a device's performance, and
in this case, other semiconducting materials can be used, such as
germanium or gallium arsenide. Iron Man's, and the military's,
night-vision capabilities make use of a semiconductor's photoconducting properties and a small energy gap that is in the infrared
portion of the electromagnetic spectrum.

All objects give off electromagnetic radiation due to the fact
that they are at a certain temperature, so their atoms oscillate at a
particular frequency that reflects their average kinetic energy. On
a dark, moonless night, the temperature of most nonliving objects
decreases (as they are not absorbing sunlight), so they emit less
radiation, and what they do give off is at lower frequencies.
Humans, on the other hand, have metabolic processes that
maintain a uniform temperature of 98.6 degrees Fahrenheit.

Consequently, we emit a fair amount of light (as much energy as a 100-Watt lightbulb) in the infrared portion of the spectrum. Our eyes are not sensitive to this part of the spectrum, but semiconductors can be chosen that have a large photoconductivity when exposed to infrared light. At night the infrared light given off by a warm-blooded person is much greater than his or her colder surroundings.

Certain night-vision goggles, such as Nite Owl II's in *Watchmen*, use "thermal imaging" to detect this light by using semiconductors, which absorb the infrared radiation given off by an object at a temperature of roughly 100 degrees. The photocurrent in the semiconductor detector is then transported to an adjacent material, which is chemically constructed to give off a flash of visible light when the photoexcited electrons and holes recombine. In this way the infrared light that our eyes cannot usually detect is shifted to the visible portion of the electromagnetic spectrum, thereby enabling us to see in the dark. These goggles also detect visible light, as well as infrared light during the daytime. All objects give off roughly the same intensity of light if they are at the same temperature (recall our discussion of light-curves from Chapter 21). When the objects around a person are warmer (due to absorbed sunlight), the contrast between the infrared light from a person and his or her inanimate surroundings is diminished, as is the utility of the goggles.

WHAT COLOR ARE THE INVISIBLE WOMAN'S EYES?

An understanding of semiconductor photoconductivity also helps to resolve a question that has long perplexed comic-book fans: Why isn't the Invisible Woman blind? When the Fantastic Four took their ill-fated rocket trip, Sue Storm (now Susan Richards) gained the ability to become completely transparent at will. How can she do this, and how can she see, if visible light passes right through her? The more basic question is: How do we see anything at all?

The molecules that make up the cells in our bodies absorb light in the visible portion of the electromagnetic spectrum. The addition of certain molecules, such as melanin, can increase this absorption, darkening the skin. As a result of her exposure to cosmic rays, the Invisible Woman gained the ability to increase the "energy gap" of all of the molecules in her body (this is presumably the nature of her "miracle exception"). If the separation between the filled lower orchestra and the empty upper balcony is increased such that it extends into the ultraviolet portion of the spectrum, then visible light will be ignored by the molecules in her body and pass right through her. This is not so far-fetched; after all, we all possess cells that are transparent to visible light. In fact, you're using them right now, reading this text through the transparent lens of your eyes.

Sunlight contains a great deal of ultraviolet light, which has more energy than visible light. We typically don't think about the ultraviolet portion of the solar spectrum until we sunburn on a bright summer day. When Sue becomes invisible, she still absorbs and reflects light in the ultraviolet region of the spectrum. We can't see her because the rods and cones in our eyes do not resonantly absorb ultraviolet light. Special UV glasses (like the ones Doctor Doom installed in his armored mask) could shift the ultraviolet light reflected from Sue down into the visible portion of the spectrum, using a similar mechanism to the one used by night-vision goggles in shifting low-energy infrared light up into the visible portion of the spectrum.

This also explains how the Invisible Woman is able to see. The rods and cones in her eyes, when she is transparent, become sensitive to the scattered ultraviolet light that bounces off us, but that we cannot see. The world Sue sees while invisible will not have the normal coloring we experience, for the shift in wavelengths of the light she detects is not associated with the colors of the rainbow. Windows appear transparent to us because they transmit visible light and absorb ultraviolet light. We can't see ultraviolet light, so we don't notice its absorption. However, when Sue is invisible, a window will appear as a large dark space, while other objects will appear transparent to her. With a little practice, she would be able to maneuver just fine.

This mechanism to account for Sue's ability to see while invisible was suggested in *Fantastic Four # 62, Vol. 3* (Dec. 2002) that corresponded to the 491st issue in the numbering scheme that began in 1961. In this issue, written by Mark Waid and drawn by Mike Wieringo, we are told that while invisible, Sue sees by detecting the scattered cosmic rays that are all around us but cannot be detected by normal vision. Right idea—wrong illumination source. Cosmic rays from outer space are not light photons but are mostly high-velocity protons that, upon striking atoms in the atmosphere, generate a shower of electrons, gamma-ray photons, muons (elementary particles related to electrons), and other elementary particles. We usually don't have to worry about radiation damage or gaining superpowers via cosmic-ray-induced mutation, at least at sea level, as the particle flux is a million trillion times less than that of sunlight. If Sue depended on cosmic rays to see at street level, she would be constantly bumping into objects and people. It is more likely that her vision makes use of the same mechanism by which she could become transparent—namely, a shift in her molecular bonding into the ultraviolet portion of the spectrum.

WHAT IS A TRANSISTOR, AND WHY SHOULD WE CARE?

Back to Tony Stark and his transistorized suit of armor. When Tony needed to increase the repelling power of his magnetic turbo-insulator, he used a top-hat transistor. How are transistors able to amplify weak signals, making radios portable and repulsor rays powerful?

While semiconductors are useful as photoconducting devices, if this were their only application, no one would think to call this era the Silicon Age. The thing about semiconductors that makes them very handy to have around the house is that you can change their ability to conduct electricity by a factor of more than a million by intentionally adding a very small amount of chemical impurities. Not only that, but, depending on the particular impurity, you can either add excess electrons to the semiconductor or remove electrons from the filled auditorium of our earlier metaphor, thereby creating additional holes that can also conduct electricity. When a material with excess electrons is placed next to a semiconductor with additional holes, you have a solar cell, and if you then add a third layer with excess electrons on top of that, you've made a transistor.

It's been known for a long time that the addition of certain chemicals can change the optical and electronic properties of insulators. After all, that's how stained glass is made. Ordinary window glass has an energy gap that is larger than the energy of visible light, which is why it is transparent. But add a small amount of manganese to the glass when it is molten, and after cooling, the glass appears violet when light passes through it. Manganese has a resonant absorption right in the middle of the glass's energy gap, as if we had parked some extra chairs on the stairways that connect the filled orchestra and the empty balcony. Particular wavelengths of visible light that would ordinarily pass through the material unmolested will now induce a transition in the manganese atoms added to the glass. In this way certain wavelengths are removed from the white light transmitted through the glass, giving the window material a color or "stain." Different chemical impurities, such as cobalt or selenium, will add different colorations (blue and red, respectively) to the normally transparent insulator.

The same principle works for semiconductors, only the chemical impurities that we choose to add can either make it very easy to promote electrons to the balcony or to take electrons out of the filled auditorium, leaving holes in their place. A semiconductor for which the chemical impurities donate electrons is termed "n-type," since the electrons are negatively charged, while those for which the impurities accept electrons from the filled lower states are called "p-type," referring to the positively charged holes created. What's special about such semiconductors with added impurities is not that their conductivity can be changed dramatically (if we wanted a more conductive material, we would just use a metal) but rather what happens when we put an n-type semiconductor next to a p-type semiconductor. The extra electrons and holes near the interface between these two different materials quickly recombine, but the chemical impurities, which also have an electrical charge, remain behind. The positively charged impurities in the n-type region and the negatively charged impurities in the p-type region create an electric field, just as exists between positive and negative charges in space. This electric field points in one direction. If I try to pass a current through the interface between the n-type and p-type semiconductors, it will move very easily in the direction of the field, and it will be very tough going opposing the field. Such a simple device is called a "diode" in the dark, and a "solar cell" when you shine light on it. When the p–n junction absorbs light, the light-induced electrons and holes create a current, even without being connected to a battery. The charges are pushed by the internal electric field just as surely as if the device were connected to an external voltage source. A solar cell, therefore, can generate an electrical current through the combination of the light-induced extra electrons and holes with the internal electric field left behind by the charged impurities. This is one of the very few ways to generate electricity that does not involve moving a wire through a magnetic field, and thus no fossil fuels need be consumed for this device to work.

A transistor takes the directionality of the electrical current of a diode and makes the internal electric field changeable. By doing this, the transistor can be viewed as a special type of valve, where an input signal determines how far the valve is opened, which in turn leads to either a large or small current flowing through the device. Returning to the water-flow analogy for electrical current from Chapter 17, a fire hose is attached to the city water supply and, as the valve connecting the hose to the faucet is opened, water flows through the hose. If the valve is barely cracked open, the flow will be very weak, and as the valve is opened wider and wider, the quantity of water exiting the hose increases. Usually I have to manually turn the handle of the valve to effect a change. Now imagine a valve that is connected to a second, smaller garden hose that brings in a small stream of water. How much or how little the valve is opened will depend on how much water the second hose brings to the valve. If I considered the water flow in the garden hose as my "signal," then the resulting water flow out of the main fire hose is an amplified version of this signal.

In this way, a small voltage can be magnified without changing any of the time-dependent information encoded within it. When Iron Man needs to increase the current to his magnetic turbo-insulator a thousandfold, or amplify the current going to the servo-motors that drive the punching force of his suit, he uses transistors to take small input currents and increase their amplitude. Despite what Tony Stark would tell you, transistors don't actually provide power, but they do enable the amplification of a small signal, increasing it many times. To do so they need a large reservoir of electrical charge, such as an external battery, just as in the water analogy the "transistor valve" would not amplify the weak input from the garden hose unless the output from the fire hose was connected to the city water supply. Consequently, rather than providing power, transistors actually use power, but the rate at which they use power in order to amplify a weak signal is much less than the old amplification technique (vacuum tubes) they replaced. This is why Iron Man would be in desperate need of a recharge after a taxing battle. Tony would frequently gasp that his transistors needed to recharge, but I'm sure that he actually meant to refer to the battery supply to his transistors. Such a slip of the tongue is forgivable—I'm sure I'd misspeak after going several rounds with the Titanium Man.

Before transistors, the amplification of a weak input current was performed by heated wires and grids that guided the motion of electrons across space. A current was run through a filament wire until it glowed white-hot, and electrons were ejected from the metal and accelerated by a positive voltage applied to a plate some distance away, pulling these free electrons toward it. Between the filament and the collector plate is a grid (that is, a screen) that can act like a valve. If the input signal is applied to this grid, it would modulate the collected current, opening and closing the valve as in the water analogy. In order to avoid collisions with air molecules that would scatter the electron beam away from the collector electrode, these wires and grids were enclosed in a glass cylinder from which nearly all the air had been removed. These so-called vacuum tubes were large, used a great deal of power to heat the wires and run the collector plate, took a while to warm up when initially started, and were very fragile. A vacuum tube–powered Iron Man would hardly be invincible, as the sound of breaking glass would accompany his first and only adventure. Semiconductor-based transistors are small, low-power devices that are instantly available to amplify current and are compact and rugged. Even so, it took years before the transistor, invented in 1947, replaced the vacuum tube in most electronic devices.

One does not accidentally discover the transistor device, but must carefully construct a semiconductor structure with high purity and low defect density so that the amplification process can be observed. The hard work and innovative experimental techniques that enabled John Bardeen, Walter Brittain, and William Shlockley at Bell Laboratories in Murray Hill, New Jersey, to construct the world's first transistor, led to their winning the Nobel Prize in physics in 1956 for this work. On the day Bardeen learned that he had been awarded his second Nobel Prize (in 1972, for his development of a theory for superconductivity), his transistorized garage-door opener malfunctioned, underscoring the need for continuing research in solid-state physics.

As manufacturing and quality-control techniques improved, and newer and smaller designs for transistors became available, another important application for this special electron valve was realized. With a small input current applied to the transistor, a small output current results. A relatively modest increase in the input current creates in turn an amplified, larger current. The output of the transistor can be either a "low current" or a "high current"; a low current is termed a "zero," whereas if there is a high current, then this state is labeled a "one." Minor adjustments to the inputs to a transistor can create either a one or zero for the output current. By combining literally millions of transistors in clever configurations, and making use of a branch of mathematics called Boolean logic (developed by a mathematician named George Boole more than ninety years before the transistor was invented and seventy years before Schrödinger's equation was developed), one has the basic building block of a microcomputer.

A full discussion of how computers manipulate "ones" and "zeroes" to represent larger numbers and carry out mathematical operations through binary code would require another, separate book. The point I want to make here is that at the heart of all microcomputers and integrated circuits is the transistor. The "chips" that underlie the commercial and recreational electronics that play a larger and larger role in society, from cell phones to laptop computers to DVD players, are all simply platforms for the clever arrangement of, and connections between, a large number of transistors. The computerized and wireless technology that surrounds us in the twenty-first century would not be possible without the transistor, which in turn could not have been invented without the insights previously gained by the pioneers of quantum physics and electromagnetism.

Schrödinger was not trying to develop a CD player, or even replace the vacuum tube, when he developed his famous equation, but without his and others' investigations into the properties of matter, the modern lifestyle we enjoy today would not be possible.

All of our lives would be very different if not for the efforts of a relatively small handful of physicists studying the behavior of the natural world. With few exceptions, these scientists were driven not by a desire to create commercial devices and practical applications, but by their curiosity, which led them, as Dr. Henry Pym put it in *Tales to Astonish # 27*, to "work only on things that appeal to [their] imagination."

THE COSTUMES ARE SUPER, TOO—
MATERIALS SCIENCE

THROUGHOUT OUR DISCUSSIONS of physics principles, we have been fairly liberal in dispensing "one-time miracle exceptions" to the laws of nature in order to account for a wide variety of superpowers—such as running at superspeed, altering temperature gradients in the atmosphere, or shrinking down to the size of an atom. Now we must take up one final suspension of disbelief, without which no superhero can reliably and safely use their powers. I speak, of course, about the miracle of superhero costumes!

Consider the Avengers and the Justice League of America, assemblies of superheroes in the Marvel and DC universes respectively. While the Avengers' membership has evolved over time, typical members include Captain America, Hawkeye (the Green Arrow of Marvel comics), Giant-Man, the Wasp, Iron Man and the mighty Thor. Similarly, the roll call of the Justice League was not fixed, but often listed as members Superman, Batman, the Flash, Green Lantern, Green Arrow (the Hawkeye of the DC universe), Aquaman, the Atom, and Wonder Woman. Before the Avengers can save the world from Kang the Conqueror, the Masters of Evil, or Ultron, or before the JLA can stop such foes as Starro, Despero, or Amazo, these heroes first have to dress for success. One can't go around avenging or forming a league for justice if one's costume is a shredded pile of rags or a loose pile on the floor after one has changed shape and size. After all, the Comics Code Authority frowns on embarrassing "wardrobe malfunctions."*

Fortunately, the question of the composition of superhero costumes that enables one to employ amazing powers and abilities— without risking an "adult warning label" on the cover of the comic book—was addressed in early issues of the first family of Marvel comics—the Fantastic Four. Recall that in the Fantastic Four, the Human Torch has a uniform that remains undamaged when he bursts into flame; the Invisible Woman's suit turns transparent when she does; Mr. Fantastic has a jumpsuit that can stretch like rubber and return to its original shape when he resumes his normal form; and Ben Grimm, the orange, rocky, superstrong Thing, has little blue shorts.

When the Fantastic Four made their comic-book debut in November 1961, they initially wore plain civilian clothes, as they needed to fly under the radar in both the comic-book and real worlds. The quartet had to sneak onto a military base in order to surreptitiously fly a rocket ship Reed had designed in order to beat the Russians in the space race—a test flight that would lead to their exposure to superpower-granting cosmic rays. Similarly, Marvel Comics kept a low profile on the costumed-hero front when the Fantastic Four first appeared. They depended at the time upon their competitor, National Periodicals (home of Superman, Batman, and the Justice League of America) for newsstand distribution. Marvel had, up till then, published Western, young-teen hijinks, and giant-monster comics. In order to not alert National that Marvel was encroaching on their superhero turf, the Fantastic Four fought the Mole Man and alien Skrull invaders in their street clothes in their first two issues. By issue # 3 they had donned blue jumpsuits that effectively became their superhero costumes. The fashion designer to the superpowered in *The Incredibles*, Edna Mode, would probably disdain the Fantastic Four's "hobo suits,"

---

* Yes, I know about how the Wasp distracted the Hulk in the Ultimate Marvel version of the Avengers (*Ultimates* # 5). And the issue that featured that particular "wardrobe malfunction" was not approved by the Comics Code Authority. And no, we won't be discussing the structural stability of strapless superheroine bodices or Black Canary's or Zantanna fishnet stockings.

but their outfits are able to mimic the FF's powers, enabling them to use their special powers without wear and tear on their clothing. How do their uniforms do it? As described in *Fantastic Four # 7*, they are composed of one of Reed Richards' miracle inventions: "unstable molecules." No doubt Reed has shared the formula for such an amazing fabric with his Avengers comrades.

Now, any veteran of a high-school chemistry lab class knows that unstable molecules do indeed exist. These are the molecules that that fall apart or explode—precisely because they are unstable! But is it physically possible for real clothing to alter its thermal and structural capabilities in response to the wearer's needs? The answer is: yes! Shape-memory materials, technically known as thermoresponsive materials, "remember" their original configuration, so that after being bent or deformed, they can return to their original shape. These materials undergo a phase transition with changing temperature, pressure, or applied electric field (as in Gotham City's crusader's cape in the film *Batman Begins*), enabling applications from the familiar (shrink-wrap) to the exotic (surgical knots that self-tighten). Unlike melting or boiling phase transitions, these materials undergo transformations between differing crystal structures.

* * *

Before we consider designing clothing using shape-memory materials, we should ask a more basic question: What is it that determines an object's "shape?" The properties of any solid are governed by two factors: its chemical composition and the arrangement of its atoms.

Consider carbon atoms, the super-flexible Mr. Fantastic or Elongated Man* of the periodic table of the elements. Unlike most atoms, which rarely deviate from the preferred number of chemical bonds they will form with other atoms, carbon is particularly malleable in the number and type of chemical bonds it can assume. As discussed in Chapter 21, chemical bonds between atoms in molecules or solids result only when the electronic

---

* A DC hero who had the same super-elasticity powers as Reed Richards—who in turn had the same powers as the Quality Comics hero of the 1940s—Plastic Man.

waves in the individual atoms are such that, when they overlap, "harmonics" of a sort arise that lower the total energy of the resulting bond, compared with the two separate atoms. When the bound atoms are in a lower energy configuration, we must supply external energy, such as heat or light, in order to separate them. When breaking apart a molecule, this energy is called the "binding energy." For phase transitions, such as when liquid turns to a vapor, the heat we must provide to effect the transformation is termed the "heat of vaporization," while before a solid melts we must add a "heat of fusion."

Carbon achieves the greatest lowering in energy when it forms four chemical bonds, but it can do so in many different ways. It can form two strong bonds with other carbon atoms in a long line, and other types of atoms at sharp angles to the linear sequence, as in long-chain polymers such as proteins or DNA. It can form three strong bonds with other carbon atoms in a single plane, forming hexagons of carbon atoms, as we saw in fig. 38 in Chapter 23. The carbon atoms then try to form a fourth bond with the atoms in the planes above and below them. As they can form a solid bond only with either the atoms above or below, but not both, the result is very strong bonds within each plane, and rather weak ones between the planes. The planes of carbon atoms then stack into thin sheets, one atop another, like phyllo dough in delicious baklava. In this case, we call the form of carbon "graphite," and the fact that this solid is mechanically soft, with the layers easy to peel apart, makes it ideal for pencil "lead." Whenever you write with a pencil, you are literally unraveling a carbon crystal, layer by layer.

Alternatively, the carbon could form four strong bonds with other atoms residing at the corners of a pyramid, with the original carbon atom at its center. If the central carbon atom binds with hydrogen atoms, then we call the resulting molecule methane gas. However, if the other atoms are also carbon atoms, the resulting solid is called "diamond." Both graphite and diamond are composed of pure carbon, but the opaque, electrically conducting and easily deformed nature of graphite results from the carbon atoms stacking in sheets, while rearranging the carbon atoms into a tetrahedral configuration produces the transparent, electrically insulating, and very hard and rigid structure of diamond.

* * *

In Chapter 15, we described how the phase of a material, whether gas, liquid, or solid, depends on the temperature and pressure. Temperature measures the average energy per atom and determines whether the atoms have enough kinetic energy to unbind into the vapor phase, for example. By squeezing on the material—that is, by increasing the pressure—we may make it harder for the atoms to vaporize, and thus a higher temperature is required to execute the phase transition. There is a balance between energy and entropy that determines at what temperature and pressure a particular material will undergo a phase transition.

Under great pressure and elevated temperatures, graphite can be transformed into diamond, as when Superman wishes to give an engagement ring to Lois Lane. Less extreme efforts are required to transform shape-memory materials from one configuration to another. Here, the phase transition involves the material shifting from one particular arrangement of atoms (called a crystalline configuration) to another. The number of atoms in the material does not change, but if the material is distorted from its previous configuration, it may require additional energy to allow the atoms to move back to their previous arrangement.

Liquid crystals are a familiar example of materials that undergo a phase transition from one crystalline configuration to another, accompanied by a variation in structural and optical properties. Liquid crystals are actually long-chain carbon molecules. Electric forces between molecules can induce them to all line up in one direction, or to self-organize into layers. The forces that lead to this ordering are weak enough that the system still flows like a liquid. The application of an external electric field can shift certain liquid crystals from one ordered configuration to another. The ability of the liquid crystal to reflect light can change dramatically depending on the arrangement of the long-chain molecules. That is, in one phase, the liquid crystal may reflect most of the light that hits it (so it will look bright), while in another phase, it will absorb light, appearing dark. Behind every liquid-crystal pixel on a flat-screen television or computer monitor are electronics, including thin film transistors, that generate a varying electric field to change the optical properties of the pixel. The procedure by which alterations on the optical properties of the pixels are translated into a moving image is the same as described in Chapter 20.

Shape-memory materials are not always long chains of molecules, nor are they necessarily composed of carbon. Rather than inducing a phase transition with an electric field and temperature as in the case of liquid crystals, shape-memory materials undergo a transformation from one crystal structure to another under stress or pressure. Flexon or Nitinol are the commercial names of nickel-titanium alloys that exhibit high elasticity and memory aspects, making them useful for eyeglass frames and other applications. Nitinol was discovered accidentally in 1961 (the same year the Fantastic Four burst upon the scene) when a research-group leader at the Naval Ordinance Laboratory passed a bent sample of a newly synthesized alloy around for inspection to the scientists at a laboratory management meeting. One of the researchers decided to warm up the twisted metal using his pipe lighter (if there's one thing I've learned from reading comic books, it's that all scientists smoke pipes, or at least they did back in the Silver Age). All were surprised to see that upon warming, the bent alloy snapped back to its original configuration, and Nitinol (the "nol" in Nitinol derives from the initials of the Naval Ordinance Lab, while the "Ni" and "ti" standing for nickel and titanium) was launched. Bending the wire changed the alloy's crystal structure, and the application of thermal energy from the pipe lighter enabled the metal to return to the lower-energy (more stable) crystal structure and regain its original shape. Varying the ratio of nickel and titanium in the alloy can alter the temperature at which the phase transition occurs. Nitinol wires are also employed in certain orthodontic applications: The wire is bent to fit inside a patient's jaw and clamped in place. When warmed by body temperature, the wire exerts a steady pressure as it tries to return to its original shape, thus aligning the teeth.

Flexible shape-memory polymers have been around longer than their metallic counterparts. A familiar example is shrink-wrap, which undergoes a structural change upon warming. Here, certain chemical groups that cross-link and connect the long-chain molecules can soften upon an external trigger, such as an electric field or temperature, allowing the molecules to reorient themselves into a lower-energy configuration. In 2002, Andreas Lendlein and

Robert Langer reported in *Science* their discovery of a biodegradable shape-memory thermoplastic polymer for surgical applications. When formed into a loose knot, threads of this material self-tighten when warmed to 104 degrees Fahrenheit, enabling endoscopic surgery with smaller incisions. Shape-memory alloys can be used in stents, catheters, and probes, following the narrow passageways in the body as easily as the Elongated Man would be able to follow these twists and turns.

While the functional jumpsuits and uniforms of the Fantastic Four and Avengers remain confined to the four-color pages of comic books, flexible shape-memory materials have recently been incorporated into real-world clothing. Certain fabrics expand in response to a lowering of temperature, so that when used as the inner layer in a winter jacket, they automatically increase the air gap, thereby improving thermal insulation. Other materials become more porous at higher temperatures, allowing body heat and water vapor to escape. Polymer-based fabrics have been developed that can be stretched to more than twice their normal length, and yet return to their original proportions when warmed. These materials may provide the answer to an age-old mystery: What holds up the Hulk's pants?

The Hulk was a founding member of the Avengers, though he wore purple shorts during his brief tenure with the team. More typically, he battles the U.S. Army, the Abomination, or the Leader while wearing only a pair of purple pants. When nuclear scientist Robert Bruce Banner was belted by gamma rays, he gained the ability to transform from a puny nuclear scientist into an eight-foot-tall, two-thousand-pound Jade Giant. This metamorphosis rends all of Banner's clothing—his shirt is shredded, his feet tear his shoes and socks apart, and the cuffs of his pants are frayed following his transformation to the Hulk, but his vast waistband remains intact. Presumably, his lavender Levi's are composed of Reed Richard's unstable molecules in the comic-book world, or shape-memory fabrics in ours.

The Hulk first appeared in his own Marvel comic in 1962, following the success of the *Fantastic Four*. Unfortunately, the Hulk might be the strongest there is, but his sales weren't, and, his comic was canceled after only six issues. You can't keep the Jade Giant down, and he returned a year and a half later in *Tales to Astonish # 60*. One can only imagine what the sales would have been if the Hulk's pants had not managed to survive his many transformations. But here ol' Greenskin was stymied by a force even he could not smash, stronger than gamma radiation: the Comics Code Authority!

Unstable molecules may do the job in the Marvel universe, but what keeps the Justice League of America's uniforms intact in DC comics? A variety of explanations have been provided for the durability of superhero costumes in the DC universe, all with a slight patina of scientific justification. Back in the Silver Age, it was explained that Superman's costume was composed of the same fabric that the baby Kal-El was swaddled with, when his birth parents sent him off in a rocket ship to Earth moments before Krypton's destruction. The fabric's extraterrestrial origin accounted for its indestructibility. In later years, it was posited that a thin aura of invulnerability extended from the Man of Steel's body, which is why his suit could survive devastating forces or being in the heart of the sun, but his cape would not escape damage. This same protective aura also accounted for the Flash's costume being able to resist shredding when the Scarlet Speedster ran at his top velocity.

Batman's costume would display the wear and tear that one might expect would result from fighting crime in Gotham City. Similarly, Spider-Man, being more of a loner in the Marvel universe (at least when he first appeared in the 1960s) was apparently not on the "unstable-molecule distribution list," and he would regularly also have to deal with a torn and damaged costume (not having Batman's advantage of a millionaire alter ego and a live-in butler, this was more of a burden on Peter Parker, living with his aged aunt in Queens, NY).

The DC hero the Atom had a unique solution to the problem of needing a costume that accommodated his shrinking power. As will be explained in detail later, the Atom's ability to change his size and weight from his normal six feet and 180 pounds down to essentially zero feet and zero pounds stemmed from his mastering miniaturization technology that employed a remnant of a white-dwarf star. This exotic material was fashioned into the Atom's blue and red costume. The miniaturization mechanism that enables the Atom to fight crime at six inches or six nanometers is built into his costume—consequently, his superhero suit is covered under the same miracle exception that accounts for the Atom's powers. In fact, the Atom never has to worry about changing out of his street clothes because his costume only appears when he is reduced in size. When he resumes his normal height, his costume stays miniature, and his street clothes grow with the rest of him. Consequently, the Atom is the only super-hero who deliberately gives himself a wedgie whenever he returns to his secret identity!*

Now that we have addressed the materials science underlying their uniforms, we come to the two main rules of superhero cos-tumes. Rule One: No capes! Rule Two: Accessorize! We next turn to the physics issues concerning some of the more famous crime-fighting accoutrements.

DOOM OF THE STAR DIAMOND

Two years after the Caped Crusader made his debut in *Detective* # 27 in 1939, another millionaire playboy took up a costumed identity, also fighting criminals without the benefit of superpow-ers and armed only with his wits and a variety of technological weapons of his own design. Green Arrow was a fairly transparent attempt to replicate Batman's earlier success. Millionaire Bruce Wayne would dress like a giant bat and, together with his teen sidekick Robin, drive from the Batcave in the Batmobile to Gotham City, where he would use the high-tech wonders in his utility belt to fight common criminals and costumed villains. Mil-lionaire Oliver Queen, on the other hand, would dress in a modi-fied Robin Hood costume and, together with his teen sidekick Speedy, drive from the Arrow Cave in the Arrow Car to Star City, where he would use a quiver full of "trick arrows" to fight com-mon criminals and costumed villains. Green Arrow's technologi-cal wonders included a boomerang arrow, an exploding arrow, a handcuff arrow, a boxing-glove arrow, an arrow that projected a net near a target, an acetylene-torch arrow, and, particularly useful for underwater adventures, an aqua-lung arrow. I understand why you might need to bring a small breathing tank along with you on

---

* Actually, the costume becomes "invisible and intangible" at his normal size—but is visible at any other size. Apparently, there is much physicists still have to learn about the properties of white-dwarf matter.

an underwater case, but what possible benefit you would derive from mounting it on the head of an arrow, I cannot begin to imagine.

Making his first appearance in the same issue of *More Fun Comics* that featured the debut of Aquaman, Green Arrow was created by writer Mort Weisinger (who also co-created the Aquatic Ace) and drawn by George Papp. Both Green Arrow and Aquaman would continue to be published through the transition period between the Golden Age of comics and the Silver Age, appearing as backup features in issues of Superboy comics. While Aquaman was one of the original set of heroes tapped for membership in the Justice League of America in 1960's *The Brave and The Bold # 28*, Green Arrow would not be offered membership until a year later, when he managed to save Wonder Woman, Aquaman, Flash, Green Lantern, and the Martian Manhunter from the "Doom of the Star Diamond" in *Justice League of America # 4*. As shown in fig. 41, the JLA were trapped within a large diamond prison, and were freed only when Green Arrow struck the crystal cage with a diamond-tipped arrow, fired with sufficient force at the only stress point of the diamond prison. The conclusion of Green Arrow's first adventure with the Justice League demonstrates some real materials science, related to crystalline defects, elastic-strain energy of bows, and the centuries-long history of trick arrows.

Let's first consider the diamond trap holding the Justice League. Diamonds are hard to break not because they are dense, but because the carbon atoms in the crystal are connected via very stiff covalent bonds. The strength of a material is primarily governed by the nature of the chemical bonds holding the atoms together in the solid phase. The strongest chemical bonds are termed "covalent bonds," where the individual atoms quantum mechanically share their outermost electrons with their atomic neighbors. In order to break these bonds, one must remove the electrons from all of the bonds connecting an atom to all of its neighbors. How the carbon atoms are configured relative to their neighbors will determine the strength of the bonds. It turns out that the overlap of the electronic matter wavefunctions that occurs when the carbon atoms all lie in a plane results in stronger bonds between the carbon atoms than those in a diamond. That is, the carbon planes that peel off on a sheet of paper when writing

Fig. 41. *Green Arrow, about to conduct an empirical test of stress fracture in covalently bonded solids, in* Justice League of America # 4 *(May, 1961).*

with a pencil have stronger bonds than those in a diamond crystal. Flaws or imperfections in a diamond occur at regions in the solid where atomic bonds, through strains or the inclusion of impurities, are broken or weakened. It is at these locations where the covalent bonding network is easiest to break, either by a jeweler cutting a diamond or by Green Arrow with his diamond-tipped arrow.

The construction of an archery bow also involves considerable materials science, as a premium is placed on substances that can be bent without breaking. A bow is essentially a spring, and the stored potential energy in the bow is converted into kinetic energy in the arrow. Recall from Chapter 12, and our discussion of the Principle of Conservation of Energy, that the definition of Work in physics is when an external force is applied across a given distance. Pulling back on the bowstring does not change the length of the string—the force supplied by the archer results in the flexing of the bow. The ideal bow will store the most potential energy per mass—that is, one wants a lightweight bow that is nevertheless strong and elastic. In materials physics, an object is termed "elastic" if, following compression or stretching, it returns to its original configuration. If the force results in a permanent alteration, then this is called a "plastic" deformation.

As it relates to Green Arrow, the larger the force necessary to reach the elastic/plastic transition, the more force he can apply without fear of breaking his bow. The more potential energy that can be stored in the bow, the more kinetic energy can be transferred to the arrow. In turn, the faster the arrow leaves the bowstring, the greater the distance it will fly through the air. With a superior bow, one could stand farther from the target and still score a bull's-eye effectively.

It has been said that the British longbow archers of the Middle Ages were the finest in the world. In the battle of Agincourt in 1415, despite having a nearly ten-to-one advantage in troop size, fifty thousand French soldiers were defeated by six thousand British soldiers, of whom roughly 4,800 were longbow archers. Nearly none of the French troops were archers, demonstrating the advantage that superior technology provides in warfare. While the British archers were no doubt accomplished, it was their longbows made of yew wood that secured their victory. Of all the woods available for constructing a bow, that of the yew tree has the greatest strength-to-weight ratio.* A cross-section of a yew tree finds an outer ring of softer sapwood, which is highly elastic, and an inner core of harder heartwood. By forming a bow from the interface of these two regions, the inner layer of heartwood resists compression upon bending, while the outer layer of sapwood's elasticity enables the bow to quickly spring back to its original configuration when released. Of course, any knots or imperfections in the wood would act the same as defect points in a crystal, where fracture would be likely to occur. Consequently, it was difficult to extract more than one functional longbow from a single yew tree. Their superior stress/strain properties ensured that this type of lumber commanded a high price by the British monarchy—with the effect that by 1600, there was massive deforestation of yew trees throughout all of Britain and Europe. Recently, yew trees and shrubs have returned to Europe and the United States, and this time they are desired for their lifesaving properties. The bark of Pacific yew trees was, up until recent advances in synthetic fabrication, the primary source of the anticancer drug Taxol (generic name paclitaxel), and once again, groves of these trees were highly valued.

Modern science and materials research have led to vast improvements in archery technology. The design of the bow itself has been subject to innovation, most notably in the 1960s, when Holless Allen invented the compound bow. By adding pulleys at the ends of the bow, a mechanical advantage is provided that increases the applied force provided by the archer pulling back on the bowstring and thereby increases the stored potential energy. The composition of the bow has also evolved from the sixteenth century. The strength-to-weight ratio of yew wood has been eclipsed by carbon fiber–reinforced polymers, which combine the high bonding strength of graphitic carbon-carbon bonds with the low-bulk density of plastics. Graphite filaments are braided into fibers, and once aligned, are epoxied into a particular form. These

---

* It has been suggested that the nutrient-conducting cells in the yew tree have a spiral thickening that acts like tiny springs. Despite being the wood of choice for archery applications for more than a thousand years, the precise origin of the superior elastomechanical properties of yew wood remains poorly understood, and is the subject of current research.

composite materials are finding applications not just in archery, but in other sporting equipment, in high-performance race cars, in helicopter blades, and even in bridge supports.

The only thing stronger and lighter than a composite of graphite fibers is a single filament of carbon, called a carbon nanotube. These materials consist of a single sheet of graphite rolled up into a hollow cylinder only three atoms in diameter and a single atom thick. Since the strongest covalent bonds are between carbon atoms in the planes of graphite, these carbon nanotubes, if synthesized free of defects, can have a strength-to-weight ratio two hundred times greater than steel, and twenty times stronger than spider silk. If these nanometer filaments can be synthesized in macroscopic lengths, a carbon nanotube cable only one millimeter in diameter could support a weight of nearly fourteen thousand pounds—a fiber no wider than the period at the end of this sentence could hold up two average-size sport-utility vehicles. An indication of how strong carbon nanotubes are can be found in Tony Stark's investigations of an alternate universe's Hulk. In an experiment designed to stress test the Hulk, Stark's researchers are stunned to discover the source of Hulk's invulnerability—there are carbon nanotubes in his skin!

One can easily imagine replacing a standard metal or flint arrowhead with one comprised of a large, single-crystal diamond, as employed by Green Arrow in *Justice League of America # 4*. Adding a small high-intensity siren to an arrowhead, producing a sonic offensive capability, is also plausible. One could incorporate an altitude monitor into the arrowhead, to indicate when the projectile is at the highest point in its trajectory, which could in turn be used to determine when a net could be ejected from the device. Less reasonable are some of the more exotic arrows Green Arrow would use to fight criminals, such as a mummy arrow (which encircled its target in a tight-fitting cloth wrapping) or handcuff arrows, the aerodynamics of which are dubious at best. But the concept of "trick arrows" is older than Green Arrow comics—by more than one thousand years! Flaming arrows and "Greek Fire" played a major role in maintaining the hegemony of the Byzantine Empire. While the exact chemical composition is still being debated, tipping the arrowheads in a mixture of petroleum and bitumen (sulfur) compounds led to an incendiary projectile that, when ignited, was difficult to extinguish and would even burn when submerged in water.

Compared with such lethal "trick arrows," a boxing-glove arrow seems rather tame.

BULLETS AND BRACELETS

In Chapter 22 we speculated that our discussion of quantum physics, coupled with a thorough knowledge of superhero comic books, would provide you, Fearless Reader, with enhanced romantic charms. Indeed, William Moulton Marston, the creator of a founding member of the Justice League of America, was, like Tony Stark and Erwin Schrödinger, a "ladies man" whose personal life would definitely not win approval by the Comics Code Authority.

In 1921, Marston had a B.A., an L.L.B. (an undergraduate law degree), and a Ph.D. is psychology, at a time when only approximately 3.3 percent of the American population had a college degree or higher. In 1917, he published a paper positing a correlation between a person's systolic blood pressure and attempts to deceive, which formed the basis of his later claim to being "the father of the lie detector." Teaching positions at American University and Tufts were followed by a year, in 1929, as the director of public services at Universal Studios in Hollywood. While he had taught at a variety of institutions, he had never obtained tenure, and after 1929, his academic publishing seems to have stopped. At this point, he set upon developing a career as a public psychologist, first in Hollywood, followed by writing a series of books, some technical while most containing popular psychology, and eventually becoming a consulting psychologist for *Family Circle* magazine.

It was in an interview in *Family Circle* in 1940, written by Olive Richard, entitled "Don't Laugh at the Comics," where Marston defended comic books as a mechanism for education, rather than dismissing them as cheap, mind-destroying juvenile entertainment. As described in Les Daniels' *Wonder Woman: The Complete History*, this article attracted the attention of M. C. Gaines and Sheldon Mayer, who offered Marston a position on the editorial advisory board of DC and All-American comics. Marston would then go on, writing under the pen name Charles Moulton, to pitch a new comic-book character that would serve to illustrate his personal convictions concerning the moral and psychological superiority of woman. In the December 1941–January 1942 issue of *All-Star Comics # 8*, Wonder Woman, was introduced, and her Amazon strength, bullet-deflecting bracelets, and golden lasso of truth (a personal and portable lie detector) were pressed into service in the four-color fight for justice.

Because Wonder Woman's superpowers were magic-based (as opposed to being derived from much more scientific means, such as light from our yellow sun, or being bitten by a radioactive spider), there isn't too much physics that relates to her exploits, with one notable exception. In her very first adventure, when she competed in Amazonian versions of track-and-field events for the right to accompany American airman Steve Trevor back to the States after his plane had crash-landed on the secret Paradise Island, the final challenge involved the potentially deadly "sport" of Bullets and Bracelets. Two women faced each other and fired pistols at their opponents. The goal of the contest was to deflect the bullets with their wrist bracelets, which were composed of the unique metal Amazonium, worn to remind the Amazons of their previous years in the captivity of men. In this "ultimate test of speed of eye and movement" (as described in 1942's *Wonder Woman # 1*), Wonder Woman's bracelets become "streaks of silver flashes of streaking light as they parry the death thrusts of the hurtling bullets." Well, assuming that Wonder Woman does indeed have the speed of Mercury and reflexes fast enough to deflect a bullet, what must the composition of her Amazonian bracelets be to withstand this barrage?

First, we must ascertain the maximum force that the bracelets must withstand. This involves a calculation of the force that the bullet exerts on the metal band when deflected by Wonder Woman. We will use the same formula from Chapter 3 that we employed to determine how much force Spider-Man's webbing must exert to stop the falling Gwen Stacy—that is, Force multiplied by time is equal to the change of momentum. In order to deflect a bullet with a mass of 20 grams traveling with a muzzle speed of 1,000 feet/sec, with a time of collision of one millisecond (a thousandth of a second), a force of 2,700 pounds is required. Given that the area of the bullet is so small, this corresponds to a pressure of seventy thousand pounds per square inch, which is more than 4,600 times greater than atmospheric pressure. What type of metal can resist this pressure without suffering plastic deformation? Nearly all of them! A typical high-strength steel alloy can easily withstand pressures from seventy-five thousand psi up to one million psi. Wonder Woman's bracelets appear to be roughly half a centimeter thick, which should be enough to deflect a speeding bullet. Amazonium seems to be nothing more exotic than cold-rolled steel.

The next basic question is if metals are so strong, why can they so easily be drawn out into long wires or fashioned into decorative jewelry? How can the atoms be so loosely bound that the resulting solid is easily manipulated, yet not fall apart under the slightest disturbance? The answer lies in electrostatics.

In the previous chapter, we introduced the metaphor of the electronic states in a solid being like the seats in the orchestra of a theater, with the empty seats in the balcony representing a band of excited states at higher energy. Metals were described as like an orchestra in which only half the "seats" are occupied with electrons, so that a current could always be induced, because there was no energy barrier separating the highest occupied state and the lowest empty state. For metals, the seats actually correspond to matter-waves that can extend over the entire solid (to explain why this is so would be too much of a digression). This is why metals make such good conductors of electricity and heat—it takes very little energy to move these "free electrons" in response to an external stimulus. In a "covalently bonded" solid like diamond, each seat in the orchestra represents a rigid, directional bond between the atoms.

If there are no direct bonds holding the metal atoms together, what keeps the metallic solid from falling apart? Electricity! Each metal atom initially is electrically neutral, with as many positively charged protons in its nucleus as there are negatively charged electrons in probability clouds surrounding it. Remove an electron from each atom to roam over the solid, and the resulting original atoms are now positively charged. Try to compress the metal, and you are pushing the atoms closer together, and the positively charged atoms (called ions) repel each other, resisting the external pressure. This makes it strong. The lack of directional rigid bonds between atoms in a metal is why they are easy to deform, to pull out into long, thin wires or to flatten into sheets. The forces holding the metal together are not as strong as in most covalently bonded solids—we can easily bend and break a metal paper clip without superpowers!—but they are much stronger than one would expect, given how malleable most metals are.

There is, obviously, a fair amount of chemistry and materials science that determines the strength of a particular metal. For example, consider Adamantium, the strongest metal in the Marvel universe, which is bonded onto the X-Man Wolverine's skeleton and claws. This miracle metal is strong not because it is dense, but apparently because it is a defect-free covalently bonded solid. It takes more energy to remove a covalently bonded atom from all of its neighbors than just localizing a free electron onto an ion in a normal metal. Adamantium must somehow combine the electrical properties of metals with the strong covalent bonds found in diamonds, while avoiding having any network-weakening defects, making the resulting material "unbreakable."

Defects weaken covalently bonded solids, but can actually make certain metals stronger.* One way to improve the strength of steel is termed "work-hardening," in which the metal is intentionally plastically deformed. This generates defects on the atomic scale within the metal. As there are no directional bonds in a metal, it is normally easy to move planes of atoms past each other when bending or drawing the material into a wire. Defects can block this atomic motion, and too many defects essentially form atomic-scale traffic jams that inhibit further deformations of the metal. This increase in strength comes at the expense of ductility. Cold-rolled steel has a tensile strength comparable to titanium alloys, but both are flimsy compared with covalently bonded carbon nanotubes.

While we're on the subject, we should address a common misconception regarding the chemical composition of Captain America's shield. Wolverine's claws are composed of pure Adamantium, but Captain America's shield is a one-of-a-kind alloy of steel and Vibranium. The steel is needed to provide rigidity, so that the

---

* And not just metals! Hold a standard 8½ × 11-inch sheet of paper at one of its narrower ends. The paper will sag under its own weight. Crumple the sheet into a tight ball, and then smooth the paper out. Holding the sheet at the same end, it will be much more rigid with all of its new wrinkles, and will be better able to resist deflecting due to gravity!

shield can ricochet off walls and supervillain minions. Vibranium is an extraterrestrial material brought to Earth when a meteorite crashed in the African nation of Wakanda, which is ruled by the superhero the Black Panther. Vibranium has the ability to absorb any and all sound and convert the energy in the sound wave into some other, not-well-specified form, making it the perfect shock absorber, a quality strongly desired in a shield. Sound waves are alternations in pressure or density, and in a solid, sound is transmitted through the vibrations of the atoms. Vibranium possibly converts the atomic vibrations from an absorbed sound wave into an optical transition (which would explain why sometimes a glow is observed when Vibranium is used), thereby conserving energy in the process.

The material that forms Captain America's shield was the result of a laboratory fluke, when a steel alloy and Vibranium were accidentally fused into an alloy. The conditions under which this metallurgical fusion occurred were not recorded, and this synthesis has never been repeated. In fact, attempts to re-create the metallic alloy used in Cap's shield led to the synthesis of Adamantium. Captain America's shield has sometimes been erroneously described as being an alloy of Vibranium and Adamantium, but this is physically impossible. When Captain America was discovered frozen alive in a block of ice by the Avengers (issue # 4, March 1964) he was found with his shield, the same disc he used for most of his superhero career in World War II. This was the same shield he used as a member of the Avengers as he fought along with and led these heroes. The same shield he had in *Avengers # 66*, when materials scientist extraordinaire Dr. Myron MacLain created Adamantium! Cap's shield can sometimes appear to defy physics, but it cannot travel backward through time.

That the shield does indeed contain a Vibranium component has been experimentally verified by no less an expert in metallurgy than Stephen Colbert. On his television show, *The Colbert Report*, Stephen was bequeathed Captain America's shield when the Sentinel of Liberty's last will and testament was read.* Rapping the shield with his knuckles and listening to the resulting "thunk," Colbert assured his audience that this was "the sound of

---

* Reports of Cap's death appear to be, fortunately, premature.

indestructible Wakandan Vibranium." He thereby resolved a long-standing mystery in physics: What sound would you hear if you struck a material that absorbed all vibrations!

Back to Wonder Woman, it turns out that her creator himself led a double life. Quite unusual for the 1930s, Marston lived with both his wife and Olive Richard, the same woman whose 1940 *Family Circle* interview with Marston led to his comic-book career. This ménage à trois lasted from the 1920s till his death in 1947. Marston's peripatetic academic career may have been a consequence of this unusual family life. The fact that at times the household's financial support was provided by the working women may have reinforced Marston's convictions concerning feminism and women's rights. Photos of Marston from this time period show a man with more than a passing resemblance to the *Daily Planet*'s Perry White. That he was able to live with two women during the depths of the Depression no doubt testifies to the irresistible allure that accompanies a mastery of both science and comic books!

While there is debate as to whether Olive or wife Elizabeth Holloway Martson served as the inspiration for Wonder Woman, there seems to be a clear paper trail crediting Olive with one of Wonder Woman's accessories. From Les Daniels' 2001 excellent history of the character, in another *Family Circle* interview less than a year after the Amazon Princess' first comic book appearance, Marston stated that the inspiration for Wonder Woman's bracelets came from Olive Richard's wrist jewelry!

DOES SIZE MATTER?

When the heroes of the Marvel universe banded together to fight Loki, the Norse god of mischief in *Avengers # 1*, Ant-Man was front and center. By *Avengers # 2*, Henry Pym had adopted a new superhero persona, as Giant-Man. When the first issue of the Avengers was published, over in *Tales to Astonish # 48*, Ant-Man was facing off against the supervillain the Porcupine. The Porcupine was Alex Gentry, an engineer who uses his technological prowess to develop a quill-coated suit that conceals a host of offensive and defensive weapons—such as tear gas, stun pellets, ammonia (presumably for clean escapes), "liquid fire" (I suspect he means a gas flamethrower), detector mines, liquid cement, and others—with which he embarks on an obligatory crime wave. Now, I must say that, as a physicist, I have worked with many engineers in my academic career, and hardly any of them dress like giant porcupines. Gentry had captured Ant-Man and placed him in a bathtub partially filled with water. Given that he cannot climb up the slick porcelain walls, Ant-Man is forced to endlessly tread water until he becomes exhausted, at which point he would drown.

In the very next issue of *Tales to Astonish # 49*, perhaps motivated by embarrassment over such pedestrian death traps as a partially filled bathtub, Henry Pym discovered a reverse version of his shrinking potion that enabled him to grow larger than his normal height of six feet. And so Giant-Man was born into the Marvel universe. Eventually Pym became a diminutive crime-fighter again, this time as the flamboyant Yellowjacket, but for the majority of his superhero career, Pym fought for justice as either the oversize Giant-Man or Goliath (same hero and same power of super-growth, just different code name and costume).

If one of the functions of comic books is wish-fulfillment for their young readers, then it must be recognized that not many kids fantasize about how cool it would be to be only a few millimeters high or even to dress like a porcupine. Now, to be able to grow to be twelve feet tall, that would be something. And yet, it turns out that being larger than normal carries with it a different, yet no less pressing, set of physical challenges. For one thing, as pointed out in *Ultimates #3* (a modern version of the Avengers, featuring a new Giant-Man), your now much-larger (super-dilated) pupils would let much more light into your eyes. You would therefore always need to wear special goggles while enlarged, to avoid overloading your optic nerves.

In addition, there is a fundamental limit to the size that someone could grow; assuming of course that one could grow far beyond his normal height in the first place, which requires just as much of a "miracle exception" as does miniaturization. This limitation is set by the strength of materials (particularly bone) and gravity. Gravity enters the situation because your mass will increase in proportion to your volume if you maintain a constant density. Density is mass divided by volume, so the bigger you are (that is, the larger your volume), the greater your mass if the ratio of the two (density) remains unchanged. You would be a less-than-imposing figure if your super-growth kept your mass fixed.

In fact, just such a situation faced Reed Richards of the Fantastic Four when he encountered the monstrous alien invader Gormuu in *Fantastic Four # 271.* This issue relates a story that took place before Richards and his three comrades had been exposed to the cosmic rays that would give them their superpowers. Gormuu, a twenty-foot-tall misshapen green creature, was a warrior invader from the planet Kraalo. This story was a nostalgic tip of the hat by writer and artist John Byrne to the Monster Invader from Space stories that dominated Marvel comics in the late 1950s until the Fantastic Four arrived to save both the universe and the financial fortunes of Stan Lee and Jack Kirby. The creatures in these comics, *Tales to Astonish, Strange Tales, Journey into Mystery,* and *Tales of Suspense,* were always at least as big as a house and all had names with adjacent double letters, such as Orrgo (the Unconquerable), Bruttu, Googam (son of Goom), and Fin Fang Foom. Gormuu's competitive advantage regarding world conquest was that his size increased whenever he was struck with any form of "broadcast energy." Richards, upon examining a ten-foot-long and inches-deep footprint left behind by the already-expanding alien, realizes that the only way to stop this menace is to continue to strike him with more and more broadcast power. Any creature large enough to leave a ten-foot-long footprint behind should also leave an impression several feet deep, if its mass were increasing at the same rate as its volume. Discovering that Gormuu's growth was at constant mass, not constant density, Richards feeds the alien so much energy that he grows larger than the Earth, and less dense than the background of space, until he becomes an insubstantial and decidedly unthreatening footnote in our nation's atomic-age history. Bearing in mind the cautionary tale of Gormuu, let us assume that Henry Pym, in his guise as Giant-Man, manages to maintain a constant density as he grows, so that his weight increases in uniform proportion to his volume.

In order to treat the Giant-Man situation mathematically, let's make a simplifying assumption—namely, that Henry Pym is a giant box. Of course, he is more naturally described as a large cylinder, but I want to keep the math as simple as possible. Well, if he's a box, then his volume is the product of his length, height, and width. A box that has a length of ten feet, a width of ten feet, and a height of ten feet will have a volume of 10 feet $\times$ 10 feet $\times$ 10 feet, or 1,000 feet cubed. This unit of volume is sometimes called "cubic feet" or feet^3, which represents the fact that we multiplied a length (feet) times another length times another length. Now assume that Giant-Man uses his Pym particles to grow twice as large in all directions. So his height is now 20 feet, and his width and length are also 20 feet each. In this case his volume is 20 feet $\times$ 20 feet $\times$ 20 feet— that is, 8,000 cubic feet. So, doubling his length in all three directions increases his volume by a factor of eight.

If Giant-Man is going to maintain a constant density as he grows, then his mass must increase at the same rate as does his volume, not his length! Doubling his height (as well as his width and breadth) requires that his weight goes up by a factor of eight to keep his density the same. Well, Hank Pym gets heavier the bigger he becomes—so what? The problem is that his weight increases faster than does the ability of his skeleton to carry his weight, so at a certain height, Giant-Man risks breaking his legs simply by standing up. The strength of an object, or its resistance to being deformed by a pull or crushed by a push, depends on how wide it is and not on its length. The technical way to say this is that an object's "tensile strength" is determined by its cross-sectional area.

Think of a fishing line rated for twenty pounds—that is, for holding up a twenty-pound fish. A heavier fish would snap the line when we tried to lift it onto the boat. If we want to keep a heavier fish suspended, changing the length of the fishing line will be absolutely no help whatsoever. To increase the strength of a fishing line, you don't increase its length, but rather, its diameter.* The wider the fishing line, the greater the area over which the force pulling it is distributed, and the less is the force applied across any given tiny element holding the line together. When a fishing line, or anything for that matter, breaks, the chemical bonds that were holding the material together snap and fly apart.

---

* Apple trees employ this principle as part of their seed-dispersal mechanism. When the fruit has matured and reached a sufficient mass, its weight exceeds the tensile strength of the narrow stem by which it is attached to the tree. When the stem snaps, the fruit falls to the ground, and following consumption by fauna, its seeds are distributed to other locations.

The larger the area that's available to support a given force, the less stress or strain is applied to any particular molecule, and the less likely it is that a catastrophic failure will occur. As discussed above, when breaks do happen, it is usually because a molecular imperfection or flaw magnifies the applied force locally, leaving the material weaker than if it were uniform and atomically perfect.

The dependence of a material's strength on its cross-sectional area also limits the distances over which Mr. Fantastic of the Fantastic Four can stretch. As explained in *Fantastic Four Annual # 1*, he cannot extend the length of his body beyond five hundred yards. A wooden two-by-four board (which has a rectangular cross-section of two inches by four inches) that is three feet long can be supported above the ground by a sawhorse at either end and remain parallel to the floor. A six-foot-long board of the same cross-sectional area would sag slightly in the middle, if still supported at the ends. A sixty-foot-long board would droop considerably, while a six-mile-long board would touch the ground at its center, even ignoring the curvature of the Earth. As Reed points out in *Fantastic Four Annual # 1*, "the further my body stretches, the weaker my muscles become, so that I cannot exert as much force stretched to a great distance as I can extending for a shorter distance." Reed Richards, whose understanding of the mass-volume relationship saved the Earth from the terror of Gormuu, is also a walking illustration of the cube-square law of tensile strength.

As our cubical Henry Pym grows into Giant-Man, his volume increases faster than his cross-sectional area. As Giant-Man grows larger, his bones naturally increase in size proportionally with the rest of his body. The strength of his femur or vertebrae grows by the square of his expansion rate. But the bigger that Giant-Man becomes, the more weight his bones must support—at constant density his weight increases by the cube of his growth factor. Suppose at his normal height, Dr. Pym is six feet tall and weighs 185 pounds. His femur at his normal height can support a weight of eighteen thousand pounds while a single vertebra can support eight hundred pounds—nature builds considerable redundancy into some crucial load-bearing structural components. At a height of sixty feet, Giant-Man's expansion factor is ten. His volume therefore grows by a factor of $10 \times 10 \times 10 = 1,000$ while the cross-sectional area of his bones increases only by a factor of $10 \times 10 = 100$. Henry Pym would now weigh 185,000 pounds, while his vertebrae could support only a weight of 80,000 pounds and his femur could carry 1.8 million pounds. At this height, his skeleton is not able to uniformly support his weight.

In order to keep Giant-Man from growing so tall that no villain, even a superpowered one, could pose a credible threat to him, Stan Lee argued in the 1960s that increasing his size placed a biological strain on Henry Pym. His optimal strength was at a height of roughly twelve feet, and if he grew to more than forty or fifty feet, he was as tall as a house but as weak as a kitten. Years later, the metabolic limitation was replaced by a physical one derived from the cube-square law. It is now recognized, as described, for example, in *Ultimates # 2*, that even if the metabolic issues involving a growth serum are resolved, gravity and physics will still provide stringent limits on your ultimate size.

The fact that volume grows faster than surface area is true even for non-cubical objects. The volume of a sphere is given by the mathematical expression of a constant $(4\pi/3)$ times the radius of the sphere cubed, or $(4\pi/3)\, r^3$, while its surface area is 4π times the radius squared, or $4\pi\, r^2$. (A volume will always have the units of a length cubed—such as cubic feet—while an area has the dimensions of a length squared, such that carpeting is measured in square yards.) Consequently, the rising bubbles in a vat of acid into which Batman and Robin are being slowly lowered provides the Caped Crusader with a textbook illustration of the physical principle of the cube-square law.

If you've ever thought that the bubbles in your beer rose faster as they neared the top of your glass, it wasn't the alcohol affecting your judgment. The bubbly drink is supersaturated with carbon dioxide (that's what also gives soda pop its fizz), which means that the pressure of the carbon dioxide gas forced into the liquid is greater than atmospheric pressure. When the top is popped on a carbonated beverage, there is a popping or hissing noise, due to some of the excess gas under pressure rapidly exiting the container. There is still additional carbon dioxide in the liquid, which collects into bubbles near small imperfections in the glass and then, being lighter than the surrounding fluid, rises to the top. The buoyancy force lifting the bubble is directly related to its spherical volume, which depends on the cube of the bubble's radius. The drag resistance force slowing down the bubble depends on its surface area (the bigger the area, the more beverage that has to be pushed out of the way of the rising bubble), which in turn grows by the square of the bubble's radius. As the bubble moves through the drink, it sweeps up additional carbon-dioxide molecules dispersed throughout the liquid, becoming bigger in the process. There is thus a net excess force raising the bubble. If there's a force, there is an acceleration (Newton's second law holds even inside a glass of beer) and the bubble will rise faster and faster.

If I had an infinitely tall glass, would the bubbles accelerate to the speed of light? No; in Chapter 4 we noted that the drag resistance depends not only on the surface area, but also on the speed of an object (it takes more effort to push the fluid out of the way of a rapidly moving object than one progressing slowly). As the bubble rises and moves faster and faster, the drag force increases and eventually balances out the upward lift, and once there is no net force, the bubble continues to move at a steady uniform speed (Newton's first law) known as the terminal (or final) velocity.

Now that this tour of the world of physics, through the agency of superheroes and their villainous foes, is completed, perhaps you should pause for some hands-on experimentation with a bubbly drink of your choice, in order to verify some of what we've have learned—but purely in the interest of science, of course!

SECTION
4

WHAT HAVE WE LEARNED?

ME AM BIZARRO!–

SUPERHERO BLOOPERS

WE STARTED THIS BOOK with a discussion of how Superman, applying Newton's laws of motion, can leap a tall building in a single bound, and ended with Iron Man's transistorized armor and the materials science of superhero costumes. Along the way we have addressed many of the major subjects that would be covered in a basic undergraduate physics curriculum, from the first topics addressed in introductory physics (such as Newton's laws of motion and the Principle of Conservation of Energy) to upper-level material (quantum mechanics and solid-state physics). However, I would be remiss if I left you, Fearless Reader, with the impression that absolutely everything in superhero comic books is 100-percent consistent with the laws of physics. I therefore would like to conclude by discussing some of those very few, rare examples where comic books actually get their physics wrong, no matter how many miracle exceptions one is willing to grant.

CYCLOPS OF THE X-MEN'S SECOND MUTANT POWER

The first young mutant that Prof. Charles Xavier recruited to join his nascent superteam the X-Men was Scott Summers, code named Cyclops. Scott's mutant gift, and also his curse, was that beams of "pure force" were emitted from his eyes. These force beams could punch a hole through a concrete wall and deflect a falling two-ton boulder. Only two materials were immune to

Scott's optic beams: his own skin (which was a good thing, or else his eye blasts would blow his eyelids off his face) and "ruby quartz." Scott was forced to either constantly wear sunglasses made of this exotic material—or a wraparound visor when he was on superhero duty. He could raise this ruby-quartz shield using buttons either on the side of the visor or in the palms of his gloves. When he wore the visor, his eye blasts were projected as a single, broad, red beam of force, from which his *nom de superhero* was derived. When the ruby-quartz shield was lowered, Scott could see the (red-tinged) world clearly, with the visor or sunglasses safely absorbing the destructive brunt of his optic-force beams.

Quartz is the name that geologists have ascribed to the crystalline form of silicon dioxide. If the silicon dioxide molecules are arranged in a disordered fashion, as in a large number of marbles randomly poured into a container, then the resulting material is called "glass," but if the molecular units are carefully stacked in an ordered array, the mineral is called "quartz." Just as there are different ways that a collection of marbles can be stacked in a regular pattern, there are different crystalline configurations of quartz. If the mineral contains a very small amount of iron and titanium, the resulting crystal will have a slight pinkish hue (as in our discussion of stained glass in Chapter 24), in which case it is called "rose quartz." A suspension of ruby dots in the quartz will result in cloudy brown and beige veins, and this dark, smoky, nearly opaque mineral is termed "ruby quartz."

As strange and inconvenient as it would be to have force beams projecting from your eyes, such that you would always have to look at the world through ruby-quartz glasses, we cannot really protest, for no matter how physically implausible this may be, it is covered by our miracle-exception policy. However, regardless of the mechanism by which Cyclops's optic blasts work, there is a key scene that is always missing in the *X-Men* comics and motion pictures whenever Scott lets loose with his mutant power. What we never see, yet know must occur by the laws of physics described earlier, is Scott's head snapping backward due to the recoil of his force beams.

Newton's third law informs us that forces always come in pairs—that is, every action is accompanied by an equal and opposite reaction. You cannot push against something if there is nothing to push against. Rockets depend on this principle when they expel hot gases at high velocities, so through Newton's third law, the recoil propels the ship in the opposite direction of the exhaust. Similarly, a large beam of force, sufficient to keep a two-ton boulder suspended in midair (from which we conclude that the force of the optic blasts must be at least 4,000 pounds), should push back on Cyclops's head with an equivalent recoil force of 4,000 pounds. From Newton's second law of motion—that is, Force equals mass times acceleration—his body (assuming a mass of 80 kilograms) would rapidly acquire an acceleration of more than twenty times that of gravity. From an initial stationary position, his head should be moving backward at several hundred miles per hour whenever he makes use of his special gift. Therefore, we must conclude that Cyclops possesses *two* mutant powers: force beams from his eyes, and exceptionally strong neck muscles.

PUT THAT BUILDING DOWN!

As mentioned in the beginning of this book, in the early days of the Golden Age of comics, Superman's powers were attributed to the fact that his home planet, Krypton, had a much stronger gravitational pull than Earth's. By using the benchmark that he is able to leap a tall building in a single bound on Earth, we calculated in Chapter 1 that the acceleration due to gravity on Krypton had to have been at least fifteen times greater than our own. The Man of Steel was therefore not really made of metal, but had muscles and a skeletal structure adapted to a much larger gravity. Imagine lifting a full gallon container of milk, which weighs nearly nine pounds. If you want to experience what life would be like on a planet with a gravity fifteen times weaker than Earth's, you should empty the gallon container and refill it with just a little more than a half pint of milk. Compared with hefting the full gallon, you would find the same container with only eight ounces of milk to be much easier to pick up. Similarly, in *Action Comics # 1*, Superman is able to raise an automobile weighing roughly 3,000 pounds over his head. A 3,000-pound weight to Superman (adapted to Krypton's heavier gravity) is similar to us lifting a 200-pound weight overhead. Difficult, but not impossible.

We noted earlier that with his rising popularity, Superman morphed from being the champion of the little guy into the star of a multimillion-dollar marketing empire. The threats that Superman faced became more extreme and his foes became more superpowered. Big Blue's strength level correspondingly increased to fantastic levels. Before long, he was lifting tanks, trucks, locomotives, ocean liners, jumbo jets, and high-rise office buildings. Similarly, Marvel Comics' hero the Incredible Hulk possessed a strength that also strained credulity. The Hulk's strength is usually, though not always, ascribed to his bloodstream's adrenaline levels, which is why emotional stress triggers his transformation from puny Bruce Banner into eight feet of thickly muscled green rage. The correlation with adrenaline also accounts for the fact that the madder the Hulk gets, the stronger he becomes. When suitably aggravated, he has been known to pick up and throw a castle; the side of a cliff; and even hold up a mountain threatening to crush him and a collection of other Marvel superheroes in the *Secret Wars* miniseries. While Reed Richards races to modify Iron Man's armor to channel both the Human Torch's nova-flame and Captain Marvel's electromagnetic energy in order to blast an escape tunnel out of the mountain, Reed deliberately insults and taunts the Hulk, knowing that their survival depends on a suitably steamed Jade Giant.

Eventually, Superman would become so strong that, as shown in fig. 42 from *World's Finest # 86*, he could carry two high-rise office buildings—one in each hand—as if he were carrying two pizza pies, while flying at the same time! An examination of this figure reveals one reason he is able to cart off these buildings from Gotham City to an open-air exhibit in Metropolis: They were not connected to any city water or electrical utilities. As astounding as this display of strength is, that's not even the strangest aspect of this scene, compared with Superman's comment: "I got permission to borrow the two Gotham City buildings you asked for." Who would you ask for "permission" to carry off two high-rise buildings? I doubt that either of those buildings' superintendents has the authority to allow Superman to borrow their building, especially if he informed them that he would carry them using only one hand each! When Superman first appeared in *Action # 1* in 1938, people fled in terror when he hefted a car over his head. But by the late 1950s, crowds would cheer as Superman flew overhead, carrying office buildings to a charity event, risking killing all!

Even if you accept that any person, whether a strange visitor from another planet or a nuclear scientist accidentally bombarded with gamma radiation, could indeed be strong enough to pick up a building, there is a separate violation of physics principles associated with these scenes: Simply put, buildings, ocean liners, and jumbo jets are not designed to be picked up. They are either intended to remain stationary, such as an office building, or to be supported at several points, such as the three wheels under an airplane on the runway, or, in the case of a battleship, uniformly buoyed by water. The problem with lifting an office tower, for example, is that any slight deviation from the vertical will result in gravity creating an unbalanced torque, trying to twist the building even further toward the horizontal.

Buildings such as high-rise towers or castles are large, so the distance from the edge of the building to its center of mass is very long (termed the "moment arm" in Chapter 8). These structures are quite heavy, so there is significant weight trying to rotate the building. The bigger the object, the larger the distance from its edge to the point where Superman or the Hulk is holding it, and the larger the moment arm of the torque trying to twist it. The torque for the buildings carried by Superman in fig. 42 is many times greater than reinforced concrete (concrete with steel rods threaded through it to increase its rigidity) can withstand before fracturing. Realistically, if you picked up a building and flew it somewhere, there would be a continuous stream of construction debris left behind you. Superman should arrive at the Gotham City charity event holding a few cinder blocks in each hand, and not two office towers with their structural integrity intact. Rather than asking for permission to borrow the buildings, Superman should practice asking for forgiveness for destroying these towers by picking them up in the first place.

Some later comic-book writers have realized that it's impossible, regardless of your level of superstrength, to pick up a building and not have it crumble under its own weight in your hands. In Marvel's *Fantastic Four # 249*, a Superman stand-in, code-named Gladiator, picks up the edge of the Baxter Building (the FF's headquarters) at its base and rocks it back and forth, but without physically harming the tower. Reed Richards, the smartest man in the Marvel universe, instantly recognizes that what Gladiator is doing is impossible. He theorizes that Gladiator must actually possess a previously unnamed superpower of tactile kinesis, defined in

Fig. 42. Panel from World's Finest # 86, where the Man of Steel demonstrates a much-improved strength level compared with his first appearance in Action *# 1, where he lifted an automobile over his head, provoking the star-tled onlookers to flee in panic. In contrast, no one in the amphitheater is overly perturbed that a flying man is carrying two high-rise office buildings over their heads.*

comic books as the ability to levitate an object that one is in physical contact with. Of course, there is no such thing as tactile kinesis, but it does reduce down to a manageable number the quantity of miracle exceptions that are needed for the story to progress.

If we compare Giant-Man to a redwood tree (and not just because his personality was sometimes a little stiff), we note that the taller the tree, the wider the trunk. In order to provide support for the large mass above it, a tree needs a very broad base. Around the time of the signing of the American Declaration of Independence, two mathematicians, Euler and LaGrange, proved that a column shorter than a certain height is stable, and will be compressed by the weight of material pressing down on its base, but above a certain height (whose value depends on the strength of the material comprising the column), the tower becomes unstable against bending. The slightest perturbation away from a perfectly vertical orientation leads to a large twisting force, that is, a "torque," as in the case of the seesaw in Chapter 8, that will cause the column to bend under its own weight. Giant-Man could, in principle, grow as tall as a redwood tree, but he would have to be just as mobile (assuming he stayed below the height limit set by the cube-square law—see Chapter 25). Any attempt to run after or fight a supervillain would inevitably lead to the upper portion of his body leaning forward over his legs. The weight of his upper trunk would then cause his body to rotate, and before you could say "Stan Lee," old High-pockets would be flat on the ground.

Just such a fate inevitably befell Stilt-Man, an early foe of Daredevil's. Stilt-Man possessed a mechanized suit that contained two hydraulic legs that, when fully extended, enabled him to stand several stories tall. As sure as summer follows spring, Daredevil would use the cable in his billy club to tangle up Stilt-Man's legs, and the resulting loss of stability would bring the issue's adventure to a rapid close.

Another mystery related to the center of mass is how Spider-Man's foe, Doctor Octopus, is able to walk. Research scientist Otto Octavius employed four robotic arms that were attached to a harness around his waist, with which he manipulated radioactive isotopes. The inevitable explosive radioactive accident caused this harness and the arms to be fused to Octavius, and Doctor Octopus was born. But these arms are very heavy, and we frequently see him standing on his two legs while all four arms move behind him! They should therefore create a large torque that would put Doc Ock flat on his back, or on his face if they're in front of him. Spidey should be able to disable (if not disarm) Doctor Octopus by simply tossing an apple at him whenever he spots the arms not anchoring him to the ground. The quick and unsatisfying resolution of these stories when the physics is taken too seriously should make abundantly clear why there is not a great demand for physics professors to write superhero comic books. Careful viewers will note that in the film *Spider-Man 2*, the filmmakers took care to ensure that Doc Ock always had his heavy robotic arms balanced around his body, or some of them tethered to the ground, in order to give Octavius at least some stability. After all, they couldn't feature it in a film if it wasn't true!

THE JUSTICE LEAGUE HAS THE MOON ON A STRING

Another unrealistic feat of strength occurs in the conclusion of a 2001 adventure of the Justice League (by this time they had dropped the "of America" part of their team name, though the comic featuring their adventures still went by the acronym JLA).

In *JLA # 58*, Superman, Wonder Woman, and Green Lantern are shown pulling the moon into Earth's atmosphere in order to defeat a group of renegade Martians. Perhaps I had better back up and explain why they considered this a good idea.

Martians were introduced into the DC universe in 1955's *Detective Comics # 225*, when a physics professor trying to develop an interstellar communication device accidentally created a transporter beam instead. He thereby forcefully brought J'onn J'onzz (the Martian Manhunter) to Earth. J'onn eventually adopted a costumed identity as a superhero crime-fighter, and was a founding member of the Justice League of America back in 1960. J'onn J'onzz possessed a dazzling array of superpowers that matched Superman's—including flight, superstrength, invulnerability, Martian-breath (equivalent to Superman's super-breath), super-hearing, Martian vision, and several that Superman could only dream of, such as mental telepathy, invisibility, and shapeshifting. Just as Superman needed Kryptonite to keep him from always solving every problem in a nanosecond, the Martian Manhunter, being more powerful, required an even more common Achilles' heel in order to justify why he would ever bother teaming up and forming a league with other superheroes. It was therefore revealed that J'onn suffered, as did all Martians, from a vulnerability to fire. Consequently, rather than having to search for an exotic meteorite from the doomed planet Krypton, all you would need is a penny book of matches to incapacitate the Martian Manhunter.

It was revealed in the pages of JLA that J'onn J'onzz is mistaken when he considers himself the last survivor of the Martian race, when the Earth is attacked by a small army of evil Martians, each one possessing J'onn's superpowers. The Justice League lures the evil Martians to the moon, where the lack of an atmosphere convinces the Martians that they have no reason to fear a fire weakening them. However, while J'onn J'onzz uses his mental telepathy to distract these villains, Superman, Wonder Woman, and Green Lantern employ an enormous cable to drag the moon into the Earth's troposphere. An array of magic-based superheroes use their mystical powers to prevent both the moon and the Earth from suffering geological catastrophes from their intense gravitational attraction. Our satellite now possesses a combustible atmosphere, and the evil Martians quickly surrender and submit to banishment to another dimension (the Phantom Zone, in fact) rather than being incinerated. Even if you grant all of the above as one major-league miracle exception, there is still a serious physics problem with this story line.

Newton's second law, $\mathbf{F} = \mathbf{ma}$, tells us that if a net force is applied to a mass, no matter how large, there will be a corresponding acceleration. By the late 1990s DC Comics had established that Superman was capable of lifting eight billion pounds. Let's assume that, given the enormous stakes, both Wonder Woman and Green Lantern exerted themselves to provide an equivalent force as they pulled on the moon. So the total force that these three heroes can supply is twenty-four billion pounds. Since the magic-based heroes are nullifying the effects of gravity, we'll assume that as the moon comes closer to the Earth there is no assist from Earth's gravitational field (this will keep the calculation at a simple level). The moon has a mass of nearly seventy billion trillion kilograms. Newton's law therefore indicates that the moon will indeed accelerate owing to this force, but the rate of change of motion will be very, very small. The acceleration of the moon will be 5 trillionths feet/sec^2 (the acceleration due to gravity on the surface of the Earth is 32 feet/sec^2), and so it will take a very long time to displace the moon a significant distance. At this acceleration, the time needed for the moon to travel roughly 240,000 miles from its normal orbit to within our upper atmosphere is more than 735 years! We can only conclude that J'onn J'onzz performed some outstanding stalling in order to keep the evil Martians from realizing what was going on for more than seven centuries!

WITH THE WINGS OF AN ANGEL, COULD YOU FLY?

Another of the original members of the mutant team the X-Men introduced in 1963 was Warren Worthington III, whose mutant gift comprised two large, feathered wings growing out of his back. None of the other members of this superhero team possessed the power of flight, and aside from Iceman's ice ramps, the Angel was the only character who could avoid walking or taking the bus when called upon to face off against the Brotherhood of Evil Mutants.* Other winged superheroes or villains, such as DC Comics' Hawkman or the Spider-Man villain the Vulture, used "antigravity" devices such as Hawkman's Nth metal, to overcome gravity. They employed their wings connected to their backs in the case of Hawkman or Hawkgirl or sprouting from his arms for the Vulture, as steering devices to help them maneuver while airborne. In contrast, the X-Men's Angel used his wings as a means of levitation. It certainly seems reasonable that having wings growing out of your back would enable you to fly, but could it really?

Birds and planes manage to slip the surly bonds of gravity through Newton's third law—that for every action there is an equal and opposite reaction. A common misconception is that the pressure change induced by a fast-moving object (termed the Bernoulli effect), underlies how airplanes fly. We encountered this pressure differential when we considered the Flash dragging Toughy Boraz behind him in his superspeed wake in Chapter 4. A fast-moving object such as the Scarlet Speedster must push the air out of his way as he runs, and consequently leaves a region of air with a lower density behind him. As the air races back to fill this partial vacuum, through the same principle as in our discussion of entropy in Chapter 13, it will push anything in its way, such as the litter swirling behind fast-moving traffic or trains. However, if the difference in wind speed above and below the wing is a result of the wing's contour, then planes should not be able to fly upside down, because the pressure difference generated by the Bernoulli effect would tend to push the plane toward the ground.

In any case, we can always rely on Newton's third law, which tells us that forces always come in pairs. To provide an upward force on the airplane wing equal to or greater than the weight of the plane, an equivalent downward force from the wing must be applied to the air moving past it. The down draft of air in the region underneath the wing results in an upward lift that carries the plane into the wild blue yonder. When Superman leaps, he pushes down on the ground so that an equal and opposite force pushes back on him, starting him up and away. Similarly, birds

---

* And too few organizations nowadays, in my opinion, are proud enough of their "evil" status to boldly incorporate it into their title and stationery.

flap their wings, pushing a quantity of air downward. The downward force of the wing on the air is matched by an upward force by the air on the wing. The greater the wingspan, the larger the volume of air displaced, and the greater the corresponding upward force. This is why it is impossible for Prince Namor the Sub-Mariner to fly using his tiny ankle wings. These petite wings are too small to provide sufficient lift to counter Namor's weight.

If Warren Worthington III weighs 150 pounds (equivalent to a mass of 68 kilograms), then his wings must provide a downward force on the air of at least 150 pounds, such that the air's reaction on his wings balances his weight and keeps him above the ground. Of course, if he wants to accelerate, then his wings have to provide a force greater than 150 pounds in order for there to be an excess force (upward lift minus downward weight due to gravity) to provide a net acceleration. If his wings provide an upward force of 200 pounds while gravity exerts a downward force of 150 pounds, then Warren experiences a net vertical force of 50 pounds. Force equals mass times acceleration, so this upward force of 50 pounds creates vertical acceleration of 11 feet/sec^2. With this acceleration, the Angel will go from 0 to 60 mph in a little more than eight seconds, neglecting the considerable air resistance that he would have to overcome. Once he stops flapping his wings, the only force acting on him is gravity pulling him back to Earth. Of course he can glide once airborne, but he must continue to apply a downward force on the air to truly fly and not coast.

Two hundred pounds is a considerable force for his wings to apply, but it is not unreasonable that a person could bench press 133 percent of his body weight. Birds such as the California condor or the wandering albatross weigh roughly thirty or twenty pounds, respectively, and yet are able to generate sufficient force to fly. But Warren Worthington III is not built like a bird. Birds do not have wings growing out of their backs—their arms have evolved into wings. They have two additional modifications that assist their arm-wings: (1) They have a keeled sternum bone—that is, birds have a hinge built into the flat bone in the center of their chests that is comparable to your rib cage. This hinge acts as an anchor point for their other adaptation, namely (2) birds have two extremely large chest muscles, the supercorocoiderus and the pectoralis, used for beating their wings. Birds have so much breast meat because they need these muscles, their pectorals, to be large, as they provide the majority of the force to the wings in flight. Recall from chapters 8 and 25 that the strength of bone or muscle increases with its cross-sectional area. Consequently, the Angel must have enormous pectorals if he is to be able to use his wings to get off the ground. With a wingspan of 16 feet and a weight of 150 pounds, Warren has a weight-to-wingspan ratio of nine pounds per foot, in contrast to a ratio of three pounds per foot for a California condor. Warren's arms do not participate in supplying a force to his wings, and he must provide an upward lift using only his chest and back muscles, making him a muscle-bound—and fairly ineffective superhero.

There are other adaptations for flight that Warren could possess that would require additional miracle exceptions. To reduce their body weight, birds have lightweight bones, with a very porous structure that yet remains remarkably strong. Birds also have very efficient respiratory systems, so that every single oxygen molecule residing in their lungs is replaced within two deep breaths. In contrast, with every breath we take, we exchange only 10 percent of the air molecules residing within our lungs. Birds need to be able to rapidly refresh their air supply, as their breast muscles are working so hard to maintain them aloft. Warren's breathing could be similarly efficient. But for all this, unless he is to be drawn with enormous pectoral muscles—more fitting for some of the female superhero characters from the 1990s—the wings on his back are more ornamental than functional.

ENTER . . . THE VISION!

When Roy Thomas took over the writing duties of the Marvel comic book *The Avengers* in the mid-1960s, he would frequently reintroduce Golden Age characters with a new Silver Age twist, just as DC Comics had done when they initiated the Silver Age. One of the more popular characters created by Thomas and artist John Buscema is the Vision. Originally a supernatural costumed crime-fighter in the 1940s, the new Vision introduced in *Avengers* # 57 is an android* created by Ultron, another android. Ultron is

---

* Technically, the Vision is a "synthezoid," and no, I don't know what the difference is.

one of the Avengers' most dangerous foes, and the Vision was initially intended to infiltrate the superteam in order to destroy them from within. Rebelling against his programming, the Vision saved the lives of the Avengers and went on to become a valued member of the team.

In addition to laser vision, the power of flight, and the mind of a computer, the Vision possessed the superpower of total independent control of his body's density. He could make his body, or any part thereof, as hard as diamond or so insubstantial that he could pass through solid objects. Kitty Pryde of the X-Men walks through walls using her mutant ability to vary her quantum-mechanical tunneling probability, but the Vision should stick to using the door when he wants to enter a room.

The density of any object is defined as the mass per volume, and can be altered either by changing the mass or varying the volume. The volume is governed by the average spacing between atoms. Any solid typically has its atoms packed fairly closely, so the atoms can be considered to be touching (they have to be this close in order to form chemical bonds, which are what hold the atoms together in a solid after all). Very roughly, all solids have the same density, within a factor of ten or so.

Even if the Vision could control his density at will and could maintain the structural integrity of his body, he could not pass through walls. A gas, such as the air in your room, is comparatively dilute, with the average spacing between atoms being roughly ten times larger than the size of an atom. Yet the fact that the air in your room is less dense than the walls does not mean that the air can pass through the solid walls. Good thing, too, otherwise the air in an airplane would leak out through the fuselage and make air travel an even more unpleasant experience. We must therefore conclude that Ultron made a second error when he constructed the density-altering Vision (the first was believing that such a noble android would betray the mighty Avengers).

CAN THE ATOM USE THE TELEPHONE TO REACH OUT AND TOUCH SOMEONE?

The DC superhero the Atom has appeared throughout this book, and his ability to reduce his size and mass independently have provided excellent illustrations of a wide range of physical phe-nomena. Of course, occasionally his shrinking would take him to ridiculous extremes, such as whenever he visited other worlds that contained civilizations, cities, and advanced technology all residing within an atom. Given that there are nearly a trillion tril-lion atoms in a cubic centimeter of a typical solid, it's amazing that the Atom ever managed to find these nanoworlds, unless they are a routine feature of every element in the periodic table. The implausibility of the Atom's powers was slyly acknowledged in 1989, in a scene in his second regular series, *The Power of the Atom # 12*. In this story, the Atom shrinks both himself and a col-league in order to escape a supervillain's death trap, and they wind up decreasing to subatomic lengths in order to pass through the empty spaces in the floor's atoms. Pausing in their miniaturiza-tion, they discuss the events of the past few issues while sitting on an electron. The Atom's friend suddenly notes that they are smaller than oxygen molecules and wonders, "How are we even breathing?" To which the Atom honestly replies, "I'm not sure."

Superman can fly, the Flash can run really fast, Hawkman has his wings and antigravity belt, Storm rides on thermally generated air currents, but how do you get around when you're very, very tiny? Ant-Man uses flying carpenter ants as his personal taxi service, the Wasp has wings that grow out of her back when she shrinks, but the Atom has Bell Telephone. In his Silver Age debut issue, *Showcase # 34*, the Atom employs a unique mode of trans-portation. In this story he needs to confront a small-time crook named Carl Ballard who is clear across town. Presumably, after looking up Ballard in the phone book, the Atom dials his number while setting up a metronome near the receiver, which creates a "tick-tock" sound. Shrinking himself smaller and smaller, the Mighty Mite jumps into one of the holes on the speaker of his tele-phone, and in the next panel we see him flying out of the receiver of Carl Ballard's phone.

The "explanation" for this trick is revealed on a text page in the back of the comic.* By dialing Ballard's phone number, the Atom

---

* In the late 1950s and early 1960s, comics always included no fewer than two pages of prose in order to qualify for second-class-postage mailing rates reserved for "magazines," which were defined as journals that contained a minimum of two pages of text.

causes an electrical impulse to travel from his phone to the central telephone exchange, which then forwards the signal to Ballard's phone. When the circuit is completed once Ballard answers the ringing phone, the signal—in this case the ticking metronome—is transmitted from the Atom's phone to Ballard's. At this point the Atom jumps into his speaker, shrinking down to the size of an electron, and rides these electrical impulses from his phone to Ballard's.

The writer of this text page, DC Comics editor Julie Schwartz, correctly describes how a telephone transfers sound into electrical impulses. A thin diaphragm vibrates when sound waves strike it, which in turn compress or dilate carbon granules that are adjacent to the membrane. The electrical conduction through the carbon grains is very sensitive to how tightly they press against one another. As you speak, the interconnections between the grains alternately contract or expand, and the electrical signal down the wire is appropriately modified. At the other end of the telephone connection, the electrical signal causes other carbon grains to undergo equivalent vibrations that are transferred to another diaphragm. The diaphragm's vibrations create pressure waves in the air that are then detected by the ear of the person receiving the call. All of this Julie Schwartz got right. Where he goofed is in assuming that the Atom could hitch a ride on the electrical impulses propagated down the wire.

When you speak, complex sound waves can convey all sorts of information. The sound waves can be detected by another membrane (such as an eardrum), causing it to vibrate in accordance with the amplitude, wavelength, and even phase information encoded in the message you spoke. But it is the wave that carries that information—not the air you expelled from your mouth. By speaking, you set up alternating regions of less-dense and more-dense air (equivalently you can think about the density variations as pressure modulations—a reasonable approximation at a constant temperature) that move away from the speaker. It is not the air coming from your mouth that reaches the listener; otherwise you would never have to worry about noisy neighbors in the apartment next door.

Similarly, the information encoded in electrical impulses in a telephone wire is transmitted by means of density waves of electrons, rather than having the electrons move down the wire. What happens is that a region of electrons of higher-than-normal density is unstable (as the negatively charged electrons repel each other) and expands into the adjacent regions, causing a buildup of electron density in the next spatial location, which in turn causes a bulge farther down the line, and so on. The speed of this transmission is determined by the electrostatic repulsion that pushes the electrons away from each other. That is, if I shake one electron, how long will it take a second electron some distance away to respond to the first electron's motion? Pretty quickly, as it turns out, as the electrical interaction between two charges in the wire is communicated at roughly one third of the speed of light. Depending on the distance, there will be a barely perceptible time lag between moving the first charge and it being noticed by the second charge. The speed of light is so fast—186,000 miles per second—that this time lag will be less than a billionth of a second over a distance of 12 inches. If the Atom were riding on one electron carrying the electrical impulse signal along the telephone wire, he would have to jump to the next bunch of electrons with the speed of light in order to "ride the wave" all the way to the receiver.

It's a good thing that the information in a telephone wire is in fact transmitted at the speed of light, as the average speed that an electron moves along a wire in response to an external electric field is less than a millimeter per second, nearly a trillion times slower. If you had to wait for the electrons to physically travel along the telephone wires before your message could be sent, it would be quicker to just walk to the house of the person you're calling and speak to her face-to-face.

EVERY PHYSICIST'S SECRET SUPERPOWER

When not fighting crime as the Atom, Ray Palmer's civilian identity is equally heroic, for he is a physics professor at Ivy University. As mentioned in Chapter 13, it was the late-night discovery of a strange meteorite that led to the research breakthrough that enabled Palmer to develop a second career as a costumed crime-fighter. As shown in fig. 43, Palmer discovers that the meteor is in fact a chunk of white-dwarf-star matter that will enable him to miniaturize himself and independently control his mass. Ray

Fig. 43. *Physics Professor Ray Palmer discovers the white-dwarf-star fragment that will turn out to be the key missing ingredient in his miniaturization device and eventually lead to his moonlighting as the superhero the Atom (from* Showcase *# 34). © 1961 National Periodical Publications, Inc. (DC)*

strains to lift and carry the meteorite, which is roughly twelve inches in diameter, over to his car. We are privy to Prof. Palmer's thoughts as he struggles with the great weight. "So heavy—I can hardly lift it! Puff! I don't know the odds against one white dwarf hitting another out in space—Puff—but it could happen—and when it did, this piece drifted until it landed in this field." (By the way, as also shown in fig. 43, in the mid-1960s, physics professors typically drove Cadillac convertibles.)

Ray's reasoning here is sound. When a low-mass star of a certain size has exhausted most of its elemental fuel, the energy released by fusion reactions is insufficient to counteract the gravitational pull of the star's core. The large force at the center of the star leads to a massive compression, until its density is three million grams per cubic centimeter, in which case we call the remnant a white dwarf. The pull of gravity on the remaining core of a white dwarf star is so great that only a cataclysmic explosion would generate sufficient energy to enable a small chunk of the core to break away from the rest of the star and float through space. Some astrophysicists have suggested that white-dwarf collisions occur roughly once a month.

As Ray reminds himself while struggling with the meteor fragment, the rock he is holding is heavy because it is composed of "degenerate" matter. The electrons are termed "degenerate" because they are all in the lowest energy quantum states, unlike in a normal star, where the electrons would be distributed over many quantum states, some at higher energies. The interior of the white dwarf is composed of carbon and oxygen nuclei and a sea of electrons packed as closely as they can be. The core of white dwarfs cannot be easily compressed further, for all the electrons are already in the lowest possible energy state. This is what Ray means when, as he nears his car, he thinks to himself that white-dwarf stars are composed of "degenerate matter from which the electrons have been stripped, greatly compressing them." The electrons are still in there, but are not associated with any particular atomic ions.

Ray is certainly correct that this "degeneracy" is why the white-dwarf star is so dense. The rock Ray is carrying appears to have a radius of 6 inches. Assuming a spherical white-dwarf fragment, the volume would be $(4\pi/3)\times(radius)^3$. In this case the volume of the rock is $(4\pi/3)\times(6\ inches)^3 = 905\ inches^3$—equivalent to nearly 15,000 cm^3 since 1 inch equals 2.54 centimeters. To find the mass of the rock we multiply the density of white-dwarf-star matter (3 million grams/cm^3) by this volume (15,000 cm^3), which gives us 45 billion grams, equal to 45 million kilograms. Converting this mass to weight, we multiply the mass by the acceleration due to gravity $(W = mg)$, and find that the meteorite in fig. 43 weighs one hundred million pounds. No wonder Prof. Palmer, physics professor at Ivy University, is huffing and puffing as he struggles with his find—that little rock weighs 50,000 tons!

But it turns out that this is, technically, not actually a blooper. Despite appearances, there is nothing wrong with the scene depicted in fig. 43. And that is because we physics professors are Just. That. Strong. Remember this the next time you're tempted to kick sand in someone's face at the beach. You never know if that seemingly ninety-eight-pound weakling actually has an advanced degree in physics.

AFTERWORD–

LO, THERE SHALL BE AN ENDING!

IT SHOULD COME AS NO SURPRISE that comic books and physics make a good match; after all, the fun underlying science is not so different from that of a good superhero comic-book story. In both situations either the scientist or the comic-book reader (in some cases they may be one and the same) are presented with a set of rules to be applied in novel, challenging situations. The rules may be Maxwell's equations of electricity and magnetism and Schrödinger's equation, and the challenging problem may be to develop a semiconductor analog of a vacuum tube. Alternatively, the rules may be that our hero can run at superspeed and has an aura that protects him from the adverse effects of air drag and electromagnetic induction, and the challenge would be that he has to capture a villain armed with a freeze gun capable of icing up any surface, while recovering the stolen bank funds and without harming any innocent bystanders. In both situations the trick is to find a solution that employs the known rules in a new way (if an old solution would work, we'd just use that), without doing anything that is deemed impossible under these guidelines. We can't design a transistor device that, in order to function, requires electrons to split into two halves or be attracted toward each other without an intervening positive charge, because the basic unit of negative charge has never been observed to behave in this fashion. Similarly, a Flash comic-book story featuring the Scarlet Speedster defeating Captain Cold by shooting heat beams from his eyes would be unsatisfying, as this is not an ability that the Flash has ever possessed.

The goal of basic scientific research is to elucidate the fundamental laws of nature, and the highest accomplishment is the discovery of a new rule or principle. Equally good is the clear
demonstration of a violation in a preexisting rule, for new physics
is discovered when we understand under what circumstances the
old rules do not apply. Similarly, there are times when an established comic-book character suddenly acquires a previously
unsuspected ability, such as when Sue Storm of the Fantastic Four
discovered in *Fantastic Four # 22* that the cosmic-ray bombardment that gave her the power of invisibility had also bestowed
upon her the ability to generate "invisible force fields."* The
dynamics between Sue and her teammates were radically altered
following the discovery of this new superpower, and over the
years, she would learn to generate her force fields in an offensive
as well as defensive capability.

But such cases are rare both in comic books and in real-world
physics. There is, however, an unending stream of exciting and
challenging problems in physics, just as there is an unlimited
source of engaging comic-book stories waiting to be told. The two
central ingredients are the same for both science and comic books:
an understanding of the basic rules of the game and a fertile imagination.

Scientists don't typically consult comic books when selecting
research topics (funding agencies tend to frown on grant proposals
that contain too many citations to superheroes), but the spirit of
"What if..." or "What would happen when..." infuses both the
best scientific research and comic-book adventures. To be sure,
there are times when comic books and science fiction anticipate
scientific discoveries, just as cutting edge research is occasionally
employed as the springboard for superhero adventures (as in the
aforementioned *JLA # 19*).

Sometimes it takes a while for the science to catch up with the
comic books. As an example, consider the magician Abra Kadabra,
a Flash villain who has plagued the Scarlet Speedster from nearly

---

* This is a bit redundant, since every force field in comic books is invisible,
save one. Given that Green Lantern's ring is powerless against anything
colored yellow, the villain the Shark in *Green Lantern # 24* had the ability
to project "yellow invisible force fields"! I've said it before, and I'll say it
again: I love Silver Age comic books!

the beginning of his crime-fighting career. Garbed in the conventional stage magician's attire of eveningwear and a top hat, he would use his "magic" to bedevil the Viceroy of Velocity, such as the time he turned him into a human marionette. However, it was revealed that Abra Kadabra was a scientist from the far future and that his "magic" in the twentieth century was actually sixty-fourth-century technology.* The creators of the Flash comics clearly subscribe to the notion that our present-day science and engineering would appear to be supernatural to those in the distant past. After all, imagine the reaction you'd receive if you could travel one thousand years into the past and display just a fraction of the appliances found in a modern home (assuming you also brought a power supply with you).

It was left deliberately vague in the Silver Age story as to how sixty-fourth-century science could transform someone into a living puppet. The "explanation" would have to wait until the late 1990s, where Kadabra informs us that he employed nanotechnology to restructure the Flash at the molecular level, demonstrating once again the trouble that a crooked ex-scientist can cause. Certainly nanometer-scale machines cannot transform people into marionettes, but who's to say what can and can't be done in another few thousand years, provided it doesn't involve a violation of established physics.

In fairness, however, the predictive ability of speculative fiction sometimes gets the technological aspects right, but widely miss other revolutions that have transformed our society. Consider, for example, the 1965 television program *Lost in Space*. This popular TV show envisioned a trip to the stars by the Robinson family, accompanied by an intelligent robot and Dr. Zachary Smith, a villainous and cowardly stowaway. The show first aired on September 15, 1965, and was imagined to take place in the distant future, all the way in October 1997. As pointed out in a *New York Times* article in 1997, discussing an anniversary

---

* Bored with the peaceful idyll that society had finally achieved, Kadabra traveled back to our time in order to wreak mischief. The desire to escape the monotony of future utopia motivated the Marvel Comics time-traveling villain Kang to visit our time, intent on world conquest. It appears that human nature—at least for some humans—will always rebel against a well-ordered, perfect society.

rebroadcast of the pilot episode, while the producers and writers of *Lost in Space* were a bit off in assuming that thirty years hence we'd have starships and robots, they goofed spectacularly regarding one very crucial aspect of modern life in the late 1990s.

A scene set in mission control as the starship is preparing to launch features a familiar bank of computer monitors manned by an array of nearly identical short-sleeved white-shirted engineers. At the elbow of each mission control engineer is a small metal disc that one would never, ever find in the NASA of today. The science-fiction writers in 1965 never imagined that in thirty years, mission control would be a smoke-free environment and, consequently, no place for ashtrays. Thereby a cautionary note, that extrapolating potential scientific and technological innovations is duck soup compared with predicting future social customs.*

*   *   *

If the study of the natural world has demonstrated anything, it is that, unlike the Hulk, the smarter we get, the stronger we become. Now that you've finished this book, perhaps you'll feel a little stronger yourself, if not in arm, then at least in mind. Which is the only type of strength that really matters. It is our intelligence that provides the competitive advantage that enabled us to become the dominant species on the planet. We are not as fast as the cougar, nor can we fly like an eagle. We are not as strong as the bear or as indestructible as the cockroach. It is our intelligence that is our superpower, if you will. As quantum mechanics pioneer Niels Bohr said, "Knowledge is in itself the basis of civilization."

The optimism at the heart of all comic-book adventures lies within the scientific endeavor as well, as they both hold out the promise that we will overcome our physical challenges and improve the world. How science is to be employed, whether to ease hunger and cure disease, or to develop an army of killer robots, is up to us. For guidance in how to use our knowledge wisely and ethically, one could do worse than look to the stories

---

* Sometimes these changes take a while to come about. If ancient galley slaves could visit a modern-day health club and see the rowing machines that wealthy (compared with themselves) free men and women employ, their heads would no doubt explode in shock!

in comic books. It is as true today as it was many years ago, at the conclusion of Spider-Man's debut in *Amazing Fantasy # 15*: "With great power there must also come—great responsibility." But responsibility to do what? One answer was provided by the Man of Tomorrow in the story "The Last Days of Superman" in *Superman # 156*. Believing that he was dying from an infection of Virus X (fortunately, a false alarm), Superman etched a farewell message to the people of Earth on the moon with his heat-vision, a message he'd intended to be discovered after his demise. His final, parting words to the people of his adopted planet were: "Do good to others and every man can be a Superman."

Face front, Fearless Reader!

RECOMMENDED READING

INTRODUCTION

There are many excellent reviews of the early history of comic books. In addition to the books explicitly cited in the text and listed below, I would recommend: *Men of Tomorrow: Geeks, Gangsters, and the Birth of the Comic Book* by Gerard Jones (Basic Books, 2004); *Tales to Astonish: Jack Kirby, Stan Lee, and the American Comic Book Revolution* by Ronin Ro (Bloomsbury, 2004); *Kirby, King of Comics* by Mark Evanier (Abrams, 2008); *The Ten Cent Plague: The Great Comic-Book Scare and How It Changed America* by David Hajdu (Farrar, Straus and Giroux, 2008), and *Great American Comic Books* by Ron Goulart (Publications International, 2001). Jim Steranko's excellent two-volume *The Steranko History of Comics* (Supergraphics, 1970, 1972) is worth searching out for his thorough and entertaining elucidation of the lineage from pulp heroes to comic-book superheroes. Les Daniels has written extensively and elegantly on the history of comic-book characters, and his *DC Comics: Sixty Years of the World's Favorite Comic Book Heroes* (Bulfinch Press, 1995); *Superman: The Complete History* (Chronicle Books, 1998); *Batman: The Complete History* (Chronicle Books, 2004); *Wonder Woman: The Complete History* (Chronicle Books, 2001); and *Marvel: Five Fabulous Decades of the World's Greatest Comics* (Harry N. Abrams, 1991) are all highly recommended, as is *Silver Age: The Second Generation of Comic Book Artists* by Daniel Herman

(Hermes Press, 2004). A historical analysis of the role of comic books in American popular culture is presented in *Comic Book Nation* by Bradford W. Wright (Johns Hopkins University Press, 2001).

While not explicitly a history of comic books, *Baby Boomer Comics: The Wild, Wacky, Wonderful Comic Books of the 1960s* by Craig Shutt (Krause Publications, 2003) is a fun overview of some of the high and low points of Silver Age comic books.

Others have explored the science underlying comic-book superheroes, and any reader disappointed that their favorite character was not sufficiently discussed here may try consulting *The Science of the X-Men* by Linc Yaco and Karen Haber (ibooks, 2000); *The Science of Superman* by Mark Wolverton (ibooks, 2002); *The Science of Superheroes* by Lois Gresh and Robert Weinberg (Wiley, 2002); and *The Science of Supervillains* by the same authors and publisher (2004). The science underlying other pop-cultural subjects has been explored in *The Physics of Star Trek* by Lawrence Krauss (Basic Books, 1995); *The Science of Star Wars* (St. Martin's Press, 1998) and *The Science of the* X-Files (Berkley, 1998), both by Jeanne Cavelos; *The Physics of Christmas* by Roger Highfield (Little, Brown & Company, 1998); as well as his *The Science of Harry Potter* (Viking, 2002), and *The Physics of the Buffyverse* by Jennifer Ouellette (Penguin, 2006).

Those readers interested in a deeper discussion of the philosophy and nature of physics investigations should consider Richard Feynman's *The Character of Physical Law* (Random House, 1994) and *The Pleasure of Finding Things Out: The Best Short Works of Richard P. Feynman* (Perseus Publishing, 2000); as well as Milton A. Rothman's *Discovering the Natural Laws: The Experimental Basis of Physics* (Dover, 1989) and *The Fermi Solution: Essays on Science* by Hans Christian von Baeyer (Dover 2001).

SECTION ONE—MECHANICS

While this book covers many of the topics treated in an introductory physics class, those readers who are gluttons for punishment and wish to consult a traditional physics textbook (or seek to verify that I am not trying to pull any fast ones) may find *Conceptual Physics* by Paul G. Hewitt (Prentice Hall, 2002) helpful. It is written as a high-school physics text, so the mathematics remains at the algebra level. An abridged version of Richard Feynman's brilliant lectures in physics, covering the basis of classical physics, *Six Easy Pieces* (Perseus Books, 1994) is highly recommended.

There are several excellent biographies of Isaac Newton. The reader interested in learning more about this towering intellect may consider *The Life of Isaac Newton* by Richard Westfall (Cambridge University Press, 1994); *Newton's Gift* by David Berlinski (Touchstone, 2000); and *Isaac Newton* by James Gleick (Pantheon Books, 2003).

The discussion of the Special Theory of Relativity in Chapter 11 went by so fast that its brevity can be attributed to Lorentz contraction. The first book anyone interested in this subject should read is *What Is Relativity* by L. D. Landau and G. B. Romer (translated by N. Kemmer) (Dover, 2003), which in only sixty-five pages (with figures!) clearly explains, without equations, the physical concepts underlying Einstein's theory. Fuller discussions of this fascinating subject can be found in *Relativity and Common Sense* by Hermann Bondi (Dover Publications, 1962); *An Introduction to the Special Theory of Relativity* by Robert Katz (D. Van Nostrand Co., 1964); *Introduction to Special Relativity* by James H. Smith (W. A. Benjamin, 1965); and *Discovering the Natural Laws: The Experimental Basis of Physics* by Milton A. Rothman (Dover, 1989). Be warned that all of these treatments deal with the mathematics underlying relativity as well as the physical concepts.

SECTION TWO—ENERGY—HEAT AND LIGHT

Excellent overviews for the nonspecialist on how energy is created and transformed, particularly at the molecular level, can be found in *The Stuff of Life* by Eric P. Widmaier (W. H. Freeman & Company, 2002); *The Machinery of Life* by David S. Goodsell (Springer-Verlag, 1992); and *Stories of the Invisible* by Philip Ball (Oxford University Press, 2001). Background information on this mysterious quantity is available in *Energies: An Illustrated Guide to the Biosphere and Civilization* by Vaclav Smil (MIT Press, 1998) and *Energy: Its Use and the Environment* by Roger A. Hinrichs and Merlin Kleinbach (Brooks Cole, 2001), Third Edition. This last is a textbook written at a practically math-free level, with abundant information concerning the environmental issues involved in energy transformation.

Excellent popular accounts of the fascinating history of thermodynamics are: *A Matter of Degrees* by Gino Segre (Viking, 2002); *Understanding Thermodynamics* by H. C. Van Ness (Dover Publications, 1969); and *Warmth Disperses and Time Passes: The History of Heat* by Hans Christian von Baeyer (Modern Library, 1998). Issues related to the measurement of temperature are considered in an accessible manner in *Temperatures Very Low and Very High* by Mark W. Zemansky (Dover Books, 1964), while phase transitions are discussed in *The Periodic Kingdom* by P. W. Atkins (Basic Books, 1995) and *Gases, Liquids and Solids* by D. Tabor (Cambridge University Press, 1979).

Popular accounts of the history of electricity and magnetism are found in: *Electric Universe: The Shocking True Story of Electricity* by David Bodanis (Crown, 2005); *The Man Who Changed Everything: The Life of James Clerk Maxwell* by Basil Mahon (John Wiley & Sons, 2003); and *A Life of Discovery: Michael Faraday, Giant of the Scientific Revolution* by James Hamilton (Random House, 2002).

SECTION THREE—MODERN PHYSICS

There are many excellent overviews of quantum physics written for the nonspecialist. Highly recommended are: *Thirty Years That Shook Physics: The Story of Quantum Theory* by George Gamow (Dover Press, 1985) and *The New World of Mr. Tompkins* by G. Gamow and R. Stannard (Cambridge University Press, 1999).

Excellent, clear discussions of cutting-edge research in string theory can be found in: *The Elegant Universe* by Brian Greene (W. W. Norton, 1999); *The Fabric of the Cosmos* by Brian Greene (Alfred A. Knopf, 2003); *The Future of Spacetime* by Stephen W. Hawking, Kip S. Thorne, Igor Novikov, Timothy Ferris, and Alan Lightman (W. W. Norton and Company, 2002); *Quintessence: The Mystery of Missing Mass in the Universe* by Lawrence Krauss (Basic Books, 2000); and *Warped Passages* by Lisa Randall (Ecco, 2005).

The solid-state physics revolution that has transformed all of our lives is documented in the highly readable *Crystal Fire: Birth of the Information Age* by Michael Riordan (Norton, 1997) and *The Chip: How Two Americans Invented the Microchip and Launched a Revolution* by T. R. Reid (Simon & Schuster, 1985).

The strength of materials, as it relates to the cube-square law in biological organisms, is discussed in the short and highly readable *Why Size Matters: From Bacteria to Blue Whales* by John Tyler Bonner (Princeton University Press, 2006).

SUMMARY

In the spirit of continuing a review of the topics addressed here, the reader should consider these fun books employing the question-and-answer approach to cover a wide range of physics for the nonexpert: *The Flying Circus of Physics with Answers* by Jearl Walker (Wiley, 1977) and *Mad About Physics: Braintwisters, Paradoxes, and Curiosities* by Christopher P. Jargodzski and Franklin Potter (John Wiley & Sons, 2000). In a similar vein, for those no longer intimidated by mathematics, is *Back-of-the-Envelope Physics* by Clifford Swartz (Johns Hopkins University Press, 2003). Those readers eager to put their physics knowledge to use are directed to *How Does It Work?* by Richard M. Koff (Signet, 1961) and Cy Tymony's *Sneaky Uses for Everyday Things* (Andrews McMeel Publishing, 2003), which contains instructions for making your own Power Ring!

Finally, some comic-book recommendations. Both DC and Marvel have comprehensive reprint lines, where comics from the Golden Age through the present are collected, frequently on better-quality paper than the originals and at a fraction of their cost if you were to buy the back-issues separately today. *The Archives* series from DC and the *Marvel Masterworks* volumes reprint Golden and Silver Age comics focusing on a given character or team in a hardcover format. In addition, Marvel has a line of paperback reprints termed *Essentials*, while DC comics has their *Showcase Presents* line, where twenty or so issues of Silver Age or later comics featuring a given character or title are reprinted on cheaper paper, in black and white, at a cost of less than a dollar per issue. Those readers whose memories of former favorites have been jogged or those who have developed a new interest will almost certainly find a reprint volume at either your favorite bookstore or your friendly neighborhood comic-book shop. To find the nearest comic-book store, dial 1-888-COMICBOOK or visit http://csls. diamondcomics.com.

There are some collections, however, that should be considered required reading as part of any well-rounded liberal education in costumed superheroes. At the top of the list would be *Watchmen* (DC Comics, 1986, 1987) by Alan Moore and Dave Gibbons, which is justifiably characterized as the *War and Peace* of comic books by the film director Terry Gilliam. For legal reasons, the characters in this story are disguised versions of Silver Age heroes originally published by Charlton Comics (such as the Question, Blue Beetle, Captain Atom, etc.) and a familiarity with them is not necessary to enjoy the story. These characters' adventures are now published by DC Comics, where they are undisturbed by the fate that their doubles met in Moore's and Gibbons' epic. Another must-read is Frank Miller's *The Dark Knight Returns* (DC Comics, 1997), which imagines a possible future fate for Batman. This miniseries is considered by most to be responsible for saving Batman from cancellation or worse—irrelevance—by returning the character to his darker, grim and gritty roots, and has set the tone for various motion-picture versions of the Caped Crusader. Continuing the concept of possible futures of superheroes, Mark Waid's and Alex Ross' Kingdom Come miniseries (DC Comics, 1998) investigates the interactions between DC Comics' superpowered heroes and villains and normal civilians. The influence of Marvel Comics superheroes on society, seen from the point of view of a non-superpowered photographer for the *Daily Bugle*, is explored in *Marvels* (Marvel Comics, 2004) by Kurt Busiek and Alex Ross. One of the best time-travel adventures can be found in the collection *Days of Future Past* (Marvel Comics, 2004), starring many of the characters from the popular X-Men films, where Kitty Pryde goes back in time to prevent a political assassination that will lead humanity to a dark, dystopian future. Finally, to cleanse the palate of all these deconstructions of the superhero myth, read Darwyn Cooke's *DC: The New Frontier Vols. 1 and 2* (DC Comics, 2004, 2005)—a brilliant reconstruction of the dawn of the Silver Age set in the Cold-War America of the late 1950s when these heroes first appeared.

KEY EQUATIONS

NEWTON'S THREE LAWS OF MOTION
Page 25

The basic principles of dynamics, as elucidated by Sir Isaac Newton, state that (1) an object at rest will remain at rest, or if in uniform straight-line motion, will remain in motion, unless acted upon by an external force; (2) if an external force does act on the object, then its change in motion (either speed or direction) is proportional to the outside force, that is $F = ma$; and (3) forces always come in pairs, commonly expressed as for every action there is an equal and opposite reaction.

DEFINITION OF ACCELERATION
Page 25

Acceleration is defined as the rate of change of velocity—either its magnitude (speed) or its direction, and has units of (distance/time)/time or distance/(time)^2.

WEIGHT = Mg
Page 27

A consequence of Newton's second law ($F = ma$) when the external force is the gravitational attraction of a planet. The force is then referred to as Weight, and the acceleration due to gravity is relabeled by the letter "g."

$$V^2 = 2gh$$
Page 30

A description of the velocity v of an object moving under the influence of gravity, whether slowing down as it rises or speeding up as it falls, through a distance h.

FORCE DUE TO GRAVITY $= (G) \times \dfrac{[\,(MASS\ 1) \times (MASS\ 2)\,]}{(DISTANCE)^2}$
Page 34

The simple expression, also elucidated by Sir Isaac Newton, for the attractive force between any two point masses. The force is proportional to the product of each mass and inversely proportional to the square of the distance that separates them.

$$g = GM/R^2$$
Page 37

A consequence of Newton's law of gravitational attraction is that the acceleration due to gravity for any large object such as a planet or moon can be expressed as a universal constant G $(G = 66.7$ trillionth of $m^3/kg\text{-}sec^2)$ multiplied by the object's mass M, divided by the square of the object's radius. This expression is correct only for spherically symmetric masses.

$$g_K/g_E = \rho_K R_K / \rho_E R_E$$
Page 37

By making use of the fact that the mass M of a planet can be written as the product of its density ρ and its volume $(4\pi R^3/3$ for a sphere), the acceleration due to gravity $g = GM/R^2$ simplifies to $[\frac{4\pi}{3}]G\rho R$, and when taking the ratio of the accelerations due to gravity for two planets, the constants G $(4\pi)/3$ cancel out.

FORCE $\times$ **TIME** $= (MASS) \times (CHANGE\ IN\ SPEED)$
Page 52

A restatement of Newton's second law $(F = ma)$ where the acceleration is the change in velocity divided by the time over which the external force acts. The momentum is defined as the product of the mass and velocity of an object.

PRESSURE = ATMOSPHERIC PRESSURE + (DENSITY × ACCELERATION DUE TO GRAVITY × DEPTH)
Page 84

The pressure under water increases linearly with depth beneath the surface. At the water's surface, the pressure is just that of the atmosphere, and the deeper one submerges, the pressure increases as the weight of the fluid above you.

CENTRIPETAL ACCELERATION $a = V^2/R$
Page 95

An object moving with velocity v in a circular arc with radius R is characterized by an acceleration for its continually changing direction. The magnitude of this acceleration is v^2/R and an external force pointing in toward the center of the circular trajectory of magnitude $F = mv^2/R$ must act on the object to account for this changing motion.

WORK = FORCE × DISTANCE
Page 144

Work in physics is another expression for energy, and any change in the kinetic energy of an object must result from an external force acting over a given distance. The expression indicates that no work is done when you hold a weight over your head, for while you do supply a force, there is no displacement of the stationary object. This is in conflict with the common usage of the term "work" but is correct physically—once you have increased the object's potential energy by raising it over your head, there is no additional change in its energy if you maintain it in this raised state indefinitely.

KINETIC ENERGY = $(1/2)mV^2$; POTENTIAL ENERGY = mgh
Page 144

Expressions for the energy associated with motion (kinetic

energy = (1/2)mv²) or for the potential of motion in a gravitational field (potential energy = mgh). Note that the expression for potential energy is the same as the Work done raising an object of weight mg by a height h.

FIRST LAW OF THERMODYNAMICS
Page 162

Essentially a restatement of the principle of conservation of energy, indicating that any change in the internal energy of a system will be the result of any Work done on or by the system, and any heat flow into or out of the system.

SECOND LAW OF THERMODYNAMICS
Page 169

In any process to convert the heat energy that flows from a hot object to a colder object into Work (defined as the product of force times distance), there will inevitably be some loss. That is, one cannot transform 100 percent of the heat flow into productive work. This is related to the entropy of the systems involved, which is a measure of the disorder of their components.

THIRD LAW OF THERMODYNAMICS
Page 172

Upon lowering the temperature of a system in equilibrium, which is a measure of the average energy of its components, the entropy also decreases. The entropy of any system is zero if there is only one configuration that it can have, and that state is only realized when the energy of each component is zero—that is, at a temperature of Absolute zero, which can never actually be reached.

COULOMB'S LAW OF ELECTROSTATIC ATTRACTION
Page 198

The mathematical expression for the force between two charged objects, indicating that the force is proportional to the product of each object's charge and divided by the square of the distance separating them. The formula is algebraically identical to Newton's expression for gravitational force. However, while gravity is always attractive, the force between two charged objects can be attractive if they have opposite signs (positive and negative) or repulsive if they have the same sign (both positive or both negative).

OHM'S LAW V = IR
Page 232

Expression relating the voltage V pushing or pulling electrical charges in a conductor of resistance R to the current I (number of charges moving past a given point per unit time). While this expression holds for most metals, not every electronic device obeys this simple linear relationship.

ENERGY = hf
Page 250

The quantum hypothesis that states that the change in energy of any atomic system characterized by a frequency f can only occur in steps of magnitude Energy = hf, where h is Planck's constant, a fundamental constant of nature. When a system lowers or raises its energy by emitting or absorbing light, it must do so through quantized packets of energy termed "photons."

DEBROGLIE RELATIONSHIP P λ = h
Page 252

The motion of any matter having a momentum p is associated with a matter-wave of wavelength λ, where the product of the momentum and the wave's wavelength is Planck's constant h.

SCHRÖDINGER EQUATION
Page 268

The fundamental wave equation for the "motion" of quantum objects. By knowing the potential V acting on the object, one can solve this equation to obtain the wavefunction ψ that characterizes its behavior. Squaring this wavefunction yields the probability density of finding the object at a given point in space and time, and from this probability density, the average or expected values for any measurable quantity (location, momentum, etc.) can be obtained.

NOTES

INTRODUCTION

Page 2 *Action # 333* (National Comics. 1966), written by Leo Dorfman, drawn by Al Plastino.

Page 4 *World's Finest # 93* (National Comics, 1958), reprinted in *World's Finest Comics Archives Volume 2* (DC Comics, 2001). Written by Edmond Hamilton, drawn by Dick Sprang.

Page 7 "middle-class sensibilities" *Comics, Comix & Graphic Novels: A History of Comics*, Roger Sabin (Phaidon Press, 1996).

Page 7 "yellow journalism" *The Classic Era of American Comics*, Nicky Wright (Contemporary Books, 2000).

Page 7 "firmly established until 1933" *Comic Book Culture: An Illustrated History*, Ron Goulart (Collectors Press Inc., 2000).

Page 8 "big money in the Depression" *The Pulps: Fifty Years of American Pop Culture*, compiled and edited by Tony Goodstone (Chelsea House, 1970).

Page 8 "Superman was the brain child" *The Illustrated History of Superhero Comics of the Golden Age*, Mike Benton (Taylor Publishing Co., 1992); *Superman. The Complete History*, Les Daniels (Chronicle Books, 1998).

Page 11 "before someone noticed and complained" *Men of Tomorrow: Geeks, Gangsters and the Birth of the Comic Book*, Gerard Jones (Basic Books, 2004).

Page 11 "Dr. Fredric Wertham's 1953..." *Seduction of the Innocent*, Fredric Wertham (Rinehart Press, 1953).

Page 12 "The U.S. Senate Subcommittee" *Seal of Approval, The History of the Comics Code*, Amy Kiste Nyberg (University of Mississippi Press, 1998).

Page 12 "Declining sales from the loss" *Comic Book Nation*, Bradford W. Wright (Johns Hopkins University Press, 2001).

Page 13 *The Atom # 21* (National Comics, Oct./Nov. 1965). Written by Gardner Fox, drawn by Gil Kane.

Page 13 "Give us back our eleven days!" *Encyclopedia Britannica* (William Benton, Chicago) vol. 4, pg. 619 (1968).

Page 14 *Brave and the Bold # 28* (National Comics, 1960), reprinted in *Justice League of America Archives Volume 1* (DC Comics, 1992). Written by Gardner Fox, drawn by Mike Sekowsky.

Page 14 "Why take the time...?" *Man of Two Worlds, My Life in Science Fiction and Comics*, Julius Schwartz with Brian M. Thomsen (Harper-Entertainment, 2000).

Page 14 "The Hugo Award winner Alfred Bester..." *Star Light, Star Bright*, Alfred Bester (Berkley Publishing Company, 1976).

Page 14 "as reflected in this joke:" Lance Smith, private communication (2001).

Page 16 "physics is not about having memorized..." Hellmut Fritszche, private communication (1979).

CHAPTER I

Page 21 *Superman # 1* (National Comics, June 1939), reprinted in *Superman Archives Volume 1* (DC Comics, 1989). Written by Jerry Siegel and drawn by Joe Shuster.

Page 21 *Superman # 330* (DC Comics, Dec. 1978). Written by Martin Pasko and Al Shroeder and drawn by Curt Swan and Frank Chiaramonte.

Page 21 *Action Comics # 262* (National Comics, 1960). Written by Robert Bernstein and drawn by Wayne Boring.

Page 23 "In his very first story..." *Action # 1* (National Comics, June 1938), reprinted in *Superman # 1* (National Comics, June 1939), reprinted in *Superman Archives Volume 1* (DC Comics, 1989). Written by Jerry Siegel and drawn by Joe Shuster.

Page 23 "In the 1940s and 1950s" FN "How a radio-active element" *Superman: The Complete History*, Les Daniels (Chronicle Books, 1998).

Page 25 "Whether we wish to describe the trajectory..." *The Principia: Mathematical Principles of Natural Philosophy*, Sir Isaac Newton, translated by I. Bernard Cohen and Anne Whitman (University of California Press, 1999); Newton's Principia for the Common Reader, S. Chandrasekhar (Oxford University Press, 1995).

Page 32 *Action # 23* (National Comics, 1940), reprinted in *Superman: The Action Comics Archives Volume 2* (DC Comics, 1998). Written by Jerry Siegel and drawn by Joe Shuster and the Superman Studio.

CHAPTER 2

Page 34 "As if describing the laws of motion..." *The Principia: Mathematical Principles of Natural Philosophy*, Sir Isaac Newton, translated by

I. Bernard Cohen and Anne Whitman (University of California Press, 1999); *Newton's Principia for the Common Reader*, S. Chandrasekhar (Oxford University Press, 1995).

Page 35 "This is the true meaning..." *The Life of Isaac Newton*, Richard Westfall (Cambridge University Press, 1994); Newton's Gift, David Berlinski (Touchstone, 2000); Isaac Newton, James Gleick (Pantheon Books, 2003).

Page 37 "cubical planets such as the home world of Bizarro" *Superman: Tales of the Bizarro World* trade paperback (DC Comics, 2000).

Page 39 "While planets in our own solar system" Astronomy. *The Solar System and Beyond* (2nd edition), Michael A. Seeds (Brooks/Cole, 2001).

Page 40 "To be precise, 73 percent of the" Just Six Numbers. *The Deep Forces that Shape the Universe*, Martin Rees (Basic Books, 2000).

Page 41 "The fusion process speeds up as the star generates..." The time necessary for iron and nickel synthesis can vary from several weeks to less than a day, depending on the star's mass. See "The Evolution and Explosion of Massive Stars," S. E. Woolsey and A. Heger, Rev. *Modern Physics* vol. 74, p. 1015 (Oct. 2002).

Page 42 "Only five years earlier..." W. Baade and F. Zwicky, *Physical Review* vol. 45, p. 138 (1934).

CHAPTER 3

Page 46 "This all changed with a golf game" *Man of Two Worlds, My Life in Science Fiction and Comics*, Julius Schwartz with Brian M. Thomsen (HarperEntertainment, 2000).

Page 46 "Instead he and Jack Kirby created a new superhero team from whole cloth." *Stan Lee and the Rise and Fall of the American Comic Book*, Jordan Raphael and Tom Spurgeon (Chicago Review Press, 2003); *Tales to Astonish: Jack Kirby, Stan Lee and the American Comic Book Revolution* by Ronin Ro (Bloomsbury, 2004).

Page 46 Footnote. "Those who were involved in publishing DC and Marvel comics..." *Alter Ego # 26*, pg. 21 (TwoMorrows Publishing, July 2003).

Page 46 *Fantastic Four # 1* (Marvel Comics, 1961), reprinted in *Marvel Masterworks: Fantastic Four Volume 1* (Marvel Comics, 2003). Written by Stan Lee and Jack Kirby.

Page 48 *Amazing Fantasy # 15* (Marvel Comics, 1962), reprinted in *Marvel Masterworks: Amazing Spider-Man Volume 1* (Marvel Comics, 2003). Written by Stan Lee and drawn by Steve Ditko.

Page 48 *Amazing Spider-Man # 44–46* (Marvel Comics, 1964), reprinted in *Marvel Masterworks: Amazing Spider-Man Volume 5* (Marvel, 2004). Written by Stan Lee and drawn by Steve Ditko.

Page 48 *Amazing Spider-Man # 121* (Marvel Comics, June 1973), reprinted in *Spider-Man: The Death of Gwen Stacy* trade paperback (Marvel Comics, 1999). Written by Gerry Conway and drawn by Gil Kane.

Page 48 *Amazing Spider-Man # 39* (Marvel Comics, Aug. 1964), reprinted in *Marvel Masterworks: Amazing Spider-Man Volume 4* (Marvel Comics, 2004). Written by Stan Lee and drawn by John Romita.

Page 51 "This question was listed…" Wizard: *The Comics Magazine # 100* (Gareb Shamus Enterprises, Jan. 2000).

Page 51 The towers of the George Washington Bridge are actually 604 feet above the water. See *The Bridges of New York*, Sharon Reier (Dover, 2000).

Page 53 "Col. John Stapp rode an experimental" *Wings & Airpower* magazine, Nick T. Spark (Republic Press, July 2003).

Page 54 *Spider-Man Unlimited # 2* (Marvel Comics, May 2004). Written by Adam Higgs and drawn by Rick Mays.

Page 55 *Wizard: The Comics Magazine # 104* (Gareb Shamus Enterprises, Apr. 2001).

Page 56 *Peter Parker: Spider-Man # 45* (Marvel Comics, Aug. 2002). Written by Paul Jenkins and drawn by Humberto Ramos.

CHAPTER 4

Page 57 "It was a dark and stormy night…" *Showcase # 4* (National Comics, Oct. 1956), reprinted in *Flash Archives Volume 1* (DC Comics, 1996). Written by Robert Kanigher and drawn by Carmine Infantino.

Page 57 Footnote. *Flash Comics # 110* (National Comics, Dec.–Jan. 1960), reprinted in *Flash Archives vol. 2* (DC Comics, 2000). Written by John Broome and drawn by Carmine Infantino.

Page 59 "Captain Cold, one of the first…" See, for example, *Showcase # 8* (National Comics, June 1957), reprinted in *Flash Archives Volume 1* (DC Comics, 1996). Written by John Broome and drawn by Carmine Infantino.

Page 59 "While friction's basic properties were…" *History of Tribology*, 2nd edition, Duncan Dowson (American Society of Mechanical Engineers, 1999).

Page 62 "This mechanism was proposed…" D. Hu, B. Chan, and J. W. M. Bush, *Nature* 424, pp. 663–666 (2003).

Page 62 Flash # 117 (National Comics, Dec. 1960), reprinted in *Flash Archives Volume 3* (DC Comics, 2002). Written by John Broome and drawn by Carmine Infantino.

Page 63 "One can, of course, move faster than the speed of sound…" "Breaking the Sound Barrier," Chuck Yeager, *Popular Mechanics* (Nov. 1987); *Yeager: An Autobiography*, Chuck Yeager (Bantam, reissue edition, 1986).

Page 64 Fig. 10 *Flash # 117* (second story) (National Comics, Dec. 1960), reprinted in *Flash Archives Volume 3* (DC Comics, 2002). Written by Gardner Fox and drawn by Carmine Infantino.

Page 67 *Flash # 124* (National Comics, Nov. 1961), reprinted in *Flash Archives Volume 3* (DC Comics, 2002). Written by John Broome and drawn by Carmine Infantino.

CHAPTER 5

Page 69 "In his first appearance..." *Tales to Astonish # 27* (Marvel Comics, Jan. 1962), reprinted in *Essential Ant-Man, Volume 1* (Marvel Comics, 2002). Written by Stan Lee and Larry Lieber and drawn by Jack Kirby.

Page 70 *The Incredible Shrinking Man: A Novel*, Richard Matheson (Tor Books, 2001).

Page 70 *Tales to Astonish # 35* (Marvel Comics, Sept. 1962), reprinted in *Essential Ant-Man, Volume 1* (Marvel Comics, 2002). Written by Stan Lee and Larry Lieber and drawn by Jack Kirby.

Page 71 "Given that ants actually communicate..." *Journey to the Ants*, Bert Holldobler and Edward O. Wilson (Belknap Press of Harvard University Press, 1994).

Page 71 "discussing the construction of 'time machines'" "Wormholes, Time Machines and the Weak Energy Condition," Michael S. Morris, Kip S. Thorne, and Ulvi Yurtsever, *Phys. Rev. Lett.* 61, 1446 (1998); "Warp Drive and Causality," Allen E. Everett, *Phys. Rev. D* 53, 7365 (1996); "Closed Timelike Curves Produced by Pairs of Moving Cosmic Strings: Exact Solutions," J. Richard Gott III, *Phys. Rev. Lett* 66, 1126 (1991); *Black Holes and Time Warps*, K. S. Thorne (Norton, 1994).

Page 72 See the novelization *Fantastic Voyage*, Isaac Asimov (based on a screenplay by Harry Kleiner) (Bantam Books, 1966).

Page 73 "As discussed in Isaac Asimov's..." Ibid, chapter 4.

Page 74 *Fantastic Voyage II: Destination Brain*, Isaac Asimov (Doubleday, 1987).

CHAPTER 6

Page 78 "Making his debut..." *More Fun Comics # 73* (National Comics, Nov. 1941), written by Mort Weisinger and drawn by artist Paul Norris.

Page 79 Footnote. *More Fun Comics # 106* (National Comics, Nov.-Dec. 1945), written by Joe Samachson and drawn by artist Louis Cazeneuve.

Page 79 Footnote "Sub-Mariner" *Marvel Comics # 1* (Timely Comics, Oct. 1939) written and drawn by Bill Everett.

Page 79 "alveoli" *Physics in Biology and Medicine*, 3rd Edition, Paul Davidovits (Academic Press, 2007) p.129.

Page 80 "water layer, gas exchange" Ibid, p. 132.

Page 80 "surface tension" *Capillarity and Wetting Phenomena: Drops, Bubbles, Pearls, Waves*, Pierre-Gilles de Gennes, Francoise Brochard-Wyart and David Quere (Springer, 2003).

Page 81 "pulmonary surfactant" *The Surgical Review*, 2nd ed., Pavan Alturi, Giorgos C. Karakousis, Paige M. Porrett and Larry R. Kaiser (Lippincott Williams & Wilkins, 2005) p. 369.

Page 82 "surfactants make water wetter..." *Fragile Objects: Soft Matter,*

Hard Science and the Thrill of Discovery, Pierre-Gilles de Gennes and Jacques Badoz (Springer, 1996).

Page 83 "a year later..." *More Fun Comics # 85* (National Comics, Nov1942), written by unknown and drawn by artist Louis Cazeneuve.

Page 85 *Justice League of America # 200* (DC Comics, Mar. 1982) written by Gerry Conway and drawn by George Perez and Brett Breeding.

Page 86 "dates back to ancient Mesopotamia" *The Sumerians: Their History, Culture and Character*, Samuel Noah Kramer (University of Chicago Press, 1971).

Page 89 "imploding tanker car" http://www.youtube.com/watch?v= E_hci9vrvfw

Page 90 "fish gas bladder" *Biology of Fishes* (2nd edition), Carl Bond (Brooks Cole, 1996).

Page 91 "chemical engineering professor Ed Cussler" "Will Humans Swim Faster or Slower in Syrup?" Brian Gettelfinger and E.L. Cussler, *AIChE Journal*, vol. 50, p. 2646 (2004).

Page 91 "dolphins are continuously shedding their skin" "The Physics of Swimming," Karen C. Fox, *Discover Magazine*, Feb. 2006.

CHAPTER 7

Page 95 "Dragline silk webbing..." "Stronger than Spider Silk," Eric J. Lerner, *The Industrial Physicist*, vol. 9, no. 5, p. 21 (Oct./Nov. 2003); Nature, vol. 423, pg. 703 (2003).

Page 95 "Spider-Man is able to alter the material properties..." *Spider-Man Annual # 1* (Marvel Comics, June 1963); reprinted in *Marvel Masterworks: Amazing Spider-Man Volume 2* (Marvel Comics, 2002). Written by Stan Lee and drawn by Steve Ditko.

Page 95 "Similarly, real spiders can control" C. L. Craig et al. *Molecular Biol. Evolution*, vol. 17, 1904 (2000); Frasier I. Bell, Iain J. McEwen and Christopher Viney, A. B. Dalton, S. Collins, E. Munoz, J. M. Razal, V. H. Ebron, J. P. Ferraris, J. N. Coleman, B. G. Kim and R. H. Baughman, Nature, vol. 416, p. 37 (2002).

Page 95 "genetic engineering experiments..." *The Goat Farmer Magazine* (May 2002) (Capricorn Publications, New Zealand).

Page 96 "other scientists have reported..." D. Huemmerich, T. Scheibel, F. Vollrath, S. Cohen, U. Gat, and S. Ittah, *Current Biology*, vol. 14, no. 22, p. 2070 (Nov. 2004).

Page 96 "The silk-producing gene..." Jurgen Scheller, Karl-Heinz Guhrs, Frank Grosse, and Udo Conrad, *Nature Biotechnology*, vol. 19, no. 6, p. 573 (June 2001); S. R. Fahnestock and S. L. Irwin, *Appl. Microbiol. Biotechnol.* Vol. 47, p. 23 (1997).

Page 96 "As Jim Robbins discussed..." "Second Nature," Jim Robbins, *Smithsonian*, vol. 33, no. 4, p. 78 (July 2002).

CHAPTER 8

Page 97 "he could ride on top of an ant..." See, for example, *Tales to Astonish # 35* (Marvel Comics, Sept. 1962), reprinted in *Essential Ant-Man Volume 1* (Marvel Comics, 2002). Written by Stan Lee and Larry Lieber and drawn by Jack Kirby.

Page 98 "to instruct hundreds of them..." *Tales to Astonish # 36* (Marvel Comics, Oct. 1962), reprinted in *Essential Ant-Man Volume 1* (Marvel Comics, 2002). Written by Stan Lee and Larry Lieber and drawn by Jack Kirby.

Page 98 *Tales to Astonish # 37* (Marvel Comics, Nov. 1962), reprinted in *Essential Ant-Man Volume 1* (Marvel Comics, 2002). Written by Stan Lee and Larry Lieber and drawn by Jack Kirby.

Page 101 "an ingenious series of levers..." *Intermediate Physics for Medicine and Biology* (3rd ed.), Russell K. Hobbie (American Institute of Physics, 2001); *Biomechanics of Human Motion*, M. Williams and H. R. Lissner (Saunders Press, 1962).

Page 101 "essentially the same as a fishing rod..." See, for example, *The Way Things Work*, David Macaulay (Houghton Mifflin, 1988) for an amusing illustration of different lever configurations.

Page 101 "The ratio of moment arms is thus 1:7," *Back-of-the-Envelope Physics*, Clifford Swartz (Johns Hopkins University Press, 2003).

Page 103 "What determines how high you can leap?" *On Growth and Form*, D'Arcy Thompson (Cambridge University Press, 1961).

Page 104 "It is an easy consequence of anthropomorphism," Ibid, page 27.

CHAPTER 9

Page 105 "Beware the Atomic Grenade" *Flash # 122* (National Comics, Aug. 1961), reprinted in *Flash Archives Volume 3* (DC Comics, 2002). Written by John Broome and drawn by Carmine Infantino.

Page 106 *Tales to Astonish # 50* (Marvel Comics, Dec. 1963), reprinted in *Essential Ant-Man Volume 1* (Marvel Comics, 2002). Written by Stan Lee and drawn by Jack Kirby.

Page 107 "dropped from the Masters of Evil by Egghead" *Avengers # 228* (Marvel Comics, Feb. 1983) written by Roger Stern and drawn by Al Milgrom.

Page 110 "innovations in gyroscope design employ electrostatically levitated micro-discs," "Micromachined rotating gyroscope with electromagnetically levitated rotor," Wu, X.-S.; Chen, W.-Y.; Zhao, X.-L.; Zhang, W.-P., *Electronics Letters* vol. 42, p. 912 (2006).

Page 111 *Avengers # 139* (Marvel Comics, Sept. 1975) reprinted in *Essential Avengers Volume 6* (Marvel Comics, 2008). Written by Steve Englehart and drawn by George Tuska.

Page 111 "Helicopters hover in the air" *Principles of Helicopter Aerodynamics* (2nd ed.), J. Gordon Leishman (Cambridge University Press, 2006).

Page 113 "When the Norse god Thor needed to travel" *Journey into Mystery # 83* (Marvel Comics, Aug. 1962) reprinted in *Essential Uncanny Thor Volume 1* (Marvel Comics, 2001) written by Stan Lee and Larry Leiber and drawn by Jack Kirby.

Page 113 *Bartman Comics # 3* (Bongo Comics, Aug. 1994) written by Steve Vance and drawn by Bill Morrison.

Page 113 "the X-Men villain the Blob" *X-Men # 3* (Marvel Comics, Jan. 1964) reprinted in *Essential Uncanny X-Men Volume 1* (Marvel Comics, 1999) written by Stan Lee and drawn by Jack Kirby.

Page 114 "Observations of the Coma cluster of galaxies" Fritz Zwicky, *Helvetica Phys. Acta* vol. 6, p. 110 (1933).

Page 115 "there must be more mass in this galaxy than we can see" *In Search of Dark Matter*, Ken Freeman and Geoff McNamara (Springer, 2006).

CHAPTER 10

Page 118 "Galileo was perhaps not the first person to notice..." *Galileo's Pendulum*, Roger G. Newton (Harvard University Press, 2004).

Page 120 "A human vocal cord is..." *Intermediate Physics for Medicine and Biology* (3rd ed.), Russell K. Hobbie (American Institute of Physics, 2001).

Page 121 "Alternatively, if one is too close to the source,..." *On Growth and Form*, D'Arcy Thompson (Cambridge University Press, 1961).

Page 122 *Atom # 4* (National Comics, Dec./Jan. 1962), reprinted in *Atom Archives Volume One* (DC Comics, 2001). Written by Gardner Fox and drawn by Gil Kane.

Page 123 "An insect's eye is very good at..." C. J. van der Horst, "The Optics of the Insect Eye," *Acta Zool*, p. 108 (1933).

CHAPTER 11

Page 125 "or of Catwoman's whip..." *JLA # 13* (DC Comics, April 1998). Written by Grant Morrison and drawn by Howard Porter. The fact that the tip of Catwoman's whip is moving at roughly twice the speed of sound (the source of the loud crack it creates) was emphasized when she snapped the villain Prometheus in a particularly vulnerable area.

Page 125 DC: *The New Frontier # 2* (DC Comics, Apr. 2004), also reprinted in *DC: The New Frontier Volume 1* (DC Comics, 2004). Written and drawn by Darwyn Cooke.

Page 125 *Flash # 202* (vol. 2) (DC Comics, Nov. 2003). Written by Geoff Johns and drawn by Alberto Dose.

Page 126 "The Special Theory of Relativity can be boiled down..." *What Is Relativity?* L. D. Landau and G. B. Romer (Translated by N. Kemmer) (Dover, 2003).

Page 127 "with a sweeping motion..." *Flash # 124* (National Comics, Nov. 1961), reprinted in *Flash Archives Volume 3* (DC Comics, 2002). Written by John Broome and drawn by Carmine Infantino.

Page 127 "In order for this to be true, Einstein argued..." *Relativity and Common Sense*, Hermann Bondi (Dover, 1980).

Page 127 "two observers can disagree...." *What Is Relativity?* L. D. Landau and G. B. Romer (Translated by N. Kemmer) (Dover, 2003).

Page 128 *Flash # 175* (National Comics, Dec. 1967), reprinted in *Superman vs. the Flash* (DC Comics, 2005). Written by E. Nelson Bridwell and drawn by Ross Andru.

Page 128 "Negative Man" *My Greatest Adventure # 80* (National Comics, June 1963), reprinted in *The Doom Patrol Archives Volume 1* (DC Comics, 2002). Written by Arnold Drake with Bob Haney and drawn by Bruno Premiani.

Page 128 "Captain Marvel" *Avengers # 227* (Marvel Comics, Jan. 1983). Written by Roger Stern and drawn by Sal Buscema.

Page 129 *Flash # 132* (National Comics, Nov. 1962), reprinted in *Flash Archives Volume 4* (DC Comics, 2006). Written by John Broome and drawn by Carmine Infantino.

Page 129 *JLA # 89* (DC Comics, late Dec. 2003). Written by Joe Kelly and drawn by Doug Mahnke.

Page 130 *Legion of Superheroes # 16* (vol. 5), (DC comics, May 2006). Written by Mark Waid and drawn by Barry Kitson.

Page 132 *Flash # 141* (National Comics, Dec. 1963), reprinted in *Flash Archives Volume 5* (DC Comics, 2009). Written by John Broome and drawn by Carmine Infantino.

Page 133 "back in the 1950s, a group of physics majors....", private communication, Hal Weisinger.

CHAPTER 12

Page 137 "As you grew and matured, you needed" See, for example, *The Stuff of Life*, Eric P. Widmaier (Henry Holt and Company, 2002); *The Machinery of Life*, David S. Goodsell (Springer-Verlag, 1998); and *Stories of the Invisible*, Philip Ball (Oxford University Press, 2001).

Page 139 "Studies of the decay of radioactive nuclei..." *The Elusive Neutrino: A Subatomic Detective Story*, Nickolas Solomey (W. H. Freeman & Company, 1997).

Page 141 "An automobile's efficiency..." *New Directions in Race Car Aerodynamics: Designing for Speed,* Joseph Katz (Bentley Publishers, 1995).

Page 142 Footnote. "Positron Production in Multiphoton Light-by-Light Scattering," D. L. Burk et al., Phys. Rev. Lett. 79, 1626 (1997).

Page 144 *Flash # 106* (DC Comics, May 1959), reprinted in *Flash Archives Volume 1* (DC Comics, 1996). Written by John Broome and drawn by Carmine Infantino.

Page 144 *Flash # 25* (vol. 2) (DC Comics, Apr. 1989). Written by William Messner-Loebs and drawn by Greg LaRocque.

Page 145 "physicists were confused about energy..." See, for example, *Warmth Disperses and Time Passes: The History of Heat,* Hans Christian von Baeyer (Modern Library, 1998).

Page 147 "At one point in Flash comics..." See, for example, *Flash # 24* (vol. 2) (DC Comics, Apr. 1989). Written by William Messner-Loebs and drawn by Greg LaRocque.

Page 147 "consider some basic chemistry" See, for example, *The Periodic Kingdom,* P. W. Atkins (Basic Books, 1995).

Page 149 "all of life is possible because the mass of a helium nucleus" *Just Six Numbers,* Martin Rees (Basic Books, 1999).

Page 151 *Watchmen* (DC Comics, 1986, 1987) Written by Alan Moore and drawn by Dave Gibbons.

Page 153 "The volume of oxygen use by a runner..." *Energies: An Illustrated Guide to the Biosphere and Civilization,* Vaclav Smil (MIT Press, 1999).

Page 153 "the Earth's atmosphere contains..." There are approximately two hundred million trillion moles of gas in the Earth's atmosphere, while each mole contains Avogadro's number (0.6 trillion trillion) molecules. The Earth's atmosphere is thus estimated to contain 0.12 billion trillion trillion trillion gas molecules.

Page 154 *Flash # 167* (DC Comics, Feb. 1967). Written by John Broome and drawn by Carmine Infantino.

Page 155 "This is one reason why golf balls have dimples." *The Physics of Golf,* Theodore P. Jorgensen (Springer, second edition, 1999); *Golf Balls, Boomerangs and Asteroids: The Impact of Missiles on Society,* Brian H. Kaye (VCH Publishers, 1996); *500 Years of Golf Balls: History and Collector's Guide,* John F. Hotchkiss (Antique Trader Books, 1997).

Page 156 Footnote. Technically, *Superman # 130* ascribed the wrong mechanism to kryptonite's resistance to air friction when it claimed "kryptonite can't combine chemically with oxygen, which causes combustion." It may indeed not be chemically reactive with oxygen, but the heat generated when an object moves at high velocity through the atmosphere is due to the work needed to push the air molecules out of the way, and is a purely physical, rather than chemical, process.

Page 156 "The first such character" *Superman: The Complete History,* Les Daniels (Chronicle Books, 1998).

Page 157 "Not to be outdone..." *DC Comics: Sixty Years of the World's Favorite Comic Book Heroes*, Les Daniels (Bulfinch Press, 1995).

Page 157 "as far as most fans of the Silver Age..." "Comics That Didn't Really Happen," by Mark Evanier, reprinted in *Comic Books and Other Necessities of Life* (TwoMorrows Publishing, 2002).

CHAPTER 13

Page 158 *Showcase # 34* (DC Comics, Sept./Oct. 1961), reprinted in *Atom Archives Volume 1* (DC Comics, 2001). Written by Gardner Fox and drawn by Gil Kane.

Page 160 Fig. *20 Atom # 4* (DC Comics, Dec./Jan. 1962), reprinted in *Atom Archives Volume One* (DC Comics, 2001). Written by Gardner Fox and drawn by Gil Kane.

Page 161 *Atom # 2* (DC Comics, Aug./Sept. 1962), reprinted in *Atom Archives Volume 1* (DC Comics, 2001). Written by Gardner Fox and drawn by Gil Kane.

Page 161 "The field of thermodynamics..." *Warmth Disperses and Time Passes: The History of Heat*, Hans Christian von Baeyer (Modern Library, 1998); *A Matter of Degrees*, Gino Segre (Penguin Books, 2002).

Page 164 "Another example:" *Energies: An Illustrated Guide to the Biosphere and Civilization*, Vaclav Smil (MIT Press, 1999).

Page 166 "This concept, called 'entropy,' is..." *Understanding Thermodynamics*, H. C. Van Ness (Dover Publications, 1969).

Page 168 *West Coast Avengers # 42* (Marvel Comics, Mar. 1989). Written and drawn by John Byrne.

Page 169 "Could I use the talents of the Atom..." *Warmth Disperses and Time Passes: The History of Heat*, Hans Christian von Baeyer (Modern Library, 1998).

Page 172 "radio-wave background radiation" *Temperatures Very Low and Very High*, Mark W. Zemansky (Dover Books, 1964).

Page 173 "Many of the elder statesmen of physics" *Philosophy of Science: The Historical Background*, Joseph J. Kockelmans (ed.) (Transaction Publishers, 1999).

Page 173 Planck quote *Scientific Autobiography and Other Papers*, Max K. Planck (translated by F. Gaynor) (Greenwood Publishing Group, 1968).

Page 173 "A key development" *An Introduction to Stochastic Processes in Physics*, Don S. Lemons (Johns Hopkins University Press, 2002).

Page 173 "it was not until 1905..." *Investigations of the Theory of the Brownian Movement*, Albert Einstein (Dover, 1956).

Page 174 "The random collisions of the air on our eardrums..." "How the Ear's Works Work," A. J. Hudspeth, *Nature* 341, 397 (1989); "Brownian Motion and the Ability to Detect Weak Auditory Signals," I. C.

Gebeshuber, A. Mladenka, F. Rattay, and W. A. Svrcek-Seiler, *Chaos and Noise in Biology and Medicine,* ed. C. Taddei-Ferretti (World Scientific, 1998). Note that this is not the high-pitch tone that many of us hear. That high-frequency sound is most likely tinnitus, resulting from damage (either from loud noises or old age) to the cilia that detect sound waves in the inner ear.

CHAPTER 14

Page 175 "Stan Lee, head writer and editor, ..." *Excelsior!: The Amazing Life of Stan Lee,* Stan Lee and George Mair (Fireside, 2002).

Page 175 *X-Men # 1* (Marvel Comics, Sept. 1963), reprinted in *Marvel Masterworks: X-Men Volume 1* (Marvel Comics, 2002). Written by Stan Lee and drawn by Jack Kirby.

Page 176 *X-Men # 47* (Marvel Comics, Aug. 1968). Written by Arnold Drake and drawn by Werner Roth.

Page 176 *X-Men # 8* (Marvel Comics, Nov. 1964), reprinted in *Marvel Masterworks: X-Men Volume 1* (Marvel Comics, 2002). Written by Stan Lee and drawn by Jack Kirby.

Page 176 "A snowflake is created when..." *The Snowflake: Winter's Secret Beauty,* Kenneth G. Libbrecht and Patricia Rasmussen (Voyageur Press, 2003).

Page 177 "Einstein's equation for how far a fluctuating atom..." *Investigations of the Theory of the Brownian Movement,* Albert Einstein (Dover, 1956).

Page 177 "The exact details..." "Instabilities and Pattern Formation in Crystal Growth," J. S. Langer, *Reviews of Modern Physics* 52, 1 (1980).

Page 179 Fig. 22 *Amazing Spider-Man # 92* (Marvel Comics, Jan. 1971), reprinted in *Spider-Man: The Death of Captain Stacy* (Marvel Comics, 2004). Written by Stan Lee and drawn by Gil Kane and John Romita.

Page 180 *All-New, All-Different X-Men # 1* (Marvel Comics, 1975), reprinted in *Marvel Masterworks: Uncanny X-Men Volume 1* (Marvel Comics, 2003). Written by Len Wein and drawn by David Cockrum.

Page 181 "At its core, the weather..." *The Essence of Chaos,* Edward Lorenz (University of Washington Press, 1996); *The Coming Storm,* Mark Masline (Barron's, 2002).

Page 183 Fig. 24 *X-Men # 145* (Marvel Comics, May 1981). Written by Chris Claremont and drawn by Dave Cockrum and Joe Rubinstein.

Page 184 "A final thought..." *Lord Kelvin and the Age of the Earth,* Joe D. Burchfield (University of Chicago Press, 1990); *Degrees Kelvin,* David Lindley (Joseph Henry Press, 2004).

CHAPTER 15

Page 186 *Tales of Suspense # 39* (Marvel Comics, Mar. 1963), reprinted in *Marvel Masterworks: The Invincible Iron Man Volume 1* (Marvel Comics, 2003). Written by Stan Lee and Larry Lieber and drawn by Don Heck.

Page 187 "When the Melter first appeared" *Tales of Suspense # 47* (Marvel Comics, Nov. 1963), reprinted in *Marvel Masterworks: The Invincible Iron Man Volume 1* (Marvel Comics, 2003). Written by Stan Lee and drawn by Steve Ditko.

Page 187 "When this happens, a chemical bond forms..." *The Periodic Kingdom*, P. W. Atkins (Basic Books, 1995).

Page 188 "What determines the exact temperature and pressure..." *Gases, Liquids and Solids*, D. Tabor (Cambridge University Press, 1979).

Page 188 "In a conventional oven..." *On Food and Cooking*, Harold McGee (Scribner, revised and updated edition, 2004); The Science of Cooking, Peter Barham (Spring, 2001).

Page 192 *Tales of Suspense # 90* (Marvel Comics, Jun. 1967), *Essential Iron Man Volume 2* (Marvel Comics, 2004). Written by Stan Lee and drawn by Gene Colan.

Page 192 "Such a microwave-based 'heat ray' that..." "Report: Raytheon 'heat beam' weapon ready for Iraq," *Boston Business Journal*, Dec. 1, 2004.

CHAPTER 16

Page 194 Footnote. *West Coast Avengers # 13* (vol. 2) (Marvel Comics, Oct. 1986). Written by Steve Englehart and drawn by Al Milgrom.

Page 195 *Adventure # 247* (National Comics, April 1958), reprinted in *Legion of Superheroes Archives Volume 1* (DC Comics, resissue edition, 1991). Written by Otto Binder and drawn by Al Plastino.

Page 196 *Adventure # 353* (National Comics, Feb. 1967) written by Jim Shooter and drawn by Curt Swan.

Page 197 *Amazing Spider-Man # 9* (Marvel Comics, Feb. 1963), reprinted in *Marvel Masterworks: Amazing Spider-Man Volume 1* (Marvel Comics, 2003). Written by Stan Lee and drawn by Steve Ditko.

Page 198 "is approximately the same size as a carbon atom..." *Back-of-the-Envelope Physics*, Clifford Swartz (Johns Hopkins University Press, 2003).

Page 200 "Wingless Wizard's anti-gravity discs" First seen in *Strange Tales # 118* (Marvel Comics, March 1964), reprinted in *Essential Human Torch Volume 1* (Marvel Comics, 2003). Written by Stan Lee and drawn by Dick Ayers.

Page 201 *Flash # 208* (vol. 2) (DC Comics, May 2004). Written by Geoff Johns and drawn by Howard Porter.

Page 202 "George de Mestral's investigations…" *Why Didn't I Think of That?*, Allyn Freeman and Bob Golden (John Wiley and Sons, 1997).

Page 202 "Evidence for Van der Waals Adhesion in Gecko Setae," K. Autumn, M. Sitti, Y. A. Liang, A. M. Peattie, W. R. Hansen, S. Sponberg, T. W. Kenny, R. Fearing, J. N. Israelachvili, and R. J. Full, *Proc. National Acad. Sciences* 99, 12,252 (2002).

Page 203 "development of 'gecko tape,'" "Microfabricated Adhesive Mimicking Gecko Foot-Hair," A. K. Geim, S. V. Dubonos, 2, I. V. Grigorieva, K. S. Novoselov, A. A. Zhukov, and S. Yu. Shapoval, *Nature Materials* 2, 461 (2003).

CHAPTER 17

Page 205 *Amazing Spider-Man # 9* (Marvel Comics, Feb. 1963), reprinted in *Marvel Masterworks: Amazing Spider-Man Volume 1* (Marvel Comics, 2003). Written by Stan Lee and drawn by Steve Ditko.

Page 207 *Superman # 1* (National Comics, June 1939), reprinted in *Superman Archives Volume 1* (DC Comics, 1989). Written by Jerry Siegel and drawn by Joe Shuster.

Page 208 *Adventure # 301* (National Comics, Oct. 1962) written by Jerry Siegel and drawn by John Forte.

Page 209 *Amazing Spider-Man Annual # 1* (Marvel Comics, Feb. 1964), reprinted in *Marvel Masterworks: Amazing Spider-Man Volume 2* (Marvel Comics, 2002). Written by Stan Lee and drawn by Steve Ditko.

Page 210 "a comic-book writer would generate a script…" *Man of Two Worlds, My Life in Science Fiction and Comics*, Julius Schwartz with Brian M. Thomsen (HarperEntertainment, New York), 2000.

Page 210 "in 1965, to pick a particular year…" *Comic Book Marketplace # 99* (Gemstone Publishing, Feb. 2003).

Page 210 "With so many stories being created every month…" *Stan Lee and the Rise and Fall of the American Comic Book*, Jordan Raphael and Tom Spurgeon (Chicago Review Press, 2003); *Tales to Astonish: Jack Kirby, Stan Lee and the American Comic Book Revolution*, Ronin Ro (Bloomsbury, 2004).

Page 212 *Daredevil # 2* (Marvel Comics, June 1964), reprinted in *Marvel Masterworks: Daredevil Volume 1* (Marvel Comics, 2004). Written by Stan Lee and drawn by Joe Orlando.

CHAPTER 18

Page 213 "a perfect illustration of one of the fundamental…" *Amazing Spider-Man # 9* (Marvel Comics, Feb. 1963), reprinted in *Marvel Master-*

works: *Amazing Spider-Man Volume 1* (Marvel Comics, 2003). Written by Stan Lee and drawn by Steve Ditko.

Page 214 "This phenomenon, termed the Ampere effect,..." *Electric Universe: The Shocking True Story of Electricity*, David Bodanis (Crown, 2005).

Page 215 *Daredevil # 2* (Marvel Comics, June 1964), reprinted in *Marvel Masterworks: Daredevil Volume 1* (Marvel Comics, 2004). Written by Stan Lee and drawn by Joe Orlando.

Page 215 "I'll use a nice argument..." *Discovering the Natural Laws: The Experimental Basis of Physics*, Milton A. Rothman (Dover Press, 1989).

Page 216 "The test charge therefore sees..." *Electricity and Magnetism— Berkeley Physics Course* Vol. 2, Edward M. Purcell (McGraw Hill, 1963).

Page 217 *Superboy # 1* (National Comics, Mar.–Apr. 1949). Written by Edmond Hamilton and drawn by John Sikela and Ed Dobrotka.

CHAPTER 19

Page 220 *X-Men # 1* (Marvel Comics, Sept. 1963), reprinted in *Marvel Masterworks: X-Men Volume 1* (Marvel Comics, 2002). Written by Stan Lee and drawn by Jack Kirby.

Page 221 Footnote. *Atom # 3* (DC Comics, Oct.–Nov. 1962), reprinted in *Atom Archives Volume 1* (DC Comics, 2001). Written by Gardner Fox and drawn by Gil Kane.

Page 222 "Hemoglobin is a very large molecule..." *The Machinery of Life*, David S. Goodsell (Springer-Verlag, 1998).

Page 222 Footnote. I thank Prof. E. Dan Dahlberg of the University of Minnesota and Dr. Roger Proksh of Asylum Research for demonstrating this low-tech "magnetic force microscope."

Page 224 "Materials that form magnetic domains..." *Magnets: The Education of a Physicist*, Francis Bitter (Doubleday, 1959).

Page 225 "It is through our diamagnetism" "Everyone's Magnetism," Andrey Geim, *Physics Today* 51, p. 36 (Sept. 1998); "Magnet levitation at your fingertips," A. K. Geim, M. D. Simon, M. I. Boamfa, and L. O. Heflinger. *Nature* 400, p. 323 (1999).

Page 225 See the web page for the High Field Magnetic Laboratory at the University of Nijmegen in the Netherlands: http://www.hfml.ru.nl/levitate.html for some great images of levitating objects.

Page 227 "Magnetism is, at its heart,..." *Discovering the Natural Laws: The Experimental Basis of Physics*, Milton A. Rothman (Dover Press, 1989).

Page 229 *The Dark Knight Strikes Again # 1* (DC Comics, 2001). Reprinted in *The Dark Knight Strikes Again* (DC Comics, 2003). Written and drawn by Frank Miller.

Page 229 "Nearly all commercial power plants..." *Energy: Its Use and the Environment*, Roger A. Hinrichs and Merlin Kleinbach (Brooks/Cole, 2002), Third Edition.

CHAPTER 20

Page 231 "help keep comic-book publishers solvent" *Seal of Approval, The History of the Comics Code*, Amy Kiste Nyberg (University of Mississippi Press, Jackson, Mississippi), 1998.

Page 231 Western comics at DC and Marvel... *Comic Book Culture: An Illustrated History*, Ron Goulart (Collectors Press Inc., 2000); *DC Comics: Sixty Years of the World's Favorite Comic Book Heroes*, Les Daniels (Bulfinch Press, 1995); *Marvel: Five Fabulous Decades of the World's Greatest Comics*, Les Daniels (Harry Abrams, 1991).

Page 232 "It was the Scottish physicist..." *The Man Who Changed Everything: The Life of James Clerk Maxwell*, Basil Mahon (John Wiley & Sons, 2003).

Page 235 "600 million tons of hydrogen nuclei every second..." "The Evolution and Explosion of Massive Stars," S. E. Woolsey and A. Heger, *Rev. Modern Physics* 74, p. 1015 (Oct. 2002).

Page 235 "light generated from a nuclear fusion reaction..." "How Long Does It Take for Heat to Flow Through the Sun?" G. Fiorentini and B. Rici, *Comments on Modern Physics* 1, p. 49 (1999).

Page 236 "While his shattered spine may have left him..." *X-Men # 20* (Marvel Comics, May 1966), reprinted in *Marvel Masterworks: X-Men Volume 2* (Marvel Comics, 2003). Written by Roy Thomas and Drawn by Jay Gavin.

Page 236 "The role of nerve cells..." *Synaptic Self: How Our Brains Become Who We Are*, Joseph LeDoux (Penguin, 2002); I of the Vortex, R. R. Llinas (MIT Press, 2001).

Page 238 "Fish use these organs as a form of radar," "The Shark's Electric Sense," R. Douglas Fields, *Scientific American*, Aug. 2007.

Page 238 *More Fun Comics # 77* (National Comics, Mar. 1942) Written by unknown and drawn by Paul Norris.

Page 238 *X-Men # 7* (Marvel Comics, Sept. 1964), reprinted in *Marvel Masterworks: X-Men Volume 1* (Marvel Comics, 2002). Written by Stan Lee and Drawn by Jack Kirby.

Page 239 "Television signals consist of..." *See The Way Things Work*, David Macaulay (Houghton Mifflin Company, 1988), for an accessible, graphical illustration of the mechanisms underlying television broadcasts and reception, and *How Does It Work?* Richard M. Koff (Signet, 1961), for a more technical discussion.

Page 240 "A sensitive antenna placed near this monitor..." "Electromagnetic Radiation from Video Display Units: An Eavesdropping Risk?," Wim Van Eck, *Computers and Security* 4, p. 269 (1985).

Page 240 Footnote. "Electromagnetic Eavesdropping Risks of Flat Panel Displays" Markus G. Kuhn, presented at the *Fourth Workshop on Privacy Enhancing Technologies*, Toronto, Canada, May 2004.

Page 241 "Neuroscientists have developed a research tool..." "Experimentation with a Transcranial Stimulation System for Functional Brain Mapping," G. J. Ettinger, W. E. L. Grimson, M. E. Leventon, R. Kikinis, V. Gugino, W. Cote et al. *Med. Image Analysis* 2, p. 133 (1998); "Transcranial Magnetic Stimulation and the Human Brain," M. Hallett, Nature 406, p. 147 (2000). A technical overview can be found in *Transcranial Magnetic Stimulation: A Neurochronometrics of Mind*, Vincent Walsh and Alvaro Pascual-Leone (MIT Press, 2003).

CHAPTER 21

Page 245 *Fantastic Four # 5* (Marvel Comics, July 1962), reprinted in *Marvel Masterworks: Fantastic Four Volume 1* (Marvel Comics, 2003). Written by Stan Lee and drawn by Jack Kirby.

Page 246 *Fantastic Four # 10* (Marvel Comics, Jan. 1963), reprinted in *Marvel Masterworks: Fantastic Four Volume 1* (Marvel Comics, 2003). Written by Stan Lee and drawn by Jack Kirby.

Page 246 *Fantastic Four # 16* (Marvel Comics, July 1963), reprinted in *Marvel Masterworks: Fantastic Four Volume 2* (Marvel Comics, 2005). Written by Stan Lee and drawn by Jack Kirby.

Page 246 *Fantastic Four # 76* (Marvel Comics, July 1968), reprinted in *Marvel Masterworks: Fantastic Four Volume 8* (Marvel Comics, 2005). Written by Stan Lee and drawn by Jack Kirby.

Page 247 *Atom # 5* (DC Comics, Feb./Mar. 1963), reprinted in *Atom Archives Volume 1* (DC Comics, 2001). Written by Gardner Fox and drawn by Gil Kane; Atom # 4 (DC Comics, Dec./Jan. 1962), reprinted in *Atom Archives Volume 1* (DC Comics, 2001). Written by Gardner Fox and drawn by Gil Kane; Atom # 19 (DC Comics, Jun./Jul. 1965), reprinted in *JLA: Zatanna's Search* (DC Comics, 2004). Written by Gardner Fox and drawn by Gil Kane; Justice League of America # 18 (DC Comics, Mar. 1963), reprinted in *Justice League of America Archives Volume 3* (DC Comics, 1994). Written by Gardner Fox and drawn by Mike Sekowsky; *Brave and the Bold # 53* (DC Comics, Apr.–May, 1964). Written by Bob Haney and drawn by Alexander Toth.

Page 248 "At the end of the nineteenth century,..." *Thirty Years That Shook Physics: The Story of Quantum Theory*, G. Gamow (Dover, 1985).

Page 249 "This is how the surface temperature of the sun..." *Temperatures Very Low and Very High*, Mark W. Zemansky (Dover, 1964).

Page 251 "The fact that the energy of electrons..." *The New World of Mr. Tompkins*, G. Gamow and R. Stannard (Cambridge University Press, 1999).

Page 254 "Imagine an electron orbiting a nucleus..." *The Quantum World*, J. C. Polkinghorne (Princeton University Press, 1984).

Page 256 "The lighter-than-air element Helium..." *Helium: Child of the Sun*, Clifford W. Seibel (University Press of Kansas, 1968).

CHAPTER 22

Page 259 *Showcase # 4* (National Comics, Oct. 1956), reprinted in *Flash Archives Volume 1* (DC Comics, 1996). Written by Robert Kanigher and drawn by Carmine Infantino.

Page 259 *Flash # 123* (DC Comics, Sept. 1961), reprinted in *Flash Archives Volume 3* (DC Comics, 2002). Written by Gardner Fox and drawn by Carmine Infantino.

Page 261 "The crossover meeting between the Silver Age and Golden Age Flash..." *Man of Two Worlds, My Life in Science Fiction and Comics*, Julius Schwartz with Brian M. Thomsen (HarperEntertainment, 2000).

Page 261 "So popular was this meeting of the two superteams..." See, for example, *Crisis on Multiple Earths* Volumes One, Two, Three (DC Comics, 2002, 2003, 2004).

Page 262 "Billy Batson, who could become a superhero by shouting 'Shazam!'" *DC Comics: Sixty Years of the World's Favorite Comic Book Heroes*, Les Daniels (Bulfinch Press, 1995).

Page 262 "The yearlong miniseries" *Crisis on Infinite Earths* (DC Comics, 2000). Written by Marv Wolfman and drawn by George Perez.

Page 263 "more like those described in the Marvel comic universe" *What If Classics* (Marvel Comics, 2004).

Page 263 "After a great deal of effort" *Thirty Years That Shook Physics: The Story of Quantum Theory*, G. Gamow (Dover Press, 1985).

Page 264 "two points about mathematics" *Euclid's Window: The Story of Geometry from Parallel Lines to Hyperspace*, Leonard Mlodinow (Touchstone, 2001).

Page 265 My *Greatest Adventures # 80* (DC Comics, June 1963), reprinted in *Doom Patrol Archives Volume 1* (DC Comics, 2002). Written by Arnold Drake and drawn by Bruno Premiani.

Page 265 *X-Men # 1* (Marvel Comics, Sept. 1963), reprinted in *Marvel Masterworks: X-Men Volume 1* (Marvel Comics, 2002). Written by Stan Lee and drawn by Jack Kirby.

Page 266 "research of comic-book historians" See, for example, *Comic Book Marketplace # 64* (Gemstone Publications, Nov. 1998).

Page 266 "Another publishing synchronicity" *Back Issue # 6* (TwoMorrows Publishing, Oct. 2004).

Page 267 "While we may not know how Schrödinger" *Schrödinger: Life and Thought*, Walter Moore (Cambridge University Press, 1989).

Page 268 "Given that the average values are the only quantities..." The notion that quantum mechanics is a complete theory that always provides accurate predictions of experimental observations, but does not necessarily describe an external reality, is not universally accepted among physicists. The growing body of experiments on macroscopic quantum behavior would, however, tend to support this interpretation (see "The Quantum Measurement Problem," A. J. Leggett, *Science* 307, p. 871 (2005)).

Page 270 "They posed the following situation..." *Schrödinger's Rabbits: The Many Worlds of Quantum*, Colin Bruce (Joseph Henry Press, 2004).

Page 271 *JLA # 19* (DC Comics, June 1998). Written by Mark Waid and drawn by Howard Porter.

Page 271 "In 1957, Hugh Everett III argued..." *The Many-Worlds Interpretation of Quantum Mechanics*, edited by Bryce S. DeWitt and Neill Graham (Princeton University Press, 1973). Contains a reprint of Everett's Ph.D. thesis and a longer discussion of his ideas, along with articles by other physicists. Interestingly, Everett referred to his theory as involving "Relative States," and it was DeWitt who coined the expression "Many Worlds."

Page 271 Footnote. *Animal Man # 32* (DC Comics, Feb. 1991). Written by Peter Milligan and drawn by Chas Troug.

Page 273 "A gross oversimplification of string theory..." An excellent introduction to String Theory can be found in *The Elegant Universe*, Brian Greene (W. W. Norton, 1999).

Page 273 Footnote. *Strange Tales # 129* (Marvel Comics, Feb. 1965), reprinted in *Essential Dr. Strange Volume 1* (Marvel Comics, 2001). Written by Don Rico and drawn by Steve Ditko.

Page 274 "This may be dangerous;..." See *The Pleasure of Finding Things Out*, Richard P. Feynman (Perseus Books, 1999).

Page 274 "Physicists investigating quantum gravity..." *The Fabric of Reality: The Science of Parallel Universes and Its Implications*, David Deutsch (Penguin, 1997); *Parallel Worlds*, Michio Kaku (Doubleday, 2005).

Page 274 "Recently some scientists have claimed that time travel..." *The Future of Spacetime*, Stephen W. Hawking, Kip S. Thorne, Igor Novikov, Timothy Ferris, and Alan Lightman (W. W. Norton and Company, 2002); *Time Travel in Einstein's Universe*, J. Richard Gott (Mariner Books, 2001).

Page 274 *Superman # 146* (DC Comics, July 1961). Written by Jerry Siegel and drawn by Al Plastino.

Page 278 *Avengers # 267* (Marvel Comics, May 1986), reprinted in *Avengers: Kang—Time and Time Again* (Marvel Comics, 2005). Written by Roger Stern and drawn by John Buscema and Tom Palmer.

CHAPTER 23

Page 279 *Flash # 116* (National Comics, Nov. 1960), reprinted in *Flash Archives Volume 2* (DC Comics, 2000). Written by John Broome and drawn by Carmine Infantino.

Page 280 Fig. 35 *X-Men # 130* (Marvel Comics, Feb. 1980). Written by Chris Claremont and drawn by John Byrne.

Page 280 "This is an intrinsically quantum mechanical phenomenon..." A mathematical discussion of this phenomenon can be found in *Quantum Theory of Tunneling*, Mohsen Razavy (World Scientific, 2003).

Page 281 Fig. 36 *Flash # 123* (DC Comics, Sept. 1961), reprinted in *Flash Archives Volume 3* (DC Comics, 2002). Written by Gardner Fox and drawn by Carmine Infantino.

Page 282 "Scanning Tunneling Microscope" "The Scanning Tunneling Microscope," G. Binnig and H. Rohrer, Scientific American 253, p. 40 (1985); "Vacuum tunneling: A new technique for microscopy." C. F. Quate, *Physics Today* 39, p. 26 (1986); *Solid State Electronic Devices* (5th ed.), Ben G. Streetman and Sanjay Banerjee (Prentice Hall, 2000).

Page 285 *X-Men # 141* (Marvel Comics, Jan. 1981), reprinted in *Days of Future Past* (Marvel Comics, 2004). Written by Chris Claremont and drawn by John Byrne.

Page 286 *Astonishing X-Men # 4* (Marvel Comics, Oct. 2004). Written by Joss Whedon and drawn by John Cassaday.

CHAPTER 24

Page 287 "The solid-state transistor is the fountainhead..." *Crystal Fire: Birth of the Information Age*, Michael Riordan (Norton, 1997).

Page 288 *Showcase # 22* (DC Comics, Oct. 1959), reprinted in *Green Lantern Archives Volume 1* (DC Comics, 1998). Written by John Broome and drawn by Gil Kane.

Page 288 *Showcase # 6* (National Comics, Jan.–Feb. 1957), reprinted in *Challengers of the Unknown Archives Volume 1* (DC Comics, 2003). Written by Dave Wood and drawn by Jack Kirby.

Page 288 "The Marvel Age of Comics began..." *Marvel Comics Presents Fantastic Firsts* (Marvel Comics, 2001).

Page 288 *Tales of Suspense # 39* (Marvel Comics, Mar. 1963), reprinted in *Marvel Masterworks: The Invincible Iron Man Volume 1* (Marvel Comics, 2003). Written by Stan Lee and Larry Lieber and drawn by Don Heck.

Page 289 *Iron Man # 144* (Marvel Comics, Mar. 1981). Written by David Michelinie and drawn by Joe Brozowski and Bob Layton.

Page 289 "And boy, did Shellhead..." See *Essential Iron Man Volume 1* and 2 (Marvel Comics, 2000, 2004).

Page 290 "The weapons that were distributed..." *Tales of Suspense # 55* (Marvel Comics, Jul. 1963), reprinted in *Essential Iron Man Volume 1* (Marvel Comics, 2000). Written by Stan Lee and Larry Lieber and drawn by Don Heck.

Page 294 *Iron Man # 132* (Marvel Comics, Mar. 1980). Written by David Michelinie and drawn by Jerry Bingham and Bob Layton.

Page 296 "the phenomenon of diamagnetic levitation..." See the Web page for the High Field Magnetic Laboratory at the University of Nijmegen in the Netherlands: http://www.hfml.ru.nl/levitate.html for some great images of levitating objects.

Page 297 "hand-held pulsed-energy weapons..." "*Star Wars* Hits the Streets," David Hambling, *New Scientist*, no. 2364 (October 12, 2002).

Page 301 "Semiconductor devices are typically constructed..." *Quantum Electronics*, John R. Pierce (Doubleday Anchor, 1966).

Page 302 *Watchmen* (DC Comics, 1986, 1987) written by Alan Moore and drawn by Dave Gibbons.

Page 303 "we all possess cells..." "Dying to See," Ralf Dahm, *Scientific American* 291, p. 83 (Oct. 2004); "Lens Organelle Degradation," Steven Bassnett, *Experimental Eye Research* 74, p. 1 (2002).

Page 303 *Fantastic Four # 62*, (vol. 3) (Marvel Comics, Dec. 2002). Written by Mark Waid and drawn by Mike Wieringo.

Page 308 "on the day Bardeen learned..." *The Chip: How Two Americans Invented the Microchip and Launched a Revolution*, T. R. Reid (Simon & Shuster, 1985).

CHAPTER 25

Page 311 Footnote. *Ultimates # 5* (Marvel Comics, Jul. 2002) written by Mark Millar and drawn by Bryan Hitch.

Page 311 "They depended at the time upon their competitor, National Periodicals..." *Men of Tomorrow: Geeks, Gangsters and the Birth of the Comic Book*, Gerard Jones (Basic Books, 2004).

Page 311 *Fantastic Four # 3* (Marvel Comics, Mar. 1962) written by Stan Lee and drawn by Jack Kirby.

Page 312 *Fantastic Four # 7* (Marvel Comics, Oct. 1962) written by Stan Lee and drawn by Jack Kirby.

Page 313 "Carbon achieves the greatest lowering in energy..." *Valence* (3rd ed.) C. A. Coulson (Oxford University Press, 1985).

Page 314 "Liquid crystals are a familiar example..." *Liquid Crystals: Nature's Delicate Phase of Matter* (2nd ed.), Peter J. Collins, (Princeton University Press, 2001).

Page 315 "shape-memory materials undergo a transformation..." *Shape Memory Materials*, K. Otsuka and C. M. Wayman eds. (Cambridge University Press, 2008); *Shape Memory Alloys: Modeling and Engineering Applications*, ed. Dimitris C. Lagoudas (Springer, 2008).

Page 316 "shape-memory thermoplastic polymer for surgical applications...", "Biodegradable, Elastic Shape-Memory Polymers for Potential Biomedical Applications," A. Lendlein and R. Langer, *Science* vol. 296, p. 1673 (2002).

Page 316 "Certain fabrics expand in response...." *Intelligent Macromolecules for Smart Devices*, Liming Dai (Springer, 2004).

Page 316 *Hulk # 1* (Marvel Comics, May 1962) written by Stan Lee and drawn by Jack Kirby.

Page 316 *Tales to Astonish # 60* (Marvel Comics, Oct. 1964) written by Stan Lee and drawn by Steve Ditko.

Page 317 "the Atom had a unique solution to the problem of needing a costume" *Showcase # 34* (National Comics, Sept./Oct. 1961) written by Gardner Fox and drawn by Gil Kane.

Page 318 *Detective Comics # 27* (National Comics, 1939) written by Bill Finger and drawn by Bob Kane.

Page 319 *More Fun Comics # 73* (National Comics, Nov. 1941) written by Mort Weisinger and drawn by George Papp.

Page 319 *Justice League of America # 4* (National Comics, May 1961) written by Gardner Fox and drawn by Mike Sekowsky.

Page 319 "Diamonds are hard to break...," *Valence* (3rd ed.) C. A. Coulson (Oxford University Press, 1985).

Page 321 "It has been said that the British longbow archers..." *The Book of Archery: Being the Complete History and Practice of the Art* George A. Hansard (Henry G. Bohn, 1841); *The History of Archery*, Edmund H. Burke (Greenwood Press Reprint, 1971).

Page 322 Footnote. "Determination of Young's and shear moduli of common yew and Norway spruce by means of ultrasonic waves," D. Kennecke, W. Sonderegger, K. Pereteann, T. Luthi and P. Niemz, *Wood Science and Technology*, vol. 41, p. 309 (2007).

Page 322 "Holless Allen invented the compound bow" "Archery Bow with Draw Force Multiplying Attachments" U.S. Patent no. 3,486,495 (Dec. 1969).

Page 323 "Flaming arrows" *A History of Greek Fire and Gunpowder*, J. R. Partington (Johns Hopkins University Press, 1998); *Greek Fire, Poison Arrows and Scorpion Bombs*, Adrienne Mayor (Overlook TP, 2004).

Page 324 "In 1921, Marston had a B.A.," *Wonder Woman: The Complete History*, Les Daniels (Chronicle Books, 2001).

Page 324 "when only approximately 3.3 percent of the American population had a college degree" *Education of the American Population*, John K. Folger and Charles B. Nam (Ayer Pub., 1976).

Page 325 *All-Star Comics # 8* (National Comics, Dec. 1941) written by Charles Moulton and drawn by Harry G. Peter.

Page 325 *Wonder Woman # 1* (National Comics, Summer 1942) written by Charles Moulton and drawn by Harry G. Peter.

Page 328 "one-of-a-kind alloy of steel and Vibranium" *Captain America # 303* (Marvel Comics, Mar. 1985) written by Mike Carlin and drawn by Paul Neary.

Page 328 "Vibranium is an extraterrestrial material" *Fantastic Four # 52* (Marvel Comics, July 1966) written by Stan Lee and drawn by Jack Kirby.

Page 328 *Avengers # 4* (Marvel Comics, 1964) written by Stan Lee and drawn by Jack Kirby.

Page 328 *Avengers # 66* (Marvel Comics, July 1969) written by Roy Thomas and drawn by Barry Smith.

Page 329 "Daniels' 2001 excellent history" *Wonder Woman: The Complete History*, Les Daniels (Chronicle Books, 2001).

Page 330 *Tales to Astonish # 48* (Marvel Comics, Oct. 1963), reprinted in *Essential Ant-Man Volume 1* (Marvel Comics, 2002). Written by Stan Lee and H. E. Huntley and drawn by Don Heck.

Page 330 *Tales to Astonish # 49* (Marvel Comics, Nov. 1963), reprinted in *Essential Ant-Man Volume 1* (Marvel Comics, 2002). Written by Stan Lee and drawn by Jack Kirby.

Page 330 "Yellowjacket" *Avengers # 59* (Marvel Comics, Dec. 1968), *Avengers # 63* (Marvel Comics, Apr. 1969), reprinted in *Essential Avengers Volume 3* (Marvel Comics, 2001). Written by Roy Thomas and drawn by John Buscema and Gene Colan.

Page 330 "Goliath" *Avengers # 28* (Marvel Comics, May 1966), reprinted in *Essential Avengers Volume 2* (Marvel Comics, 2000). Written by Stan Lee and drawn by Don Heck.

Page 330 *Ultimates # 3* (Marvel Comics, May 2002). Written by Mark Millar and drawn by Bryan Hitch.

Page 331 *Fantastic Four # 271* (Marvel Comics, Oct. 1984). Written and drawn by John Byrne.

Page 331 "Orrgo" *Strange Tales # 90* (Marvel Comics, Nov. 1961). Written by Stan Lee and drawn by Jack Kirby.

Page 331 "Bruttu" *Tales of Suspense # 22* (Marvel Comics, Oct. 1961). Written by Stan Lee and drawn by Jack Kirby.

Page 331 "Googam (son of Goom)" *Tales of Suspense # 17* (Marvel Comics, May 1961). Written by Stan Lee and drawn by Jack Kirby.

Page 331 "Fin Fang Foom" *Strange Tales # 89* (Marvel Comics, Oct. 1961). Written by Stan Lee and drawn by Jack Kirby.

Page 332 Footnote. *On Growth and Form*, D'Arcy Thompson (Cambridge University Press, 1961).

Page 333 *Fantastic Four Annual # 1* (Marvel Comics, 1963), reprinted in *Marvel Masterworks: Fantastic Four Volume 2* (Marvel Comics, 2005). Written by Stan Lee and drawn by Jack Kirby.

Page 334 *Ultimates # 2* (Marvel Comics, May 2002). Written by Mark Millar and drawn by Bryan Hitch.

Page 334 "If you've ever thought that the bubbles..." *200 Puzzling Physics Problems*, Peter Gnädig, Gyula Honyek, and Ken Riley (Cambridge University Press, 2001).

CHAPTER 26

Page 339 "The first young mutant..." See, "Call Him...Cyclops!" in *X-Men # 43* (Marvel Comics, Apr. 1968). Written by Roy Thomas and drawn by Werner Roth.

Page 341 "in the early days of the Golden Age..." *Superman # 1* (National Comics, June 1939), reprinted in *Superman Archives Volume 1* (DC Comics, 1989). Written by Jerry Siegel and drawn by Joe Shuster.

Page 342 "Before long he was lifting..." See *Superman: The Man of Tomorrow Archives Volume 1* (DC Comics, 2004) for a selection of feats of superstrength.

Page 342 "even hold up a mountain..." *Secret Wars # 4* (Marvel Comics, Aug. 1984), reprinted in *Marvel Super Heroes Secret Wars* (Marvel Comics, 2005). Written by Jim Shooter and drawn by Bob Layton.

Page 342 *World's Finest # 86* (National Comics, Jan.–Feb. 1957), reprinted in *World's Finest Comics Archives Volume 2* (DC Comics, 2001). Written by Edmond Hamilton and drawn by Dick Sprang.

Page 343 *Fantastic Four # 249* (Marvel Comics, Dec. 1982), reprinted in *Essential John Byrne Volume 2* (Marvel Comics, 2004). Written and drawn by John Byrne.

Page 344 "two mathematicians, Euler and LaGrange, proved..." Euler, Acta Acad. Sci. Imp. Petropol., pp. 163–193 (1778); G. Greenhill, Proc. Camb. Phil. Soc. 4, p. 65 (1881); *On Growth and Form*, D'Arcy Thompson (Cambridge University Press, 1961).

Page 345 "Just such a fate inevitably befell Stilt-Man..." *Daredevil # 8* (Marvel Comics, June 1965), reprinted in *Essential Daredevil Volume 1* (Marvel Comics, 2002). Written by Stan Lee and drawn by Wallace Wood. *Daredevil # 26* (Marvel Comics, Mar. 1967), reprinted in *Essential Daredevil Volume 2* (Marvel Comics, 2004). Written by Stan Lee and drawn by Gene Colan. *Daredevil # 48* (Marvel Comics, Jan. 1969), reprinted in *Essential Daredevil Volume 2* (Marvel Comics, 2004). Written by Stan Lee and drawn by Gene Colan.

Page 345 "Doctor Octopus, is able to walk." See *Amazing Spider-Man # 3, 11, 12* (Marvel Comics, July 1963, Apr. 1964, May 1964), reprinted in *Essential Spider-Man Volume 1* (Marvel Comics, 2002). Written by Stan Lee and drawn by Steve Ditko.

Page 346 *JLA # 58* (DC Comics, Nov. 2001). Written by Mark Waid and drawn by Mike Miller.

Page 346 *Detective Comics # 225* (National Comics, Nov. 1955) written by Joe Samachson and Jack Miller and drawn by Joe Certa.

Page 348 "A common misconception is that the pressure change..." K. Weltner, *American Journal of Physics* 55, pp. 50–54 (1987).

Page 350 "Birds such as the California condor..." "The Simple Science of Flight: From Insects to Jumbo Jets," Henk Tennekes (MIT Press, 1997).

Page 350 *Avengers # 57* (Marvel Comics, Oct. 1967), reprinted in *Essential Avengers Volume 3* (Marvel Comics, 2001). Written by Roy Thomas and drawn by John Buscema.

Page 352 *Power of the Atom # 12* (DC Comics, May 1989). Written by William Messner-Loebs and drawn by Graham Nolan.

Page 352 *Showcase # 34* (DC Comics, Sept./Oct. 1961), reprinted in

Atom Archives Volume 1 (DC Comics, 2001). Written by Gardner Fox and drawn by Gil Kane.

Page 353 "When you speak, complex sound waves..." *The Way Things Work*, David Macaulay (Houghton Mifflin Company, 1988) has an accessible description of the physics underlying telephones. See also *How Does It Work?* by Richard Mikoff (Signet, 1961).

Page 355 "When a low mass star..." *Astronomy. The Solar System and Beyond* (2nd ed.), Michael A. Seeds (Brooks/Cole, 2001).

Page 356 "Some astrophysicists have suggested..." "Disks of Destruction," Robert Irion, *Science* 307, pp. 66–67 (2005).

AFTERWORD

Page 358 *Fantastic Four # 22* (Marvel Comics, Jan. 1964). Reprinted in *Marvel Masterworks: Fantastic Four Volume 3* (Marvel Comics, 2003). Written by Stan Lee and drawn by Jack Kirby.

Page 358 Footnote. *Green Lantern # 24 (*DC Comics, Oct. 1963), reprinted in *Green Lantern Archives Volume 4* (DC Comics, 2002). Written by John Broome and drawn by Gil Kane.

Page 358 *JLA # 19* (DC Comics, June 1998). Written by Mark Waid and drawn by Howard Porter.

Page 359 "turned him into a human marionette." *Flash # 133* (DC Comics, Dec. 1962). Written by John Broome and drawn by Carmine Infantino.

Page 359 "if you could travel one thousand years into the past..." In *What If Tale # 33* (Marvel Comics, Jun. 1982), written by Steven Grant and drawn by Don Perlin, Iron Man is trapped back in the days of King Arthur by a double-crossing Doctor Doom. With no way to return to the present, he employs his twentieth-century knowledge of science and engineering to usher in a millennium of worldwide peace and prosperity.

Page 360 "Knowledge is itself the basis of civilization." "To the United Nations," *Niels Bohr, Impact of Science on Society* 1, p. 68 (1950).

Page 361 "A New Model Army Soldier Rolls Closer to the Battlefield," Tim Weiner, *New York Times*, Feb. 16, 2005; "Who Do You Trust: G.I. Joe or A.I. Joe?," George Johnson, *New York Times*, Ideas and Trends, Feb. 20, 2005.

Page 361 *Superman # 156* (DC Comics, Oct. 1962), reprinted in *Superman in the Sixties* (DC Comics, 1999). Written by Edmond Hamilton and drawn by Curt Swan.

Page 361 "Face Front," à la Stan Lee, in practically every Marvel Comic in the 1960s.

ACKNOWLEDGMENTS

THE SEED OF THE IDEA for the freshman seminar on the physics of superheroes that preceded the writing of this book was unintentionally planted by Prof. Terry Jones of the astronomy department at the University of Minnesota during a student's preliminary oral exam. A traditional academic hurdle in many physics graduate programs involves an examination of the students' general knowledge of physics, administered by faculty members who direct their questions to a student, who is armed only with chalk and a blackboard, and must answer in real time. Terry's question: "How much energy would it take for the Death Star to blow up the planet Alderaan in *Star Wars (Episode IV—A New Hope)*?" led me to think of other exploding planets, and inspired me to develop my freshman seminar on superhero science.

Speaking of inspiration, I owe a debt to my many physics teachers and mentors at the college, graduate, and postgraduate levels. In particular I would like to thank Steve Cotsalas, John Jacobson, Peter Tea, Robert Alfano, Narkis Tzoar, Timothy Boyer, Frederick W. Smith, Kenneth Rubin, Sidney R. Nagel, Robert A. Street, and Hellmut Fritzsche. They taught me physics and, by their example, how to be a physicist.

I am also grateful to the creators of the many comic book adventures I have enjoyed over these many years. There are too many names to list them all, but I am especially appreciative of the efforts of the comic book creators of my youth: Gardner Fox, John Broome, Carmine Infantino, Gil Kane, Gene Colan, John

Romita, Robert Kanigher, Steve Ditko, and the big three, Julius Schwartz, Stan Lee, and Jack Kirby. Their stories, featuring heroes that employed their intelligence as well as their superpowers to save the day, taught me an early lesson on the importance of "brain power," though a power ring can be handy also.

I would like to thank Prof. Lawrence M. Krauss for graciously agreeing to write the foreword for this book. I would also like to take this opportunity to acknowledge Craig Shutt (Mr. Silver Age), from whose book *Baby Boomer Comics* I shamelessly lifted the joke in the afterword's title (quoting a Stan Lee Fantastic Four story, who in turn got it from the Bible).

I am grateful to my mom for instilling in me a love of reading and setting an example of lifelong learning and critical thinking. My own children, Thomas, Laura, and David, have served as willing test subjects for many of the arguments presented here, and I benefited from their feedback. I also thank Laura Adams and Allen Goldman for the STM image in fig. 38. My thanks to friends, family, and the employees at Dreamhaven Books and Comics for support and advice.

I could not have written this book without the input from the students in my freshman seminar course from 2001 through 2003, and the evening Compleat Scholar class I taught in 2003. They enriched the class with their insightful comments, clever ideas, and unique perspectives on superhero physics. In particular, questions raised by Eric Caron, Kristin Barbieri, Matt Bialick, Drew Goebel, and Christopher Brummund directly inspired the discussion of some of the topics covered in this book.

One of the bonuses of writing a book on the physics of superheroes is that it has provided a new perspective from which old problems can be viewed. However, a drawback to seriously considering fantastic topics such as running at superspeed or being able to adjust one's tunneling probability at will is that one cannot rely on experiment as a check on analysis. This handicap was mitigated by the thoughtful consideration provided by my colleagues in the School of Physics and Astronomy at the University of Minnesota. I am extremely grateful to Prof. E. Dan Dahlberg, who generously volunteered his time to carefully read the entire manuscript in draft form, and caught many mistakes or oversimplifications. In addition, Professors Benjamin Bayman, Charles E. Campbell, Michel Janssen, Russell Hobbie, Marco Peloso, and

John Broadhurst reviewed selected chapters of this book—the book is much improved for their input. I also benefited from the helpful comments and suggestions of Mark Waid, Gerard Jones, and Kurt Busiek. The paperback version has benefited from the close reading and comments of Markus Poessel. Any errors or confusion that remain are solely my responsibility, though I reserve the right to claim that they have been deliberately left in the text as "Easter eggs" for the attentive reader to discover.

Finally, I would like to offer thanks and acknowledgment for the contributions of the following individuals—without their assistance this book would have been much poorer: my agent, Jay Mandel, for wondering whether there might be a book in this superhero approach to teaching physics, for crucial guidance in the early drafts, and for helping to set the tone of the book. I am very fortunate to have Brendan Cahill of Gotham Books for an editor. His technical advice for structuring the book greatly improved the manuscript and saved this first-time author from several "freshman mistakes." Brendan's vision for this book complemented my own, and—just as importantly—his own knowledge of superhero comic books enabled him to suggest examples that I hadn't considered. In addition, the copy editor, Rachelle Nashner, played a large role in improving the readability of the final version of the text. Jenny Allen went above and beyond the call of friendship in scanning all of the figures used here. The occasional technical crisis was always ably resolved at William Morris by Tali Rosenblat (present at the beginning) and Liza Gennatiempo (there at the end) and by Patrick Mulligan at Gotham Books.

My wife, Therese, has been a constant source of encouragement. She has, from its inception, been more supportive of this project than I could ever have hoped. She has read through all of the many drafts of the manuscript, and I cannot imagine this book without her editorial advice and general counsel. I am a lucky man.

INDEX

Page numbers in *italics* refer to illustrations